# REVERSE ENGINEERING

## TECHNOLOGY OF REINVENTION

# REVERSE ENGINEERING

## TECHNOLOGY OF REINVENTION

WEGO WANG

CRC Press is an imprint of the
Taylor & Francis Group, an **informa** business

CRC Press
Taylor & Francis Group
6000 Broken Sound Parkway NW, Suite 300
Boca Raton, FL 33487-2742



International Standard Book Number: 978-1-4398-0630-2 (Hardback)



**Library of Congress Cataloging-in-Publication Data**

Wang, Wego.
Reverse engineering : technology of reinvention / Wego Wang.
p. cm.
Includes bibliographical references and index.
ISBN 978-1-4398-0630-2 (hardcover : alk. paper)
1. Reverse engineering. I. Title.

TA168.5.W36 2011
620--dc22 2010026751

**Visit the Taylor & Francis Web site at**
**http://www.taylorandfrancis.com**

**and the CRC Press Web site at**
**http://www.crcpress.com**

# *Dedication and Acknowledgments*

The author is deeply indebted to his parents for their inspiration and encouragement all these years. He expresses his whole-hearted appreciation to his wife for her support throughout the course of writing this book. His son and daughter have also been invaluable advisors and proofreaders for the book. His family provides him with both spiritual and professional support and deserves a lot of credit for the completion of this book.

The author is grateful for the support of Charlie Yongpravat for his preparation of several figures and plots, Dr. Indu M. Anand and Robert J. Sayre for their advice on subjects relating to patent and copyright laws, and James G. Serdy of the MIT Laboratory for Manufacturing and Productivity for his advice on three-dimensional printing. The author also thanks Innovmetric, Capture 3D, 3DScanCo/GKS Global Services, ReliaSoft, SEMTech Solutions, Advanced Heat Treat, Metallurgical Technologies, and Dissemination of IT for the Promotion of Materials Science (DoITPoMS) for the photos, images, micrographs, and information they provided, as well as Howard W. Ferris and the Automotive Technology Center at Massachusetts Bay Community College for their support. Roger Oldfield and Jarek Adamowski also kindly granted their copyrighted photo or micrograph for this book. Additionally, several photographs in this book were taken at Instron Corporation in Norwood, Massachusetts; the New England Air Museum in Windsor Locks, Connecticut; the Museum of Flight in Seattle, Washington; the MTU-Museum of MTU Aero Engines in Munich, Germany; the Department of Plastics Engineering of the University of Massachusetts–Lowell in Lowell, Massachusetts; laboratories at Massachusetts Institute of Technology in Cambridge, Massachusetts; and the Bingham Canyon Mine Visitors Center of Kennecott Utah Copper in Bingham Canyon, Utah.

This book is dedicated to Shie-Chih Wong, Yung Tsung Tung Wong, Tsai-Hui Chang, Andrew F. Wang, and Eileen F. Wang.

# Contents

# *Preface*

This book was written with two primary objectives: to advance the technology of reinvention through reverse engineering, and to improve the competitiveness of commercial parts in the aftermarket. While achieving these goals, this book will also show the readers the skills, knowledge, and abilities necessary to succeed in their reverse engineering endeavors by:

1. Enriching the readers' professional knowledge of reverse engineering and empowering them with alternative options in part creation
2. Providing the readers with information on the latest emerging technologies in reverse engineering
3. Familiarizing the readers with current practices and regulations related to reverse engineering
4. Enabling the readers to apply reverse engineering in all disciplines, including the aerospace, automotive, and medical device industries, academic research, accident investigation, and legal and forensic analyses

Reverse engineering has been used to study and replicate previously made parts for years. Modern technology makes this replication easier, and the evolving industry makes it more acceptable today. Legally, reverse engineering is deemed as "a fair and honest means of starting with the known product and working backwards to divine the process which aided in its development or manufacture" (U.S. Supreme Court, 1974). This book introduces the fundamental principles of reverse engineering and discusses the advanced methodologies applicable to reverse engineering with real-world examples. It also discusses relevant regulations and rules that govern industrial practice in reverse engineering.

This book defines the critical elements of reverse engineering and discusses the proper measurements and analyses required to duplicate, reproduce, or repair an original equipment manufacturer (OEM) part using reverse engineering. This knowledge will help readers determine if an OEM part can be duplicated, reproduced, or repaired with reverse engineering. The information in this book will help readers judge if a duplicated or repaired part will meet the design functionality of the OEM part and will assist the readers in evaluating the feasibility of a reverse engineering proposal or project. It covers all areas of reverse engineering. It compares machine design with reverse engineering and introduces the applications of modern metrologies, which make dimensional and geometric measurement easy. It discusses how to analyze the relevant properties for materials identification. It explains the necessary data required for manufacturing process verification. It explains

statistical applications in data processing for reverse engineering. The book also cites legal precedents on intellectual property and proprietary data, and discusses their effects on reverse engineering practice. The economic driving force of the market and its effect on reverse engineering are also briefly discussed. This book enhances the readers' ability to describe and implement a process to duplicate, reproduce, or repair a part using reverse engineering.

Currently there is no universally accepted set of terms used in reverse engineering. All terms are clearly defined before they are used in this book. For the purposes of this book, the International System of Units (SI) is used. In some instances, the U.S. customary units are also included for reference.

# Author

Dr. Wego Wang has been a technical instructor and a researcher in mechanical engineering and materials science for three decades. He is currently adjunct faculty, teaching machine design in the engineering technology department at the University of Massachusetts–Lowell, and he previously taught at Northeastern University and Boston University.

He was elected an ASM International fellow in 2009, and has received many awards, commendations, and recognitions from the Army Research Laboratory, the Federal Aviation Administration, and TMS International. In addition to this book, Dr. Wang has authored and co-authored a number of technical and professional articles and presented lectures and reports at numerous seminars and conferences. He was the 2005–2006 chairman of the ASM International Boston Chapter and is currently on the executive committee of this professional organization. He also served on the executive committee of TMS Boston Section, where he was president from 1993 to 1995.

Dr. Wang earned a bachelor of science degree in mechanical engineering from National Cheng-Kung University, a master of science degree (MS) in mechanical engineering from National Taiwan University, as well as a second MS and a doctor of science degree in materials science and engineering from the Massachusetts Institute of Technology. He works at the Federal Aviation Administration (FAA), primarily on parts manufacturer approval and engine certification programs.

# *Declaration and Disclaimer*

The data and information presented herein are the author's personal perspective and for informational purposes only. They do not represent the positions of any institution or office. Great efforts were made to ensure their accuracy. The readers are responsible for the confirmation and suitability of applying these principles for their own purposes. They hold harmless the author with respect to any and all claims to the fullest extent permitted by law.

Any commercial products or services mentioned herein are solely for technical purposes and do not imply any endorsement. Some regulatory guidance relating to reverse engineering is cited in this book for information only. This book holds no position on any of this guidance. Any discussion on this subject does not constitute either support or opposition to the reference regulatory guidance.

# 1

# *Introduction*

Reverse engineering (RE) is a process of measuring, analyzing, and testing to reconstruct the mirror image of an object or retrieve a past event. It is a technology of reinvention, a road map leading to reconstruction and reproduction. It is also the art of applied science for preservation of the design intent of the original part.

Reverse engineering can be applied to re-create either the high-value commercial parts for business profits or the valueless legacy parts for historical restoration. To accomplish this task, the engineer needs an understanding of the functionality of the original part and the skills to replicate its characteristic details. Though it roots back to ancient times in history, the recent advancement in reverse engineering has elevated this technology to one of the primary methodologies utilized in many industries, including aerospace, automotive, consumer electronics, medical device, sports equipment, toy, and jewelry. It is also applied in forensic science and accident investigations. Manufacturers all over the world have practiced reverse engineering in their product development. The new analytical technologies, such as three-dimensional (3D) laser scanning and high-resolution microscopy, have made reverse engineering easier, but there is still much more to be learned.

Several professional organizations have provided the definitions of reverse engineering from their perspectives. The Society of Manufacturing Engineers (SME) states that the practice of reverse engineering "starting with a finished product or process and working backward in logocal fashion to discover the underlying new technology" (Francis, 1988). This statement highlights that reverse engineering focuses on process and analysis of reinvention in contrast to creation and innovation, which play more prominent roles in invention. Reverse engineering is a process to figure out how a part is produced, not to explain why this part is so designed. The functionality of the original part has already been demonstrated in most cases. The Military Handbook MIL-HDBK-115A defines reverse engineering in a broader perspective to include the product's economic value as "the process of duplicating an item functionally and dimensionally by physically examining and measuring existing parts to develop the technical data (physical and material characteristics) required for competitive procurement" (MIL-HDBK-115A, 2006). This definition casts light on the primary driving force of reverse engineering: competitiveness.

This book concentrates on reverse engineering undertaken for the purpose of making a competing or alternate product because this is the most common reason to reverse engineer in industries. In this context, reverse

engineering is used primarily for three functions: (1) to duplicate or produce original equipment manufacturer (OEM) parts whose design data are not available, (2) to repair or replace worn-out parts without knowledge of the original design data, and (3) to generate a model or prototype based on an existing part for analysis. Reverse engineering has been used to produce many mechanical parts, such as seals, O-rings, bolts and nuts, gaskets, and engine parts, and is widely used in many industries.

Reverse engineering is a practice of invention based on knowledge and data acquired from earlier work. It incorporates appropriate engineering standards and multiple realistic constraints. The part produced through reverse engineering should be in compliance with the requirements contained in applicable program criteria. To accomplish a successful reverse engineering project requires broad knowledge in multiple disciplines. This book aims at further enhancing readers' abilities in

1. Applying knowledge of mathematics, engineering, and science in data analysis and interpretation
2. Using techniques, instruments, and tools in reverse engineering applications
3. Conducting appropriate experiments and tests to obtain the necessary data in reverse engineering
4. Identifying, formulating, and solving issues related to reverse engineering
5. Understanding legal and ethical responsibilities pertinent to reverse engineering
6. Assessing and evaluating documents and fostering attainment of objectives of a reverse engineering project

## 1.1 Historical Background

### 1.1.1 Industrial Evolution

The impact of reverse engineering on today's industry is beyond just introducing less expensive products and stimulating more competition. It also plays a significant role in promoting industrial evolution. The life cycle of a new invention usually lasted for centuries in ancient times. It took thousands of years to invent the electric light bulb for the replacement of the lantern. Both industry and society accepted this slow pace. However, the average life cycle of modern inventions is much shorter. It has only taken a few decades for the invention of the digital camera to replace the film camera and instant camera. This has led to a swift evolution of the photo industry.

To accommodate this rapid rate of reinvention of modern machinery and instruments, reverse engineering provides a high-tech tool to speed up the reinvention process for future industrial evolution.

Reverse engineering plays a significant role in the aviation industry primarily because of the following reasons: maturity of the industry, advancement of modern technologies, and market demands. From the dawn of the aviation industry in the early 1900s to its hardware maturity with the development of jet aircraft in the 1950s, the aviation industry revolutionized the modes of transportation in about 50 years. The early airport is unpaved and looks like a bus stop in the countryside (as shown in Figure 1.1a, also posted at the Automotive Hall of Fame in Dearborn, Michigan). It is a sharp contrast to today's open-field, paved-runway airport vested with modern technologies, as shown in Figure 1.1b. A similar analogy is also found in the aircraft engine and airframe. Figure 1.2 shows an early radial reciprocal aircraft engine that could generate a thrust up to 2,500 horsepower. It is exhibited in the New England Air Museum, Windsor Locks, Connecticut. An advanced turbine engine can

(a)

(b)

**FIGURE 1.1**
(a) An unpaved airport in early days. (From the Henry Ford Museum.) (b) A typical modern open-field, paved-runway airport.

**FIGURE 1.2**
An early radial reciprocal aircraft engine.

generate a thrust of more than 100,000 horsepower today. However, this revolution of the flying machine has slowed down significantly since the invention of the jet engine. The fundamental principles of propulsion and aerodynamics have not changed for decades. Despite that the flight and air traffic control systems have continuously made striking advances with the integration of computer technology into the twenty-first century, the basic hardware designs of jet engine and airframe structures remain virtually the same. The maturity of the aviation industry, hardware in particular, gradually shifted the gravity of this industry from a technology-driven to an economic-driven business. This shift provides a potential market for reverse engineering. During the same period, the advancement of modern metrology introduced many new tools for precision measurements of geometric form and accurate analysis of material composition and process. The fact that the aviation industry is a safety industry subject to rigorous regulations further augments the role of reverse engineering in this industry because certification requirements lead to an inevitable boost in part costs. Consequently, the market demands the least expensive certificated spare parts that are best provided by reverse engineering. Similarly, great potentials of reverse engineering exist in the medical device field.

### 1.1.2 Reinvention of Engineering Marvels from Nature

Many modern machines were invented with inspiration from nature, or reinvented through reverse engineering based on what was observed in nature.

(a)

(b)

**FIGURE 1.3**
(a) Movement of flying birds. (b) The model plane of the Wright brothers' historical first flight in 1903.

The airplane is one of the most noticeable examples. The first self-powered airplane invented by the Wright brothers was designed partially based on their observations, and imitations of flying birds. Figure 1.3a shows the maneuver and movement of flying birds. Figure 1.3b is a photo of the model plane that carried Orville Wright at the beach in Kitty Hawk, North Carolina, in his historical first flight on December 17, 1903. This model plane is exhibited in the Museum of Flight in Seattle, Washington. The first flight lasted about 1 minute and over a distance of approximately 260 m (~850 ft). Today, the cruise altitude of a commercial jet is about 10,000 m (~33,000 ft). However, the altitude of the first flight was about the same altitude of the flying birds.

The Wright bothers tried to reinvent a manmade "bird" by reverse engineering the functionality of a flying creature in nature. A century later, we find ourselves still far behind when it comes to catching the maneuverability of most birds, bats, or even bugs. A hawk moth can easily put up an aerial show flying up, down, sideways, and backwards with rapid acceleration or deceleration. Bats are capable of agile flight, rolling 180°, and changing directions

in less than half a wingspan length; today's aerospace engineers can only dream of an airplane being so maneuverable. Various lengthy mathematic formulas with complex scientific variables and parameters are introduced to decode these myths. Nonetheless, today even the most intelligent aerospace engineers can only wish they could design a flying machine that remotely resembles these features. Reverse engineering is the key for scientists and engineers to deconstruct the basic skills of flying animals and reinvent the next-generation aircraft with better maneuverability and stability.

The human body is a beautiful piece of engineering work in nature. Reverse engineering is the most effective way to reinvent the component parts of this engineering marvel due to lack of the original design data. The production of an artificial knee for implementation in the medical field is a good example. It also reflects one of the major purposes of reverse engineering: replacing the original part. The reinvention process of an artificial knee highlights the key elements of typical reverse engineering practice. It requires accurate dimensional measurement and proper material for suitability and durability. It is also very critical for the substitute new part to meet the performance requirements and demonstrate system compatibility with the surrounding original parts.

### 1.1.3 Reverse Engineering in Modern Industries

The distinction between an OEM and a supplier has been blurred in recent years in today's dynamic and competitive global market. The three major OEMs for aircraft engines—General Electric (GE), Pratt & Whitney (PW), and Rolls Royce (RR)—all just manufactured approximately one-quarter of the components in their respective brand engines. The identities of both the OEM and supplier are disappearing. On February 15, 2006, Pratt & Whitney launched its Global Material Solutions (GMS) program, a new business that will provide CFM56 engine operators with new spare engine parts through the Federal Aviation Administration (FAA) Parts Manufacturer Approval (PMA) and Supplemental Type Certificate (STC) processes. CFM56 is an aircraft engine manufactured through the cooperation of GE in the United States and Snecma Moteurs of France; PMA parts are those developed by companies other than the OEM, and are approved by the FAA based on either identicality or test and computation. The GMS program made Pratt & Whitney the first engine OEM to produce PMA/STC parts by reverse engineering for its rival products, GE engines. Upon the establishment of PW's GMS program, United Airlines immediately signed on as a prospective customer, with a potential long-term parts agreement for its fleet of ninety-eight CFM56-3-powered Boeing 737 aircraft. It brings the application of reverse engineering to a new era. The reverse engineering endeavor, such as reinventing PMA/STC parts, usually is market driven. From 1996 to 2006, more than half the new aircraft with 100 or more passengers were powered by CFM engines. The introduction of CFM engine spare parts produced by

reverse engineering for the repair and replacement of worn-out components will have significant economic impact on the aviation industry and its customers, who will have more options in their maintenance programs.

Advancements in technology have dramatically changed the landscape of reverse engineering. In the 1970s, to reverse engineer a single-crystal high-pressure turbine (HPT) blade was a challenge due to the need to decode highly guarded industry proprietary information. In the 2000s, reverse engineering a single-crystal HPT blade might be just a textbook exercise. Not only have technical innovations changed the reverse engineering process, but the practice itself is also more widely accepted.

The PMA is rooted in the aviation industry. It is both a design approval and a product approval for the reproduction of OEM parts. Each PMA part is issued with an FAA certificate document referred to as a supplement. The criteria of PMA approval are constantly updated along with the advancement of reverse engineering technology. The fact that a supplement requires signatures from both the certification office and the manufacturing office highlights the dual aspects of a reverse engineered PMA part: engineering design and part manufacturing. The production of quality reverse engineered parts does require the full reinvention of engineering design and manufacturing process.

To obtain precise geometric information for the aftermarket automobile parts, many companies also resort to the technology of digital scanning and reverse engineering. United Covers, Inc. is an automobile aftermarket manufacturer. It provides a variety of auto parts, including spoilers, running boards, fenders, and wheel covers. The company is not always able to take advantage of the OEM CAD data, partially because the as-built parts are often slightly different from the CAD data. As a result, real-life data acquisition is required to produce a high-quality replicate part to satisfy customers' expectations. Eventually United Covers contracted with 3DScanCo, a company specializing in 3D scanning and reverse engineering, to help it obtain accurate CAD data and modeling.

The genuine parts manufactured by reverse engineering have been used in automobile repairs and maintenance for years. In contrast to the PMA parts that are certificated by the U.S. federal government, the reverse engineered automotive parts are certified by the industry itself. The Certified Automotive Parts Association (CAPA) was established in 1987. This nonprofit organization develops and oversees a test program ensuring the suitability and quality of automotive parts to meet the standards for fit, form, and function in terms of component materials and corrosion resistance. CAPA encourages price and quality competition in the marketplace so that customer expenses are reduced while still maintaining part quality. It provides consumers with an objective method for evaluating the quality of certified parts and their functional equivalency to similar parts manufactured by automotive companies.

One of the widely cited reverse engineering examples in the military is the Soviet Tupolve Tu-4 (Bull) bomber. During World War II, three battle-damaged

U.S. B-29 Superfortress bombers made emergency landings in then Soviet Union territory after missions to Japan. Most airplanes can be distinguished from one another by their respective characteristics. However, the similarity between the general characteristics of the U.S. B-29 Superfortress bomber and the Soviet Tupolev Tu-4 bomber, illustrated in Figure 1.4a and b and Table 1.1, has led many people to believe that the Tupolev Tu-4 was a replica of the B-29

**FIGURE 1.4**
(a) B-29 Superfortress bomber. (b) Tupolev Tu-4 bomber. (Reprinted from Oldfield, R., http://www.airliners.net/photo/Russia—Air/Tupolev-Tu-4/1297549/&sid=53544687ba303b720943707110073baf, accessed January 12, 2010. With permission.)

**TABLE 1.1**

Characteristics of the B-29 and the Tu-4

| Characteristics | B-29 (Model 345) | Tu-4 |
|---|---|---|
| Maiden flight | September 21, 1942 | May 19, 1947 |
| Wingspan | 43.1 m (141 ft 3 in.) | 43 m (141 ft) |
| Length | 30.18 m (99 ft) | 30.18 m (99 ft) |
| Height | 8.46 m (27 ft 9 in.) | 8.46 m (27 ft 9 in.) |
| Cruising speed | 220 mph (190 knots, 350 km/h) | 220 mph (190 knots, 350 km/h) |
| Service ceiling | 10,241 m (33,600 ft) | 11,200 m (36,750 ft) |
| Power for takeoff | 2,200 HP | 2,200 HP |

*Source:* National Museum of the U.S. Air Force, Boeing B-29 fact sheets, http://www.nationalmuseum.af.mil/factsheets/factsheet.asp?fsID=2527; Wikipedia, Tupolev Tu-4, http://en.wikipedia.org/wiki/Tupolev_Tu-4 (accessed September 25, 2009).

*Note:* mph = miles per hour, HP = horsepower.

Superfortress. It is also widely believed that the U.S. fighter F-86 was reverse engineered for modification from a defected Mikoyan-Gurevich MiG-15 fighter during the Korean War. An F-86F (on loan from the National Museum of the U.S. Air Force) and a MiG-15 aircraft are exhibited in the New England Air Museum, as shown in Figure 1.5a and b. The F-86F was first introduced in 1951. It was a variant of the original North American Sabre, and later evolved into an all-weather jet interceptor and fighter. The F-86F aircraft was powered by a General Electric J47 turbojet engine that was exhibited in front of the aircraft. The MiG-15 fighter first flew in 1947. It was a superior fighter and extensively used during the Korean War. The exhibited MiG-15 fighter was manufactured under license by the People's Republic of China and later obtained by the New England Air Museum in 1990. The general characteristics of the MiG-15b is that debuted in early 1950 and the F-86F-30 are compared in Table 1.2.

A successful reverse engineering program requires great attention to the miniature details and accuracy of all measurements, in addition to a thorough understanding of the functionality of the original part. Not all reverse engineering projects are successful. For example, a reproduction of the 1903 Wright Flyer fell into a puddle after attempting flight on December 15, 2003, at the 100th anniversary of the feat of powered flight. This ill-fated flight attempt brought out another risk factor in reverse engineering. Even though we might have produced a seemingly identical replica of the original part, the operability of the reverse engineered part also depends on the operating environment, such as wind speed in this case, and system compatibility in more sophisticated operations.

(a)

(b)

**FIGURE 1.5**
(a) F-86 fighter. (b) MiG-15 fighter.

**TABLE 1.2**

Characteristics of MiG-15 and F-86F-30

| Characteristics | MiG-15bis | F-86F-30 |
|---|---|---|
| Wingspan | 10.06 m (33 ft 0.75 in.) | 11.91 m (39 ft 1 in.) |
| Length | 11.05 m (36 ft 3 in.) | 11.27 m (37 ft) |
| Height | 3.4 m (11 ft 2 in.) | 4.26 m (14 ft) |
| Cruising speed | 947 kph (589 mph) | 826 kph (513 mph) |
| Maximum speed | 1,075 kph (668 mph) at sea level | 1,107 kph (688 mph) at sea level |
| Service ceiling | 15,514 m (50,900 ft) | 14,630 m (48,000 ft) |

*Source:* Swinhart, E., The Mikoyan-Murevich MiG-15, Aviation History On-line Museum, http://www.aviation-history.com/mikoyan/mig15.html, and North American F-86 Sabre Aviation History On-line Museum, http://www.aviation-history.com/north-american/f86.html (accessed September 25, 2009).

*Note:* kph = kilometers per hour, mph = miles per hour.

## 1.2 Reverse Engineering vs. Machine Design

Engineering design is the process of devising a system, component, or process to satisfy engineering challenges and desired needs. It focuses on creativity and originality. However, reverse engineering focuses on assessment and analysis to reinvent the original parts, complementing realistic constraints with alternative engineering solutions. Reverse engineering has become a standard practice for engineers who need to replicate or repair a worn component when original data or specifications are unavailable. The reverse engineering technology is also applicable to new designs of old parts. Reverse engineering is a top-down reinvention process, while machine design is a bottom-up creation process. In the reverse engineering process an existing and sometimes worn-out part is measured and analyzed with proper methodology to re-create a design drawing for future production. In a machine design process, the design drawing is first created from a new idea or innovation, and the production of the part follows. The first step of reverse engineering is measurement and data acquisition of an existing part. This collected information is then analyzed and interpreted. During data acquisition, the engineer should obtain as much relevant information as possible, including available documentation, existing technical data, and nonproprietary drawings. It is also important to identify any missing engineering data as early as possible. A successful reverse engineering practice requires sufficient familiarity and adequate knowledge of the part being reverse engineered.

Although the primary purpose of reverse engineering an OEM part is to imitate the original part and duplicate it, usually the reproduced part is not identical to the original piece. It may be comparable, but it is unlikely to duplicate the identical dimensional tolerances and manufacturing processes. However, reverse engineered parts should resemble OEM parts as much as possible. In the aviation industry, PMA parts are preferred to be the identical twins, whenever possible, to OEM parts to ensure the same functionality and safety. Occasionally the PMA parts intend to integrate some improvement. It is always challenging to determine how much "improvement" is acceptable for a PMA part that is created using reverse engineering.

Under some circumstances, reverse engineering is one of the few options engineers have to accomplish a task; for example, when the OEM design data are not available but repair to the original part is required, or the original designer is now out of business but more parts are needed.

### 1.2.1 Motivation and Challenge

Another difference between machine design and reverse engineering is their respective economic driving forces. To develop a new innovative part or an improved old part is often the primary motivation in machine design;

the market acceptance of this part is yet to be tested. In contrast, the market acceptance of a reverse engineered part has already been proven. In fact, the best candidate for reverse engineering is often determined by the market demand of this part. The challenge for reverse engineering is to reproduce this "same" part with better or equivalent functionality at lower costs. In 2000, the parts of Pratt Whitney JT8D engines were among the most popular candidates for reverse engineering in aviation industry. However, in 2002, the market changed when these engines retired from service with plenty of surplus around; the interest in reverse engineering JT8D engine parts also noticeably decreased. Later, the parts installable on CFM56 series engines became the most popular candidates in aviation industry because of the unprecedented market shares of these engines, and the high demand of spare parts for maintenance and repairs. Due to the unique financial consideration, a successful reverse engineering project often integrates the legal, economic, environmental, and other realistic constraints into consideration early on. For instance, it is advisable before launching a reverse engineering project to be in close consultation with all the stakeholders, including the reverse engineering practitioner, the prospective customer, and the governmental agency that regulates and approves the final product. The prospective customer might have specific market demand on the part that can dictate the project planning, for example, the product quantity and schedule. The regulatory agency might require some specific demonstration to show the part's compliance to certain environmental regulations before its approval, which can affect the product test plan. All these requirements can significantly impact a reverse engineering project, costs in particular.

From time to time reverse engineering faces the following tough challenges to replicate an original part, that usually do not apply to machine design. First, the information might be lost during the part fabrication. For instance, the filler alloy will be consumed during a welding process. The original composition of the filler alloy is theoretically intractable because it is completely melted and usually metallurgically reacted with the base alloy during the welding process. In other words, the original alloy composition information is lost in the process. Second, the data might be altered during the process. For example, the melting points of lithium and aluminum are approximately 180 and 660°C, respectively. During casting, an Al–Li alloy will be heated up to above 660°C for a period of time. More lithium will evaporate than aluminum during this process. The alloy composition of the final ingot will be different from the original composition of the raw material. The reverse engineering based on the part made of cast ingot has to consider the composition alteration during casting. Third, the details of intermediate processes might have been destroyed to produce the final product. Analysis can easily confirm that a part is manufactured by forging. However, how many cycles of reheat and what presses are used at each cycle are much more difficult to verify because most the evidence has been destroyed before the final cycle.

Reverse engineering does not duplicate an identical twin to the original part because it is technically impossible. The primary objective of reverse engineering is to reinvent a part that possesses equivalent form, fit, and function of the original part based on engineering analysis of the original part. Reverse engineering is an ultimate art of applied science. It uses scientific data to recreate a piece of art that resembles the original one as much as technically possible. Engineering judgment calls based on the best available data play a much more significant role in reverse engineering than machine design.

## 1.3 Analysis and Verification

It is essential to meet the form, fit, and function requirements, and other design details. In a reverse engineering process, the part's physical features are determined by measuring its geometric dimensions, and the tolerance has to be verified. Two other key elements in reverse engineering are material identification and processes verification, including material specification conformity. The material properties to be evaluated are contingent on the service environment and expected functional performance. The material properties at room temperature, high temperature, and sometimes even at cryogenic temperatures may be required. It is worth noting that the material property depends not only on its chemical composition, but also on its manufacturing process. It is critical in reverse engineering to verify the manufacturing process to ensure that the reinvented component will meet the functional and performance requirements of the original design.

Theoretically each individual part requires its own specific analysis or test to demonstrate its functional performance. However, this book will focus on generic comparative analysis and universal scientific methods applicable to reverse engineering. As such, part-specific tests and subjects will only be discussed in case studies. For instance, specific tests to demonstrate a reverse engineered crankshaft meeting the original design functionality will not be discussed in this book. Instead, the discussion will focus on whether the aforementioned reverse engineered crankshaft can be verified as equivalent to the OEM part by demonstrating that it has the same geometric shape, dimensions within the same tolerance, and is made of the same alloy by the same process. If additional tests are required, this book will focus on the rationales for these tests.

In light of part performance verification, communication among all stakeholders and documentation of engineering data often are among the most important factors for a successful reverse engineering project. It is advisable to keep all relevant documents and records in order, and get all stakeholders to buy in as early as possible. It is also highly recommended to justify any technical modifications to the part, including alterations to the design.

The following two examples of modification are usually acceptable in reverse engineering: (1) the use of a new material to substitute an obsolete material that is no longer available, and (2) using an alternate manufacturing process that is commercially available to substitute an OEM-patented process, provided that they are comparable with each other, and both will produce similar products.

### 1.3.1 Accreditation

Both professional competence and data reliability are essential to reverse engineering. Engineering judgment is often called upon for the discrepancy between measurements due to instrumental and human inconsistency in reverse engineering practice. To ensure data reliability, all the tests and evaluations should be conducted at accredited laboratories and facilities. The following will present a brief introduction of several organizations providing quality accreditation services. The Nadcap program (formerly National Aerospace and Defense Contractors Accreditation Program) is one of the most widely recognized accreditation programs in the aviation industry. The Nadcap program, as part of Performance Review Institute (PRI), was created in 1990 by the Society of Automotive Engineers (SAE). It is a global cooperative program of major companies, designed to manage a cost-effective consensus approach to engineering processes and products and provide continuous improvement within the aerospace and automotive industries. Through the PRI, Nadcap provides independent certification of engineering processes for the industry. All the following aerospace companies require their affiliates to obtain and maintain Nadcap accreditation: Boeing, Bombardier, Cessna, GEAE (short for General Electric Aircraft Engine), Hamilton Sundstrand, Honeywell, Lockheed Martin, MTU, Northrop Grumman, Pratt & Whitney, Raytheon, Rolls-Royce, Sikorsky, and Vought. It is reasonable to expect that reverse engineering a part manufactured by these OEMs should hold up to similar accreditation requirements.

The International Organization of Standardization (ISO) is another internationally recognized quality certification organization. The ISO 9001 is a series of documents that define the requirements for the Quality Management System (QMS) standard. It is intended for use in an organization that designs, develops, manufactures, installs any product, or provides any form of service. An organization must comply with these requirements to become ISO 9001 registered. Many facilities and companies are ISO 9001 registered. For instance, Wencor West, a commercial aircraft part distributor and leading PMA manufacturer, is ISO 9001 certificated. Certification to the ISO 9001 standard does not guarantee the quality of end products; rather, it certifies that consistent engineering processes are being applied.

Instead of obtaining accreditations or certifications independently from various organizations, an association can provide a universal certification service acceptable by many regulatory agencies and companies worldwide.

The International Accreditation Forum (IAF) is an association of conformity assessment accreditation bodies. It provides a single worldwide program of conformity assessment that has multilateral recognition arrangements (MLAs) between the members.

The American Association for Laboratory Accreditation (A2LA) is a nonprofit, nongovernmental, public service, membership society. It provides laboratory accreditations based on internationally accepted criteria for competence in accordance with ISO and International Electrotechnical Commission (IEC) specifications, such as ISO/IEC 17025: *General Requirements for the Competence of Testing and Calibration Laboratories.* A2LA is a signatory to several bilateral and multilateral recognition agreements. These agreements facilitate the acceptance of test and calibration data between A2LA-accredited laboratories around the globe. A2LA is recognized by many federal, state, and local government agencies, companies, and associations.

Several accreditation organizations are associated with institutes representing standards and quality, for example, the Registrar Accreditation Board (RAB), which was first established in 1989. In 1991 the American National Standards Institute (ANSI) and the RAB jointly established the American National Accreditation Program (NAP) for Registrars of Quality Systems. In 1996, the ANSI-RAB NAP was formed, replacing the original joint program. On January 1, 2005, ANSI and the American Society for Quality (ASQ) established the ANSI-ASQ National Accreditation Board (ANAB), which is a member of the IAF. ANAB later expanded its conformity assessment services to include accreditation of testing and calibration laboratories.

### 1.3.2 Part Criticality

One of the driving engines propelling the advancement of modern reverse engineering is its ability to provide competitive alternatives to OEM parts. The rigorousness of a reverse engineering project depends on the criticality of the part and cost-benefit consideration. The criticality of a part depends primarily on how it is used in the product. A fastener such as a bolt will be a less critical component if it is used to assemble a non-load-bearing bracket only for division. However, when a bolt is used with glue to hold a 2-ton concrete ceiling in an underground tunnel, it can be a very critical component. The fasteners are among the most popular candidates for reverse engineering. It is also estimated that approximately 70% of all mechanical failures are related to fastener failures. Fortunately, most times the failures are not devastating, and proper corrective actions can be taken to avoid further damage. For example, the utilization of SAE class H11 bolts in aeronautic structures was attributed to a "higher than normal" failure rate due to stress corrosion cracking. FAA Advisory Circular 20-127 discourages the use of H11 bolts in primary aeronautic structures to avoid more incidents.

The precision and tolerance required to reverse engineer a part are often determined by the criticality of the part. From operation safety point of view, the criticality of a part is determined by checking the impact of safety if the part fails. A critical aeronautic part is deemed a part that, if failed, omitted, or nonconforming, may cause significantly degraded airworthiness of the product during takeoff, flight, or landing. However, in different fields and services the definition of criticality varies significantly. When analyzing a load-bearing critical component, the critical strength varies from tension, compression, torsion to fatigue or creep when it is subject to different types of load. The service environment also plays an important role in determining the essential characteristics of the part. High-temperature properties such as creep and oxidation resistance are the determining factors for a turbine blade operating in a high-temperature gas generator. The tensile strength is critical for a static load-bearing component, and also used to determine if a turbine disk will burst out at high rotating speed. However, for a part subject to cyclic stress, such as the automobile axle, fatigue strength is more relevant than the tensile strength. The corrosion resistance becomes a key material property for a part used in the marine industry. In other words, the critical property for a critical part in reverse engineering depends on its functionality and operating condition. For a critical part, higher-dimensional accuracy and tighter tolerance along with higher evaluation costs are expected, and it can become prohibitively expensive for a reverse engineering project.

To best meet the form, fit, and function compliance, and maximize the exchangeability, many commercial parts commonly used in industries, many of them are standardized by individual companies, government agencies, professional societies, or trade associations. Reverse engineering rarely applies to these standard parts because they are readily available on the shelf, and therefore lack financial sensitivity. However, a standard part set by one organization is not always a standard part according to the criteria of another organization. An FAA standard part needs to provide the public with all the relevant information of the part, while a Boeing standard part does not need to provide the public with all the relevant information of the part; as a result, a Boeing standard part is not necessarily an FAA standard part. Globalization also adds a new dimension to the business of part supply. When the Boeing 727 was first introduced in 1964, all seventeen of its major components were made in the United States. By contrast, thirteen of the similar seventeen components of the Boeing 787, which had its first test flight in 2008, are made exclusively or partially overseas. Beyond standardization and globalization, technology advancement definitely has made it easier to reinvent the OEM part with little knowledge of original design details. More and more high-quality spare parts are manufactured through reverse engineering to substitute OEM counterparts at a competitive price.

## 1.4 Applications of Reverse Engineering

Reverse engineering is a multidisciplinary generic science and virtually can be applied to every field universally. The primary applications of reverse engineering are either to re-create a mirror image of the original part, decode the mechanism of a function, or retrace the events of what happened. It is widely used in software and information technology industries, from software code development to Internet network security. It is also used to reconstruct the events just before and immediately after accidents in the aviation, automobile, and other transportation industries. Forensic science is another area where reverse engineering is used to help resolve the myth. Other fields, such as medical systems, architecture and civil engineering, shipbuilding, and art galleries, also find a lot of reverse engineering applications. This book will focus on its applications in hardware, and mechanical components in particular, which itself is a broad area with great potential. In this aspect the utilization of reverse engineering is beyond just reproducing mechanical components. It is used in prototype production for new design and repairs for used parts as well. Thousands of parts are reinvented every year using reverse engineering to satisfy the aftermarket demands that are worth billions of dollars.

The invention of digital technology has fundamentally revolutionized reverse engineering. Compared to the aviation and automobile industries, the applications of digitalized reverse engineering in the life science and medical device industries have faced more challenges and advanced at a more moderate pace. This is partially attributable to human organs' delicate function and unique geometric form. The rigorous regulatory requirements in life science also demand a thorough test before any reverse engineered medical device can be put into production. The fact that we have yet to fully understand the engineering originality of the human body has put reverse engineering in a unique place in the life science and medical device industries, particularly in implementing artificial parts into the human body. The lack of original design drawing often makes reverse engineering one of the few options to rebuild the best replacement part, such as a spinal implement. Applying scanned images with finite element analysis in reverse engineering helps engineers in precisely modeling customized parts that best fit individual patients.

The fundamental principles and basic limitations of reverse engineering are similar in most industries. The general practice of reverse engineering, such as data collection, detailed analysis at a microscale, modeling, prototyping, performance evaluation, and regulation compliance, are the same in principle for all industries. The success of this endeavor is usually subject to the general limitations of modern technologies. However, the specific methodologies used in different fields can be vastly different. Later in this book the discussion of geometric form in Chapter 2 is primarily on hardware

dimensional measurement. The discussions on materials characteristics and analysis in Chapter 3, part durability and life limitation in Chapter 4, and material identification and process verification in Chapter 5 will all focus on the reverse engineering applications to hardware. The principles of data process and analysis discussed in Chapter 6 are applicable to all reverse engineering applications. Part performance and system compatibility are the basic requirements for all reverse engineering applications. Most of the examples in Chapter 7 are based on hardware. Each industry has its own specific regulatory requirements, standards, and certification process, if applicable. A brief general discussion on this subject with a focus on aerospace and automotive industries is presented in Chapter 7. Acceptability and legality are very sensitive and critically vital issues to reverse engineering. Most legal precedents are related to software and information technology industries. However, the discussions in Chapter 8 on intellectual properties and proprietary information are generic and applicable to all industries.

The applications of reverse engineering in software and information technology, in the life science and medical device industries, are a significant part in the overall reverse engineering applications. Though these applications are not the focal points of this book, a brief discussion on these subjects below will help present a broad picture of reverse engineering applications. These discussions also provide a high-level comparison in terms of objective, methodology, and the final product among the applications of reverse engineering in different industries.

### 1.4.1 Software Reverse Engineering

Software reverse engineering is defined as "the process of analyzing a subject system to create representations of the system at a higher level of abstraction" (Chikofsky and Cross, 1990, p. 13). Abstraction is a concept or idea without affiliation with any specific instance. In software development, the higher abstraction levels typically deal with concept and requirement, while the lower levels accentuate design and implementation. Generally speaking, reverse engineering performs transformations from a lower abstraction level to a higher one, restructuring transformations within the same abstraction level; while forward engineering performs transformations from a higher abstraction level to a lower one.

Several levels of abstraction are labeled in Figure 1.6, which illustrates the building blocks in software development. A standard software development model can be represented as a waterfall, starting with concept at the top, then requirement, followed by design, and finally implement. The requirement and design levels are separated by a validation vs. verification division. The reverse engineering process moves upward, analyzing the implementation of the existing system, extracting the design details, recapturing the requirements, and facilitating the original concept. Reverse engineering will, step-by-step, represent the system at a gradually higher level of abstraction,

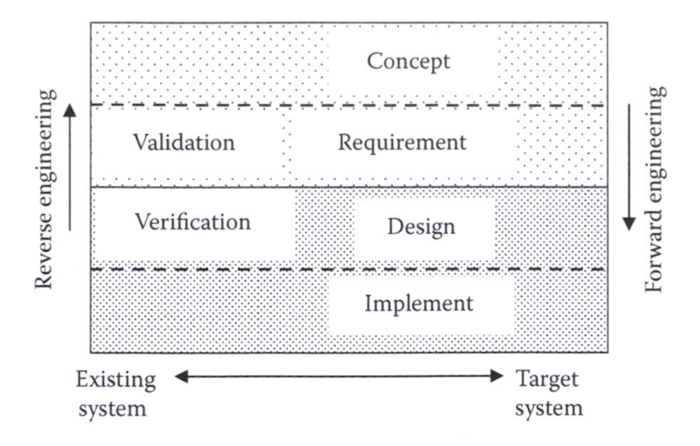


**FIGURE 1.6**
Level of abstraction of software development.

from implement level through design, requirement, and finally reaching the concept level. The key objectives of reverse engineering are to recover the information, extract the artifacts, and synthesize higher abstractions. Reverse engineering will not change the software functionality or alter the system. Any alteration is made only at the completion of reverse engineering in a reengineering process. Reverse engineering builds the foundation that can be used by the subsequent forward engineering to complete the software maintenance or revision, when applied. The software forward engineering process is similar to the typical software code development process. Any refinement will only be made in the forward engineering process to reach the goals of the target system.

There are two commonly acknowledged aspects of software reverse engineering. First, it is a coding process to rewrite a source code that is either not accessible or not available in the field of software development. In this case, great caution has to be taken to avoid potential infringement of any proprietary information or intellectual properties. Second, it is a decoding process to dissolve (or debug) an intrusion in the software security arena. In this aspect, reverse engineering plays an increasingly important role in modern information technology.

Software reverse engineering is a backward process starting with a known functionality to produce a code that can deliver or dissolve this given functionality. There are many potential applications of software reverse engineering. One is to provide an open and fair use option for the maintenance and revision of the ever-growing large volume of software by reengineering, both reverse and forward (Boehm, 1979). Reverse engineering is the first leg of software reengineering. The essential tasks in software reverse engineering are to understand the structure and behavior of the legacy software code, and to process and redescribe the information on an abstract level.

As illustrated in Figure 1.6, there are two primary activities, validation and verification, during a typical software code development and in

the subsequent software life cycle. A software code development usually starts with the establishment of the requirement baseline. This refers to the requirement specifications, which are developed and validated during the plan and requirement phase, accepted by the customer and developer at the plan and requirement review as the basis for the software development contract, and formally change controlled thereafter. The subsequent verification activities involve the comparison between the requirement baseline and the successive refinement descending from it, such as the product design and coding, in order to keep the refinement consistent with the requirement baseline. Thus, the verification activities begin in the product design phase and conclude with the acceptance test. They do not lead to changes in the requirement baseline—only to changes in refinements descending from it. In the context of validation and verification of software code development, software reverse engineering usually will get involved with the following activities (Freerisks, 2004):

- Determining the user demand
- Realization of software-related improvements
- Restoring technical aspects
- Restoring user-level aspects
- Mapping the user-level aspects on the technical aspects
- Software integration
- System integration
- Reintroduction of the system
- Finding software items that can be reused

Detailed elaboration on the above nine activities is beyond the scope of an introductory discussion on these subjects. Interested readers are urged to reference other publications on these subjects for more details.

Software reverse engineering defines the system architecture with the elements of the generic product structure, and identifies the technical requirements for the overall system. In the end, software reverse engineering will generate sufficient data on system interfaces among various units, and provide an integration plan containing the regulations governing the technical aspects for the assembly of the system. Software reverse engineering usually also identifies the user requirements and the application environment.

### 1.4.2 Applications of Reverse Engineering in the Life Science and Medical Device Industries

The physiological characteristics of living cells, human organs, and the interactions among them form the baseline requirements for reverse engineering in life science and medical devices. Some success has been reported from

time to time in identifying the biological components of the control systems and their interactions. However, a fully comprehensive understanding of the complex network of the interacting human body is still beyond today's science and modern technology. In fact, engineers and scientists often work in the reverse direction with the belief that between the observed body behaviors and the biological elements there must underlie the mechanisms that can reproduce these biological functions. This is the typical reverse engineering approach, similar to trying to figure out how a complex piece of electromechanical equipment works without having access to the original design documentation.

To reverse engineer a medical device, engineers first have to identify the materials that are used for this part and their characteristics, then the part geometric form has to be precisely measured, and the manufacturing process has to be verified. Also, more frequently than most other industries, a medical device is operated with sophisticated software for proper function. The operating software has to be fully decoded. For example, the software compatibility of a reverse engineered implantable cardiac pacemaker is one of the most critical elements of the device. In another example, to reverse engineer a blood glucose monitoring device that can be used to measure the glucose level of a diabetes patient, compatible software is a mandatory requirement for the proper transfer of the test results to a computer, and any communication between this meter and the host computer.

Reverse engineering is used in several medical fields: dentistry, hearing aids, artificial knees, and heart (Fu, 2008). Two medical models produced by prototyping are shown in Figure 1.7, including a dental model that illustrates a detailed teeth configuration. The different and unique shape of each individual's teeth configuration provides an excellent application opportunity of reverse engineering in orthodontics. The three-dimensional high-resolution scanner used in reverse engineering can be utilized to accurately measure and model the dental impression of a patient's upper and lower arches. Based on the input digital data, advanced computer-aided manufacturing processes can build customized orthodontic devices for individual patients. Modern computer graphics technology also allows the close examination of teeth movement during follow-ups and the necessary adjustment, if required. Traditional braces with wires and brackets are no longer needed. The application of reverse engineering offers a less expensive and more comfortable treatment in orthodontics. It is worth noting that this new treatment is possible only because of the recent advancement of the modern digital process and computer technology.

High-tech computer hardware, sophisticated software, feature-rich laser scanners, advanced digital processes, and rapid prototype manufacturing have also made more effective applications of reverse engineering to other medical devices, such as the hearing aid, possible since the early 2000s. The digital technology processes sound mathematically, bit by bit, in binary code, and provides a much cleaner, crisper, and more stable sound than that from

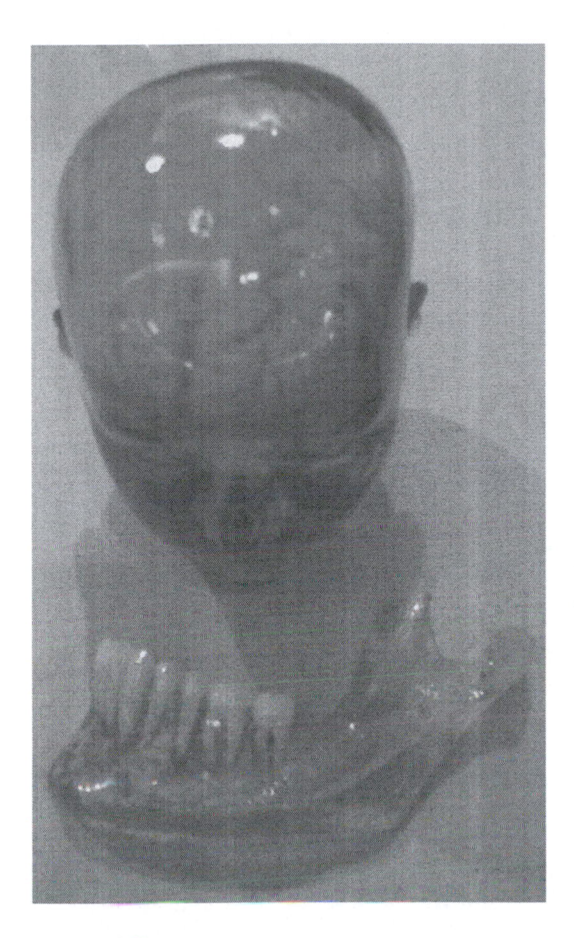

**FIGURE 1.7** **(See color insert following p. 142.)**
Prototype models in the medical field.

analog processing. It offers better overall performance and is relatively easy to update, modify, and revise, thereby providing superior consumer satisfaction in hearing aids. The further growth of reverse engineering applications in this field is mostly dependent on technology evolution to make the wireless hearing aid smaller, more sophisticated, and more efficient, while easier to manufacture and at lower cost.

The applications of reverse engineering to orthopedics, such as the knee, hip, or spine implantation, are very challenging, partially due to the complex motions of the knees, hips, or spine. A proper function of these implants manufactured by reverse engineering requires them to sustain multiaxial statistic stresses and various modes of dynamic loads. They are also expected to have sufficient wear and impact resistance. Several institutes, such as ASTM International, originally known as American Society for Testing and Materials, have published various standards on the testing of these implants. For

instance, ASTM F1717-04 provides guidance on the standard test methods for spinal implant constructs in a vertebrectomy model (ASTM, 2004). The ASTM standards are issued under an established designation system, such as F1717, and are frequently updated. The numerical suffix immediately following the designation, such as 04, indicates the year of adoption or last revision. It is critical to understand the purpose of these standard tests and correctly interpret the test results. The complex loading condition of a spine is difficult to mimic with the limitations of a laboratory testing environment. The test conducted in a dry laboratory environment at ambient temperature might follow all the guidelines of ASTM F1717 and still not accurately predict the fatigue strength of a spinal assembly exposed in the body fluid. The biological environment effects can be significant. The body fluid may lubricate the interconnections of various components in a spinal assembly; it can also have serious adverse effects, such as fretting and corrosion. Therefore, the test results are primarily aimed at a comparison among different spinal implant assembly designs, instead of providing direct evaluation of the performance of a spinal implant. A simulated fatigue test applied with real-life walking and running profiles is often desirable to ensure the high quality of these orthopedic implants.

Medical devices, biomedical materials, and orthopedic implants are usually thoroughly tested to satisfy the rigorous regulatory requirements. U.S. Food and Drug Administration (FDA) regulations require them to get premarket approval (PMA) before they can be put on the market, no matter whether they are brand-name products produced by the original inventors or genuine products produced by reverse engineering. The European Union and many other countries often accept FDA test data and approval in accordance with specific agreements. In an interesting coincidence, the acronym PMA is also used in the aviation industry, where it stands for Parts Manufacturer Approval. U.S. Federal Aviation Administration regulations require all the parts approved under PMA procedures to satisfy the relevant airworthiness requirements before they can be put on the market as well. However, most aviation PMA parts are either produced through licensee agreement with the OEM or reinvented by reverse engineering. The European Union and many other countries also accept FAA PMA approvals with discretion in accordance with specific bilateral agreements.

## References

ASTM. 2004. *Standard test methods for spinal implant constructs in a vertebretomy model.* ASTM F1717-04. West Conshohocken, PA: ASTM International.

Boehm, B. W. 1979. Guidelines for verifying and validating software requirements and design specifications. In *Euro IFIP 79*, ed. P. A. Samet, 711–719. Amsterdam: North-Holland Publishing Company.

Chikofsky, E. J., and Cross, J. H., II. 1990, January. *Reverse engineering and design recovery: A taxonomy*. IEEE software. Washington, DC: IEEE Computer Society, 7:13–17.

Francis, P. H. 1988. Project Management. In *Tool and Manufacturing Engineers Handbook volume 5–Manufacturing Management*, ed. R. F. Veilleux and L. W. Petro, 17–20. Dearborn: SME.

Freerisks, C. 2004. GD250: Lifecycle process mode "V-model" in the World Wide Web. http://www.informatik.uni-bremen.de/uniform/gdpa/part3/p3re.htm (accessed September 25, 2009).

Fu, P. 2008. Reverse engineering in the medical device industry. In *Reverse engineering: An industry perspective*, ed. V. Raja and K. J. Fernandes, 177–93. Berlin: Springer.

MIL-HDBK-115A. 2006. US Army Reverse Engineering Handbook (Guidelines and Procedures). 6. Redstone Arsenal: US Army Aviation and Missle Command.

# 2

# *Geometrical Form*

In recent years the part geometric form has been very accurately measured and replicated by the advanced technology of metrology. The precision hardware and sophisticated software allow engineers to visualize, meter, and analyze the part geometric details. They also allow the transformation of raw data to be intelligently reconstructed into computer modeling. The revolutionary advancement in software algorithm and hardware infrastructure offers a set of new tools for rapid prototype in reverse engineering. All the miniature geometrical details of a part can be captured and retained. The development and deployment of the interchangeable operating systems and data transformability further accelerate today's reverse engineering capability in geometric form analysis and reproduction. These new technologies have a huge impact on modern reverse engineering and have been ubiquitously deployed in this field. This chapter will discuss these technologies and their applications in reverse engineering.

## 2.1 Surface and Solid Model Reconstruction

One of the first steps in reverse engineering is to reconstruct the subject of interest from the data obtained by scanners or probes. The process can be divided into four phases: data acquisition, polygonization, refinement, and model generation. The details and quality of the final models depend on the data collected, the mathematical methods utilized, and the intended application. New data acquisition is accomplished with various measurement instruments, such as a three-dimensional (3D) scanner or a direct-contact probe. The accuracy of the data largely depends on the reliability and precision of these instruments. The polygonization process is completed using the software installed with these instruments. This process is often followed up with a refinement phase such as segmentation to separate and group data point sets. The segmentation methods vary from completely automatic approaches to techniques that rely heavily on the user. Related mathematical techniques include automatic surface fitting and constrained fitting of multiple surfaces. These techniques are also used for computer model refinement.

Figure 2.1 illustrates the flowchart of a reverse engineering process. A typical reverse engineering process starts with the selection of the part of

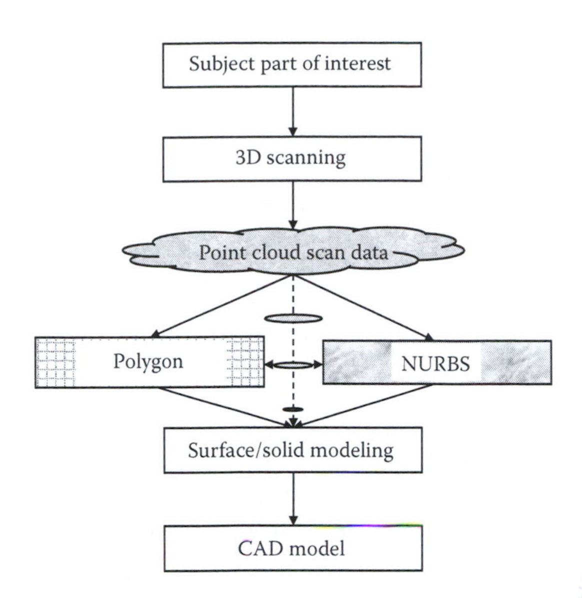


**FIGURE 2.1**
Reverse engineering process flowchart.

interest. Proper measurement devices for data acquisition are then used to generate raw data, usually a point cloud data file. The point cloud is a set of 3D points or data coordinates that appear as a cloud or cluster. Point clouds are not directly usable in most engineering applications until they are converted to a proper format, such as a polygon mesh, nonuniform rational B-spline (NURBS) surface models, or computer-aided design (CAD) models, as input for design, modeling, and measuring through a process referred to as reverse engineering. Figure 2.2a is a polygonal model in the wireframe view, while Figure 2.2b shows the surface format of the same polygonal model. Figure 2.2c is a NURBS model that is ready for export to the CAD system.

The primary technologies to transform a point cloud data set obtained by scanning into a CAD modeling are based on the formation of either a triangular polyhedral mesh or pieces of segments that fit in the model. The method of triangular polyhedral mesh is to first construct a triangular mesh to capture the part topological features based on the point cloud data. It is an approximation presentation of surfaces and other geometric features with triangles. Increasing the number of triangles will yield a better presentation of the surface, but will increase the file size at the same time. The software file for triangulation is usually written in the Standard Triangulation Language (STL), frequently referred to as STL format. It is worth noting that the acronym STL is originally derived from the rapid prototyping process stereolithography, although this process is now usually abbreviated as SLA,

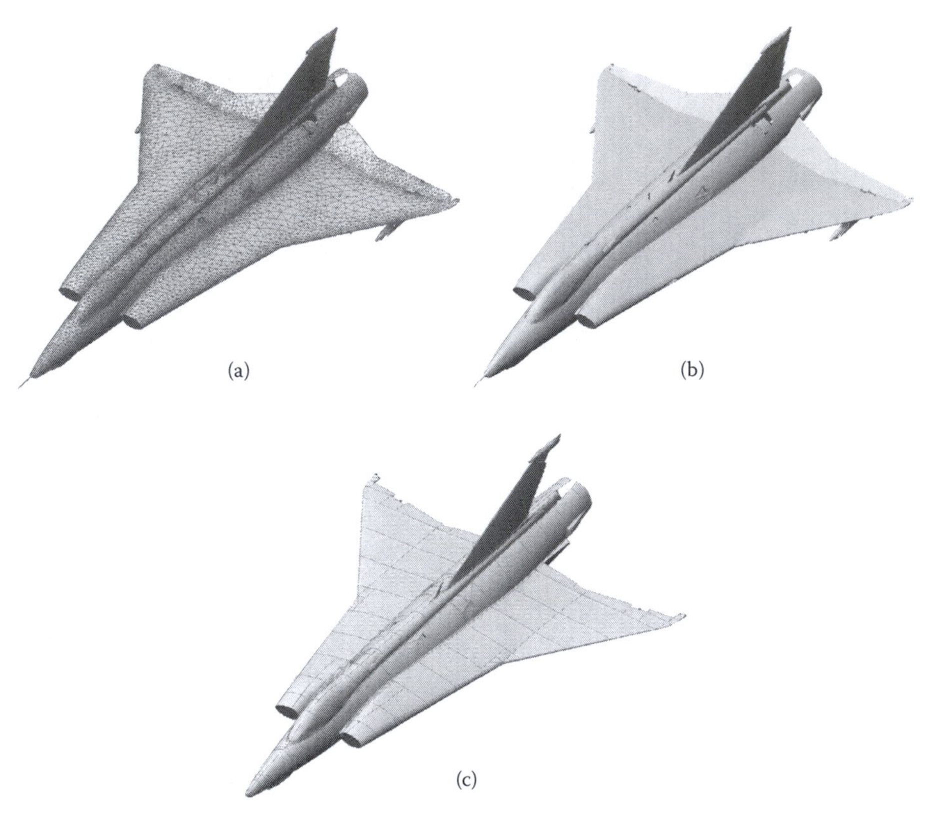


**FIGURE 2.2**
(a) Wireframe polygonal model. (b) Polygonal surface model. (c) NURBS model. (All reprinted from InnovMetric. With permission.)

as detailed later in this chapter. The optimum number of triangles is determined either automatically by the software or manually by the designer to balance between the part precision and data file size. The triangular mesh will subsequently be polished up to reduce the redundant vertices (connected points) and smooth the surface curvatures to meet the design requirements. In the initial point cloud data collection, redundant and intense data are often overlapped to ensure complete coverage of the subject part. An appropriate processing of these raw data by reorientation, realignment, removal, and addition of patch is essential. Various data process methods are developed and applied in different reverse engineering software packages. In the segment approach, the initial point cloud data are segmented into patches with defined boundaries. These discrete surface patches will subsequently be smoothed by appropriate mathematical modeling, such as parametric modeling, quadric functions, or NURBS. Each patch will then be fit into a region of the part surface to build the simulated model.

### 2.1.1 Scanning Instruments and Technology

One of the biggest challenges of reconstructing a mechanical part is to capture its geometric details. Fortunately, advanced devices have been developed to image the three-dimensional features of a physical object and translate them into a 3D model with high accuracy. Data can be obtained directly using a digitizer that is connected to a computer installed with reverse engineering software. The two most commonly used digitizing devices are probes and scanners. They both measure the part external features to obtain its geometrical and dimensional information. Probes obtain data by either a direct-contact or noncontact imaging process. The contact probe is an arm with a tiny ball attached to the end that comes into direct contact with the part being digitized. The noncontact probe is tipped with a small laser probe that never makes direct contact with the subject, and is usually used for more delicate or complex parts. The contact probe is the most economical 3D digitizer. It measures a limited number of points across a target part, and feeds the data back to a computer where the information is processed by software to build an electronic image of the part. It works best for small parts, up to the size of a book, in simple geometric shapes, and it usually provides high accuracy. The user can simply move the stylus, tracing over the contours of a physical object to capture data points and recreate complex models.

A scanner usually does not contact the object and obtains the data by a digital camera. To scan a physical part in a reverse engineering practice, sometimes the only required manual actions are just to point and shoot. All other actions, such as focusing and topographic features imaging, will be processed automatically by the scanning instrument itself. The liquid crystal display (LCD) viewfinder and autofocusing technology are used in modern scanning instruments to frame the object being digitized. Figure 2.3 illustrates the schematic of the scanning process by a 3D non-contact scanner. The laser beam is projected through an emitting lens and reflected by a mirror that is rotated by a galvanometer to sweep the laser light across the entire target object. The reflected laser light from the surface of the scanned object passes through a receiving lens and a filter, and then is collected by a video camera located at a given triangulation distance. The captured images are saved to flash memory. Most scanning instruments are also bundled with digitizing software to help engineers modify and scale the data. Its imaging process is based on the principle of laser triangulation. The Konica Minolta vivid 9i scanner measures $640 \times 480$ points with one scan, and can capture the entire object image in a few seconds, then convert the surface shape to a lattice of over 300,000 vertices. A file beyond just point cloud, such as a polygonal mesh, can be created with all connectivity information retained, thereby eliminating geometric ambiguities and improving detail capture. A photo image is also captured at the same time by the same camera.

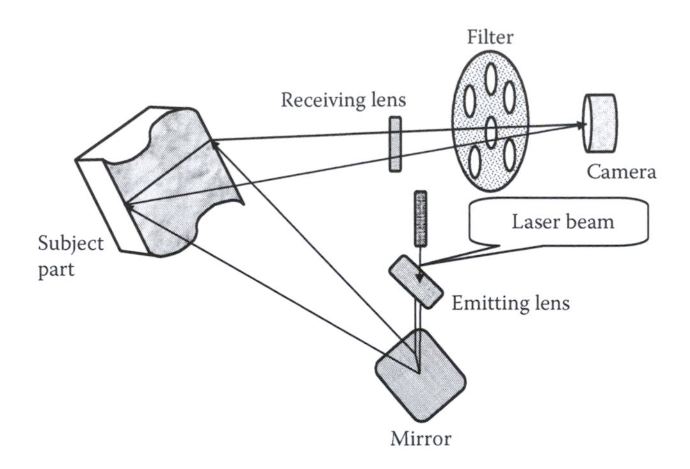


**FIGURE 2.3**
Schematic of scanning process.

Digitized scanning is a very dynamic and rapidly evolving field. More compact new instruments with still higher resolution and more functionality are introduced to the industry every year. Within just a few years, ATOS III was introduced over ATOS II and ATOS IIe by GOM mbH, and Konica Minolta Range7 was introduced over Range5, vivid 9i and 910 scanners. Figure 2.4a is a photo of ATOS II in operation, and Figure 2.4b shows a photo of Konica Minolta Range7. The Range7 scanner is a lightweight box digitizer; the laser beam is emitted from the right and a camera is installed on the left when facing the object. It provides an accuracy up to ±40 μm with a 1.31-million-pixel sensor. The installed autofocus functionality can automatically shift the focus position to provide sharp, high-accuracy 3D measurement data. The implemented sensor and measurement algorithm provides an expanded dynamic range up to 800 mm, and can measure the objects with a wide range of surface reflections, from shining glass and metallic surfaces to dark surfaces with a reflectance as low as 2.5%. The 3D digitizer Range7 can be used with various software packages, such as Geometric, PolyWorks, and Rapidform. The data output of Range7 is in the format of ASCII or binary, including normal vectors, and can be imported to various CAD systems.

The improvement in scanning rate and processing speed, and the advancement of graphical user interface have made real-time scanning possible. The users can instantaneously check the scanned data on the preview screen to see if any data are missing. This allows them to make timely adjustments and take a sequential scanning for the missing data if necessary. Several sources of illumination are available for 3D scanning. The white-light digitizing system is an optical 3D digitizing system that measures the subject surface geometry using a white light. Because white light covers a spectrum of frequency, it usually provides the best-quality data, compared with other measurement technologies, such as infrared or X-ray.

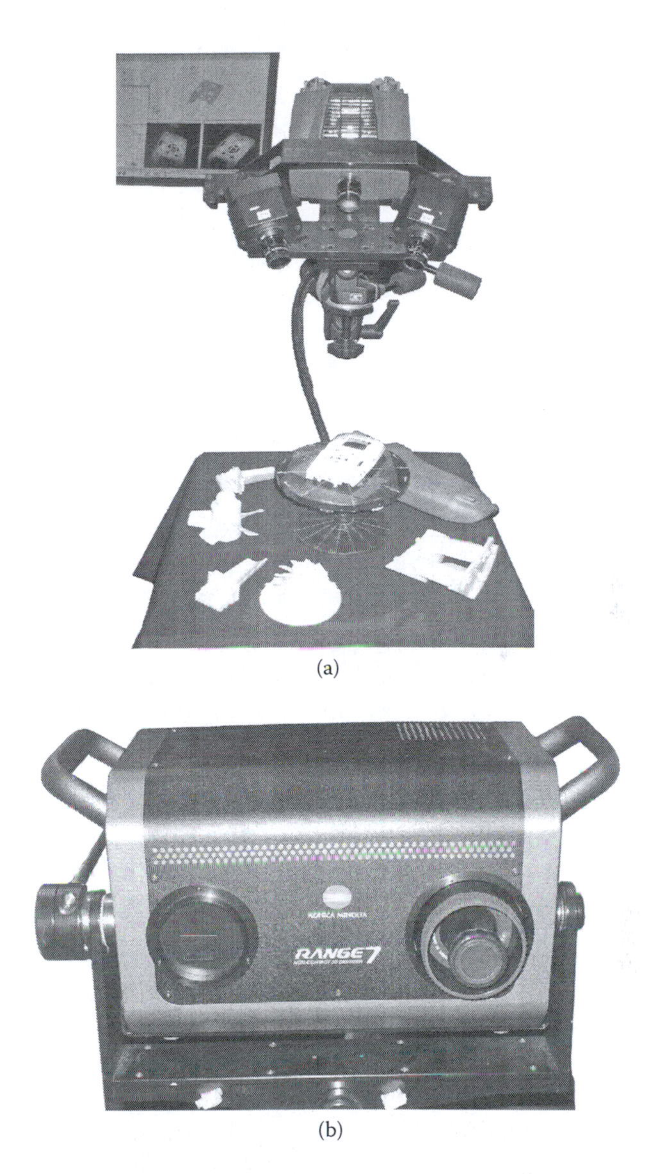


(a)

(b)

**FIGURE 2.4** **(See color insert following p. 142.)**
(a) ATOS II in operation. (b) Konica Minolta Range7 scanner.

The probes and scanners are available in a variety of forms and brands, provided by many manufacturers in various models. Some examples include the Faro laser probe, Konica Minolta laser scanner, and Leica T-scan/tracker. Scanning is often conducted using a coordinate measuring machine (CMM) that is structured like an extended arm with various degrees of freedom to provide the necessary flexibility for digitizing. For example, the Faro Laser ScanArm V3 is a seven-axis, fully integrated contact/noncontact digitizer

with an accuracy of about 35 μm. The Leica T-Scanner can digitize all types of surfaces by projecting a laser beam onto them.

Compared to the contact probe, the noncontact laser scanner is more sophisticated and expensive. It captures millions of data points to better define larger parts with free-form shapes and contours. It is usually used for large parts with complex or curved lines. A large number of data points are needed to accurately capture the intricacies of the design to create digital representations of these objects. The automatic process and feature-rich software requires less training and supervision, and provides faster feedback. The metrology technology and devices are becoming more environmentally friendly: less and less hardware has to be prepared with coating or paint for proper measurements, and thus eliminating the adverse environmental effects of coating and paint due to their chemical contents. The quality of these tools has also improved with every new generation, allowing for higher precision and the measurement of smaller dimensions. The latest scanning technology is more intelligent and can integrate data collection and feature identification together. For instance, the reverse engineering software Rapidform XOR does not just capture the geometric shape of the scanned object; it also captures the original design intent. It automatically detects features, such as revolves, extrusions, sweeps, and fillets, on the scanned object.

Both the contact probe and laser scanner work by digitizing an object into a discrete set of points. In contrast to analog signals that are continuous, digital signals are discrete. Therefore, a digital image can only be an approximation of the object it represents. The image resolution depends on the area and rate of scanning. To obtain the required information within a reasonable amount of time, the engineer usually tracks the sharp edges of the object with a direct-contact probe for better-defined details and combines this information with the results of a laser scan.

Small and economically affordable contact probes, as well as sophisticated laser scanners, are available for reverse engineering applications. This affordability has eased many engineers' reliance on outside services, allowing them to purchase their own devices and keep their digitizing design work in-house. Work in-house means shorter turnaround times, better control of the design process, and better security of proprietary information. New digitizers are also packaged with feature-rich software that makes it easier to turn static raw physical data into dynamic computer images.

### 2.1.2 Principles of Imaging

For a small part, the scanning can be completed with reference to a single coordinate system. For a large part, such as the fuselage of an aircraft or an automobile body, the subject part is usually divided into several regions. One or even multiple scans for each region are performed to capture all the geometric details. The quality of the final combined image heavily depends

on the accuracy of the alignment of these scans. The operator will then photograph this subject setting with a digital camera. Based on these pictures, the coordinates of the reference markers are determined using photogrammetric technology. The 3D spatial coordinates ($x$, $y$, $z$) of a surface point are calculated and formatted in a point cloud file, and a 3D "constellation" reflecting the shape of the subject will then be created. Point clouds collected with laser-based measurement devices may also include characteristics such as intensity and color. These scanned images will be automatically aligned to the 3D constellation created earlier to establish the final configuration for the reverse engineering process.

Photogrammetry is a three-dimensional coordinate measuring technology that uses photographs as the principal medium. In reverse engineering it is used to determine the geometric characteristics of an object and reconstruct it. The fundamental principle of photogrammetry is triangulation; however, many other disciplines, including optics and projective geometry, are also used. By taking photographs from two different locations, common points are identified on each image. A line of sight (also referred to as a ray) can be constructed from the camera location to the point on the object to produce the three-dimensional coordinates of the point of interest using the principle of triangulation. It is a stereoscopic technique and uses the law of sines to find the coordinates and distance of an unknown point by forming a triangle with it and two known reference points. In Figure 2.5, A and B are the two reference locations given by the camera locations, and C is the location of the object point of interest. The distance from A to B can be measured as $c$, and the angles $\alpha$ and $\beta$ can also be measured. The angle $\theta = 180° - (\alpha + \beta)$ because the sum of three angles in any triangle equals 180°. Following the law of sines, as described in Equation 2.1, the distances $a$ and $b$ can be calculated. If the coordinates of A and B are known, then the coordinate of C can also be calculated. Triangulation is also the way our human eyes work together to gauge distance. In addition to reverse engineering, photogrammetry is used in many other fields, including topographic mapping, architecture, and manufacturing.

$$\frac{a}{\sin\alpha} = \frac{b}{\sin\beta} = \frac{c}{\sin\theta} \tag{2.1}$$

Photogrammetry was used in a National Aeronautics and Space Administration (NASA) program: Airborne Research Integrated Experiments System. Over time, a Boeing 757-200 aircraft, as shown in Figure 2.6, has gone through numerous customizations and modifications in this program. NASA needed to further modify the fairings of this aircraft, and therefore required much higher quality CAD data in certain sections than the CAD data on file. The fairing is an airframe structure whose primary function is to produce a smooth outline and reduce drag. The 757-200 aircraft has canoe-

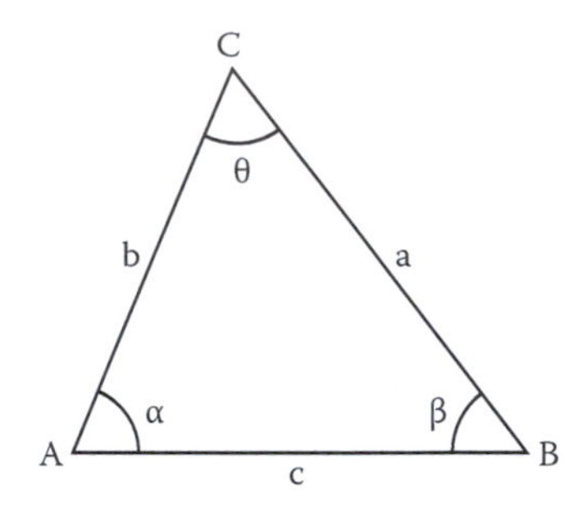


**FIGURE 2.5**
Schematic of law of sines.

**FIGURE 2.6**
A Boeing 757-200 aircraft showing fairings under the wing. (Reprinted from 3DScanCo/GKS Global Services. With permission.)

shaped flap track fairings connecting the wing to the flaps. They protect and streamline the flap-deploying mechanisms. 3DScanCo/GKS Global Services was contracted to perform on-site 3D scanning on this aircraft at Langley Air Force Base in Virginia. Using bolt hole locations and existing features of the plane to align the data, 3DScanCo/GKS Global Services scanned five sections of the 757 for fairing attachment placement. Each section was aligned with the existing scan data based on known distances in rivet placement on the plane, and the rivet placements in the scan area. Photogrammetry was used to ensure that the scan data, and the rivet distances in particular, were accurate for the critical alignment. 3DScanCo/GKS Global Services used this scan data to reverse engineer the aircraft surfaces that were then incorporated into the existing CAD data. Figure 2.7a illustrates photogrammetry, and Figure 2.7b shows the scan compilation (3DScanCo/GKS Global Services, 2009a).

Several portable 3D measurement systems have been developed that can measure very large objects, such as cars or jet engines. They use laser light

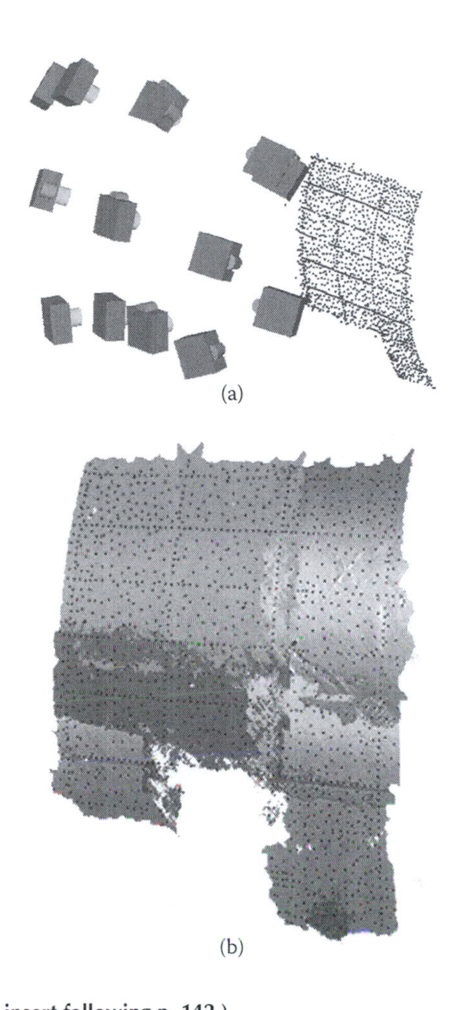

(a)

(b)

**FIGURE 2.7** **(See color insert following p. 142.)**
(a) Photogrammetry. (b) Scan compilation. (Both reprinted from 3DScanCo/GKS Global Services. With permission.)

to illuminate the targets with a 3D grid of interferometric waves for high-accuracy surface measurement, and are capable of acquiring up to 4 million data points in 1 minute, with better than 50 μm of accuracy. Interferometry is the technique of superimposing two or more waves together to detect differences between them. It is based on the physical principle that two waves with the same frequency and the same phase will add to each other (constructive interference), while two waves with the same frequency but opposite phases will subtract from each other (destructive inference). It is applied in a wide variety of fields, from astronomy to metrology. Typically in an interferometer, a wave is split into two or more coherent component waves that travel along different paths. These component waves are later combined to create interference. When the paths differ by an even number

of half-wavelengths, the superposed waves are in phase and interfere constructively, increasing the amplitude of the output wave. When they differ by an odd number of half-wavelengths, the combined waves are 180° out of phase and interfere destructively, decreasing the amplitude of the output. This makes interferometers sensitive measuring instruments for anything that changes the phase of a wave, such as path length. Optical interferometry is a technique of interferometry combining light from multiple sources in an optical instrument in order to make various precision measurements. Optical interferometry might use white light, monochromatic light such as a sodium lamp, or coherent monochromatic light such as a laser light. The main difference between these types of light is their coherence lengths: for the white light, the coherence length is only a few microns, but for the laser light it can be decimeters or more. Therefore, they show different formability of interference fringes.

### 2.1.3 Cross-Sectional Scanning

Most metrologies utilized in reverse engineering are nondestructive surface scanning technologies using CMMs, laser scanners, or white-light scanners. Scanning only on the outside surface provides a challenge to precisely determine the dimensions of internal details, such as internal cavities or deep channels. A cross-sectional scanning (CSS) technique was developed by CGI, an acronym for Capture Geometry Internally. As the company's name implies, this CSS technique can capture the internal geometrical details of the part. It is particularly effective for complex injection-molded or die-cast parts to obtain data on hidden features or critical dimensions where conventional surface scanning cannot reach. The CSS technique is widely used in the biomedical device field.

The three primary steps of the CSS technology are mounting, milling, and scanning. The subject part of interest is first placed within a mold. The mold is then filled with potting material to completely cover all the internal and external features of the part. A proper selection and subsequent hardening of the potting material will be able to provide high-contrast contours between the part and its surroundings. Afterwards, the mold and the potted part are mounted on a base, usually made of aluminum, and secured to the mill table. The part will be milled away layer by layer, usually at ultra-thin increments. After each layer is machined away, the features of the newly exposed surface of the part will be captured through scanning by a digital camera. The obtained data are then sent and filed in the system software for further processing. A simulated 3D part will be built up along the $z$-axis with the input data. Later it can be output in various formats, such as IGES, ASCII, or binary points, for reverse engineering applications. The primary limitations of this technique are that it is usually restricted to parts made of plastics or soft metals, and the original part will be consumed during the process.

### 2.1.4 Digital Data

Three-dimensional scanning, also often referred to as 3D digitizing, is the utilization of a 3D data acquisition device to acquire a multitude of $x$, $y$, $z$ Cartesian coordinates on the surface of a physical object. Each discrete $x$, $y$, $z$ coordinate is referred to as a point. The conglomeration of all these points is referred to as a point cloud. Typical formats for point cloud data are either an American Standard Code for Information Interchange (ASCII) text file containing the $x$, $y$, $z$ values for each point or a polygonal mesh representation of the point cloud in what is known as an STL file format that presents a model part in triangular mesh. The STL file format was first created by the 3D Systems company for stereolithography application. It describes the geometry of a three-dimensional object by triangulated surfaces. Each of these triangles is defined by the coordinates ($x$, $y$, $z$) of the three vertices and the normal vector to the surface. STL files describe only the surface geometry of the object, without any specifications of color, texture, or other common CAD model attributes. It is supported by many software packages, widely used for rapid prototyping, computer-aided manufacturing (CAM); and is also a common file format in point cloud processing for reverse engineering and inspection tasks.

Multiple scans are usually required to obtain sufficient data because the scanned area is restricted by the width of the beam. To capture the three-dimensional features, the data are obtained by scanning the front, back, and all other sides of the part. Each set of scanned data can be color coded and subsequently integrated together. The point cloud data can be collected in different formats, as shown in Figure 2.8a to d. A planar point cloud, as shown in Figure 2.8a, is a uniform grid of points, and is generated from a digitizer that captures the points with respect to a viewing plane. A linear point cloud, as shown in Figure 2.8b, is produced by a line scanner. It generally requires several scan passes; each one composes of a set of line scans. A spherical point cloud, as shown in Figure 2.8c, is a spherical grid produced by a spherical grid scanner. An unorganized point cloud, as shown in Figure 2.8d, is a group of random points without any ordering or connectivity information. These digital data can be used to create a polygon mesh model, NURBS surface model, solid model, color-coded inspection report, cross section, or spline analysis. These data can be further converted into a compatible format for a CAD system.

An image can be created by interpolating a grid of points using the raw scanned data. Alignment of multiple images is often necessary to best fit all the "partial" images into one complete image by loading, aligning, optimizing, analyzing, and reducing overlap of duplicated data. In computer graphics, image polygonization is a method of visualization. It converts a point cloud database to triangles by generating interconnecting mesh, and displays a polygonal approximation of the implicit surface.

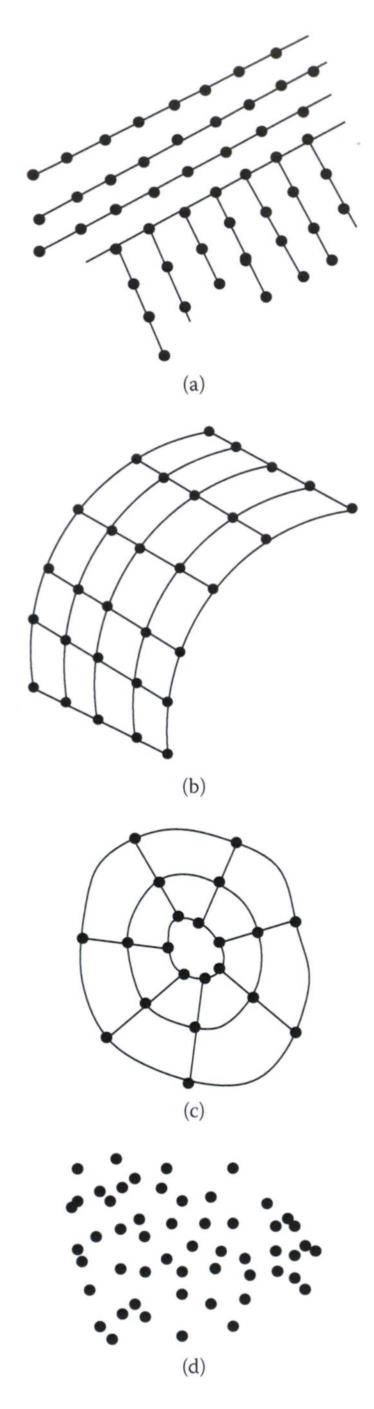


**FIGURE 2.8**
Cloud point data formats: (a) planar point cloud, (b) linear point cloud, (c) spherical point cloud, and (d) random point cloud.

The digital data can be obtained by either surface scanning, probing, or cross-sectional scanning. The quality of the digital data and the subsequent precision of the reinvented part are critical to reverse engineering. They are determined by the density and accuracy of the point clouds formed by the data acquisition device. At the heart of reverse engineering, the proper conversion of these raw data into meaningful information for further modeling and part reproduction is an equally critical concern. The capability of importing and transferring point cloud files from one format to another among digitizing and analytical systems plays an essential role in reverse engineering.

### 2.1.5 Computational Graphics and Modeling

In the 1950s, to develop a mathematical representation for the autobody surface, Pierre Bézier, at Renault in France, first published his work on spline that is represented with control points on the curve, which is now commonly referred to as the Bézier spline. Figure 2.9a illustrates a Bézier curve in solid line with four control points, 1, 2, 3, and 4, and its control polygon in dashed line, and Figure 2.9b illustrates two B-spline curves, each with multiple Bézier arcs, in solid, dash, or dot line, with a unified mechanism defining continuity at the joints.

A nonuniform rational B-spline (NURBS) surface is a surface generated by a mathematical model to represent the surface of a model. It can accurately describe any shape, from a simple two-dimensional (2D) line, circle, arc, or curve to the most complex 3D free-form surface or solid. It is continuous, as opposed to the discrete polygon model composed of triangles and vertices. In the 1960s, it became clear that NURBS was just a generalization of the Bézier spline, which could be regarded as a uniform nonrational B-spline. In 1989, the real-time, interactive rendering of NURBS curves and surfaces was first made available on workstations, and the first interactive NURBS modeler for personal computers became available in 1993. Today, most professional desktop computer graphics offer NURBS technology. Because of their flexibility and accuracy, NURBS models can be used in any process, from illustration and animation to manufacturing. NURBS surfaces are the standard method for importing and exporting data to CAD, CAM, and computer-aided

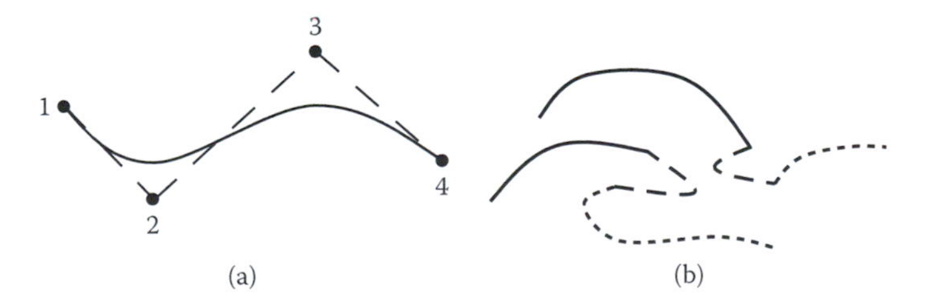


**FIGURE 2.9**
(a) A Bézier curve. (b) Two B-spline curves with multiple Bézier arcs.

engineering (CAE) applications. The Initial Graphics Exchange Specification (IGES) and Standard for the Exchange of Product Model Data (STEP) are two of the most common file interchange formats. Some applications can also accept polygon models, often using the STL format. A NURBS surface can be created from a polygon, and the deviation or fitting errors of a NURBS surface can be verified by a color code or different gray level. It is much easier to transfer a NURBS surface generated from a polygonal model to a CAD system than to create a free surface directly from a CAO program.

Parametric modeling is a modeling technology that employs parametric equations to represent geometric curves, surfaces, and solids. From the reverse engineering perspective, parametric equations are a set of mathematics equations that explicitly express the geometric parameters, such as the $x$ and $y$ locations of a circle in a Cartesian coordinate. Equation 2.2a to c is a set of example parametric equations of a circle, where $r$ is the radius of the circle and $\theta$ is the measurement of the angle from the zero reference.

$$r^2 = x^2 + y^2 \tag{2.2a}$$

$$x = r\cos\theta \tag{2.2b}$$

$$y = \sin\theta \tag{2.2c}$$

A quadratic surface is a second-order algebraic surface that can be represented by a general polynominal equation, as described by Equation 2.3, with the highest exponent power up to 2.

$$ax^2 + by^2 + cz^2 + 2fyz + 2gzx + 2hxy + 2px + 2qy + 2rz + d = 0 \tag{2.3}$$

Many common geometric surfaces, such as sphere, cone, elliptic cylinder, and paraboloid, can be represented as quadric surfaces. The quadric surfaces have been employed by engineers for solid model generation from measured point data (Chivate and Jablokow, 1993), and reverse engineering physical modeling (Weir et al., 1996).

In the 1990s, various techniques were developed to reconstruct implicit surfaces from laser data and other mathematical approaches. Implicit surfaces are two-dimensional, infinitesimally thin geometric contours that exist in three-dimensional space. They are defined by a mathematical function of specific measurable quantity, such as distance. This quantity varies within the space but is constant along the surface. For example, a spherical surface can be represented by an implicit function as

$$r_x^2 + r_y^2 + r_z^2 - r^2 = 0 \tag{2.4}$$

where $r_x$, $r_y$, and $r_z$ are the $x$, $y$, and $z$ coordinates of a point in space and $r$ is the radius of the sphere. Mathematically, those points are considered "inside" the spherical surface if $r_x^2 + r_y^2 + r_z^2 - r^2 < 0$, and "outside" the spherical surface if $r_x^2 + r_y^2 + r_z^2 - r^2 > 0$. The implicit surface is determined by implicitly distinguishing whether the points in space are either inside, outside, or on the surface. Alternatively, a spherical surface can also be represented by a parametric mathematical expression that simply calculates the point location on the spherical surface at a given angle. The points on the parametrically defined sphere can be readily specified by trigonometric equations. Both parametric and implicit methods are well developed and widely used for their respective advantages, often complementary to each other. The parametric surfaces are generally easier to draw and more convenient for geometric operations, such as computing curvature and controlling position and tangency. Nonetheless, the offset surface (a surface at a fixed distance from the base surface) from an implicit surface remains an implicit surface, whereas the offset from a parametric surface is, in general, not parametric.

The implicitly defined surface can be bounded with finite size, such as a sphere; or unbounded, such as an irregular plane. Implicit surfaces are widely used in computer-aided design, modeling, and graphics. It is particularly effective when the geometric surface cannot be explicitly expressed by a simple mathematical expression. In many computer graphics and image process applications, it is useful to approximate an implicit surface with a mesh of triangles or polygons, a popular conversion of visualization that is often referred to as polygonization. The polygonization usually involves partitioning space into convex cells and is processed with graphic software.

### 2.1.6 Data Refinement and Exchangeability

In reverse engineering, refinement is used to fine-tune a coarsely polygonized surface. If the center of a triangle is too far off the surface, the triangle may be split into two or three new triangles at its center, to bring the centers of the new triangles down to the surface. Similarly, a triangle may be divided along its edges if the divergence between surface normals at the triangle vertices is too far off. The final formation of mesh is controlled by several parameters, such as the maximum distance between two reference points. Some reverse engineering software, for example, Polyworks, uses the maximum distance parameter as a primary factor in its operation of connecting dots. The degree of polygonization depends on the complexity of the surface of interest. A higher degree of polygonization is required for a surface of greater curvature to better represent its details. Figure 2.10 illustrates that a much more dense polygonization is applied to a curved surface than to a flat surface.

To meet the quality requirements of a reconstructed surface, follow-up editing and refinement of the data play an essential role. When a model is first generated from the raw data, a hole might exist on the surface, as shown in Figure 2.11, that needs to be filled. Several options are usually provided by

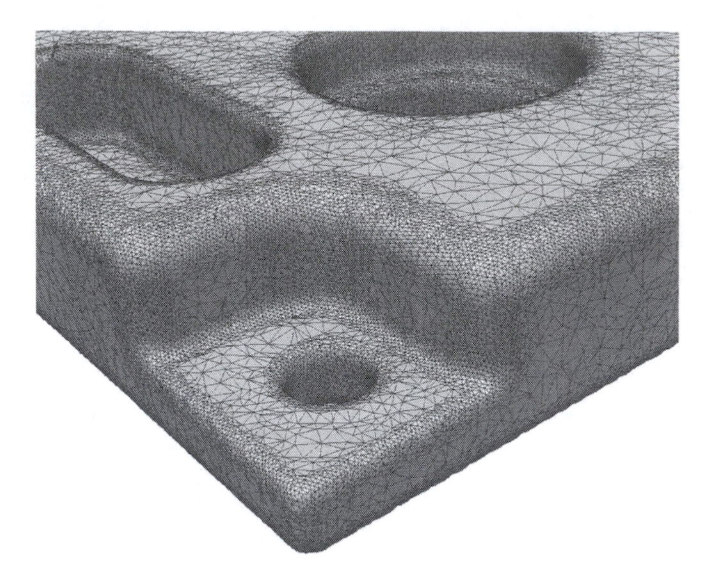

**FIGURE 2.10**
Polygonization of curved surfaces. (Reprinted from InnovMetric. With permission.)

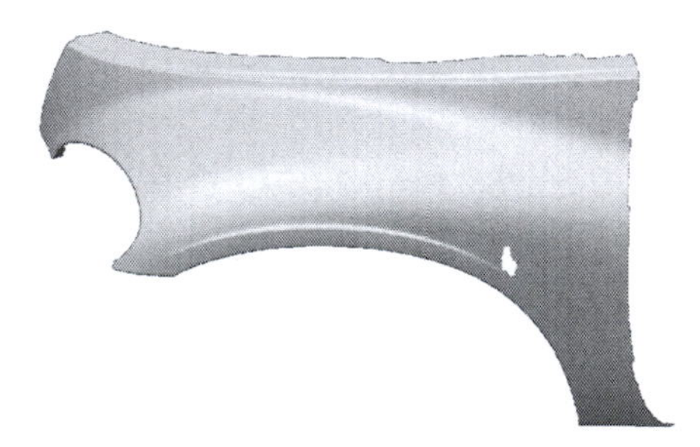

**FIGURE 2.11**
A hole in the model first generated from raw data. (Reprinted from InnovMetric. With permission.)

the software package to paste this hole. The edge at the intersection of two faces might also need to be refined. Figure 2.12 shows the densification of polygons around the intersection area to sharpen the edge.

For reverse engineering applications, a variety of CAD software programs with various analytical and modeling capabilities are available, such as AutoCAD and Inventor® by Autodesk, Solidworks® by Dassault, Pro-Engineer by Parametric Technology, and I-DEAS by Siemens. CAD software is a very dynamic and competitive field with a short product life cycle and quick business turnaround. All the software packages are constantly being revised, and annual updates are very common. The current publisher is not necessarily the original producer. For example, the computer-aided design software I-DEAS, short for Integrated Design and Engineering Analysis Software, was originally produced by Structural Dynamic Research Corporation, which

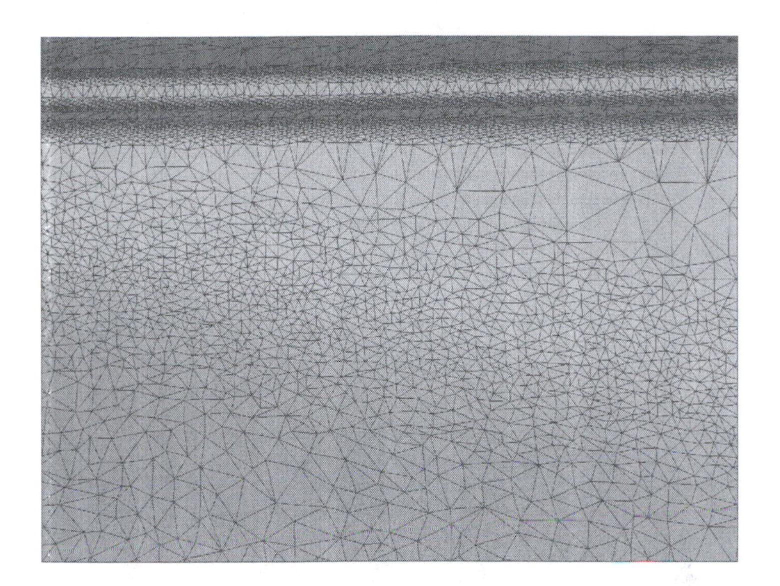

FIGURE 2.12
Refinement of sharp edge. (Reprinted from InnovMetric. With permission.)

was bought by Electronic Data Systems in 2001, then sold and restructured again in 2007, and is currently owned by Siemens PLM Software.

Many solutions have been proposed to resolve the data exchangeability and compatibility problems in design and manufacturing, and various standards have been developed. They include SET in France, VDAFS in Germany, and IGES in the United States, which has been the ANSI standard since 1980. The IGES defines a neutral data format that allows the digital exchange of information among CAD systems. Using IGES, a CAD user can exchange product data models in the form of circuit diagrams, wireframe, free-form surface, or solid model representations. The applications supported by IGES include traditional engineering drawings, models for analysis, and other manufacturing functions. In 1994, under ISO's effort, an international standard named STEP (ISO 10303) for the Product Data Representation and Exchange was released. However, the overall objective of STEP is beyond just for the exchange of product data; it is to provide a means of describing product data throughout the life cycle of a product, and independent from any particular computer system.

## 2.2 Dimensional Measurement

Today the geometrical dimension of a part can be precisely measured easily with a modern high-tech instrument. Figure 2.13 shows a measuring

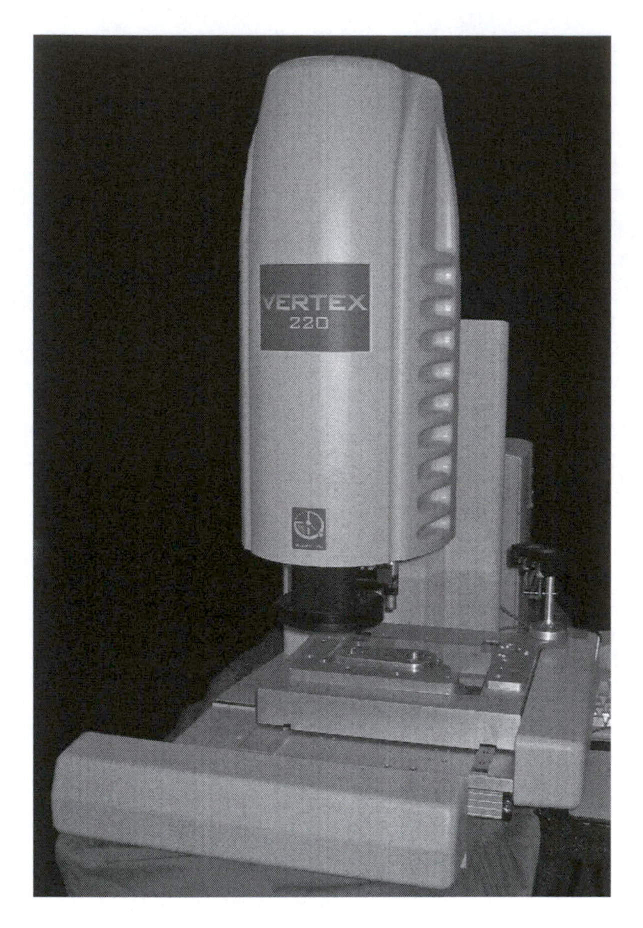


**FIGURE 2.13**
A measuring machine, Model VERTEX 200.

machine Model VERTEX 200 manufactured by Micro-Vu Corporation. It has an accuracy range in micrometers. However, the challenge of obtaining precise geometric dimension measurements is beyond just using precision measurement instruments and employing skilled technicians or highly qualified engineers. A new part is always required for dimensional measurement in reverse engineering, but it is not always available, particularly for repairs. The inherent variables of a used part can adversely affect the accuracy of an otherwise perfect measurement. Figure 2.14 is a photo of two Schick shavers; one is brand new (top) and the other is used (bottom). Besides minor differences in appearance due to design, the used one also shows subtle deformation due to usage that can introduce erroneous dimensional measurement. Dimensional changes can also be introduced by deformation or alteration resulting from repair and welding. Certain surface treatment, such as coating or plating, might also make the precise measurement of the base difficult.

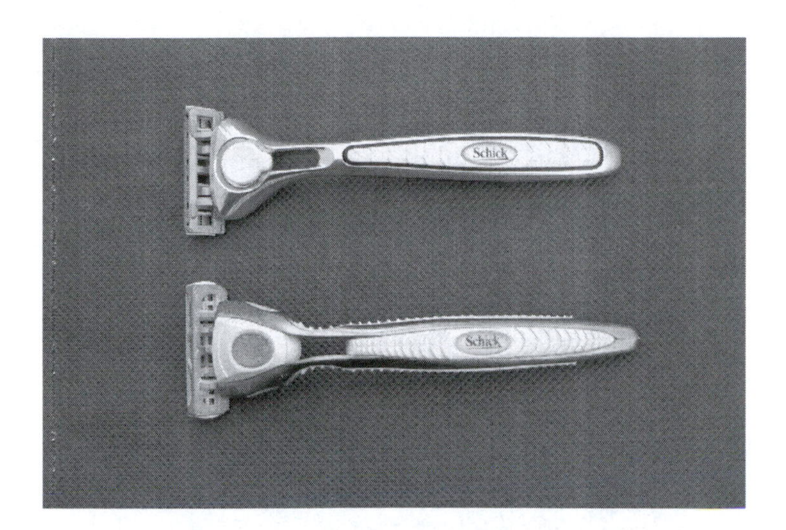


**FIGURE 2.14**
Subtle dimensional differences between the new (top) and used (bottom) Schick shavers.

The following scenario shows a constant challenge for engineers. The reverse engineering data from a true measurement show a bolt hole diameter as 0.510 in., while the OEM data that are not available to the reverse engineering shop show 0.500 in. Which diameter value should prevail? The issue can become even more complex if a maximum 0.510 in. diameter is allowed, as documented in the OEM repair manual for a part with a diameter of 0.500 in. in its original design. The best practice in this situation usually depends on regulatory policies, part criticality, and tolerance requirements.

Many optical, laser, and video precision devices have been utilized for part dimension measurement in reverse engineering. They need to be calibrated periodically and traceable to the National Institute of Standards and Technology (NIST) or other standards. The number of samples required for reliable data depend on the specified accuracy and part complexity; it can vary from a single measurement to four, five, even ten measurements. It is worth noting that the reference OEM part dimensions are sometimes available in the OEM design drawings, repair manuals, or service bulletins, and provide a good base for comparison.

## 2.3 Case Studies

Computational fluid dynamics (CFD) has been used in several programs to analyze the airworthiness of airplanes. The aerodynamic characteristics of a massive aircraft can be accurately analyzed only if its detailed geometry

and shape can be precisely modeled. The digital 3D scanning technology is the most effective method to ensure this accuracy. 3DScanCo/GKS Global Services once scanned an entire Airbus A319 with the Trimble GS200, a scanner for capturing large-scale objects. The scan data were used to generate a CAD model of the aircraft by reverse engineering. Figure 2.15a shows the scanning of the aircraft. Figure 2.15b illustrates the raw scan data, showing holes and other imperfections on the model surface. Figure 2.15c and d depicts the polygonal wireframe model and the CAD rendering, respectively. Computational Methods, an aerodynamic analysis company, later applied this CAD data and CFD analysis to evaluate the performance of an Airbus A319 aircraft installed with custom-built parts (3DScanCo/GKS Global Services, 2009b).

The 1954 Chevy 3100 is an American classic and vintage truck. Southern Motor Company partnered with Panoz Automotive to bring this legacy car back into production. Panoz contracted 3DScanCo/GKS Global Services to scan and reverse engineer the body of this automobile and to capture the bolt hole locations on the chassis. 3DScanCo/GKS Global Services scanned and generated the point cloud data for the auto body. In order to actually capture the location and size of the bolt hole locations on the chassis, 3DScanCo/GKS Global Services used the Konica Minolta vivid 9i scanner along with photogrammetry to establish a data file within 0.002 in. in precision. Applying reverse engineering with these scan data, 3DScanCo/GKS Global Services modeled the entire truck body in smooth CAD surfaces, which could then be incorporated into Southern Motor Company's manufacturing process. Figure 2.16a and b shows the STL polymesh used as a basis for reverse engineering and the CAD rendered in this project, respectively (3DScanCo/GKS Global Services, 2009c).

In another case, Capture 3D, Inc. utilized two complementary noncontact data acquisition devices to capture the full exterior surfaces of a Falcon-20 aircraft that has a span of 16.3 m (53 ft 6 in.), length of 17.15 m (56 ft 3 in.), and height of 5.32 m (17 ft 5 in.). Despite the large size of the airplane and its complex geometric surface features, the measurement of the full aircraft was done in one coordinate system. This project was commissioned by the Aerodynamics Laboratory of the National Research Council (NRC) Institute for Aerospace Research in Ottawa, Canada, for simulated CFD analysis with computer-generated models. To obtain the actual surface data of the aircraft as built, reverse engineering played a key role in linking the physical and digital model environments. The reverse engineering devices used and the fundamental principles applied in this study will be briefly discussed below.

In the early 1990s, a digitizing system, Advanced Topometric Sensor (ATOS), was developed primarily for automotive industry applications. The system was utilized to capture the geometric information from automobiles and their components to generate CAD models. Today, ATOS is used for many industrial measuring applications. The first device used in the Capture 3D/NRC Falcon-20 project is an ATOS II digitizer that is equipped with structured white-

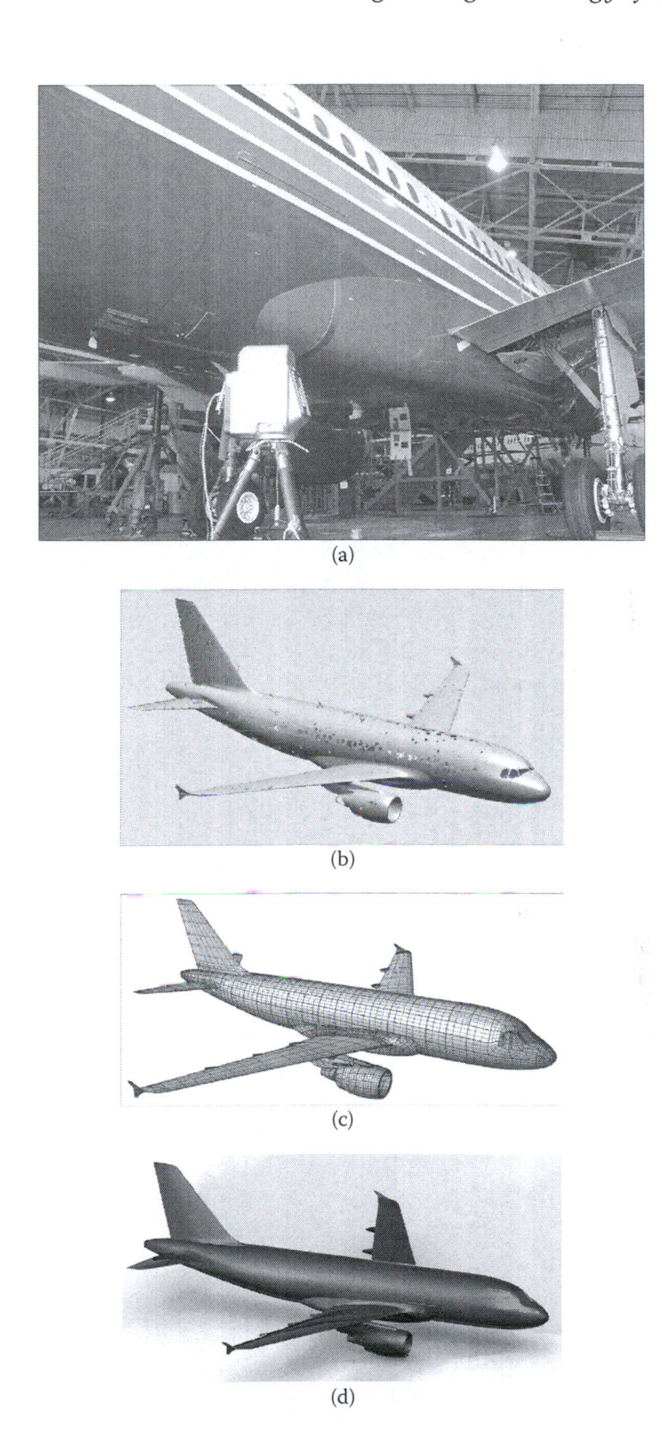

(a)

(b)

(c)

(d)

**FIGURE 2.15**
(a) Scanning of the aircraft. (b) Surface model from raw scan data. (c) Wireframe model of A319. (d) CAD rendering. (All reprinted from 3DScanCo/GKS Global Services. With permission.)

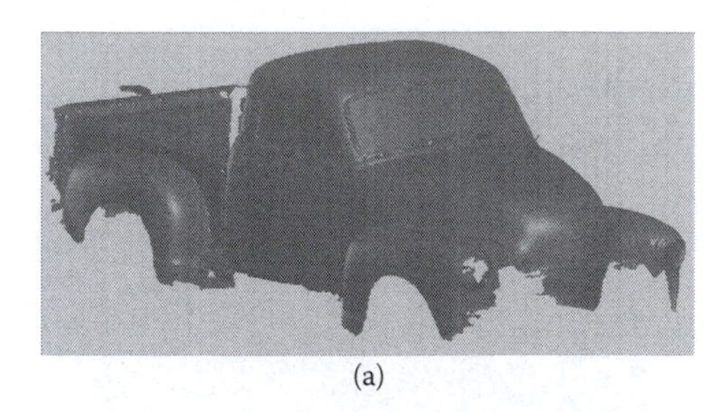

(a)

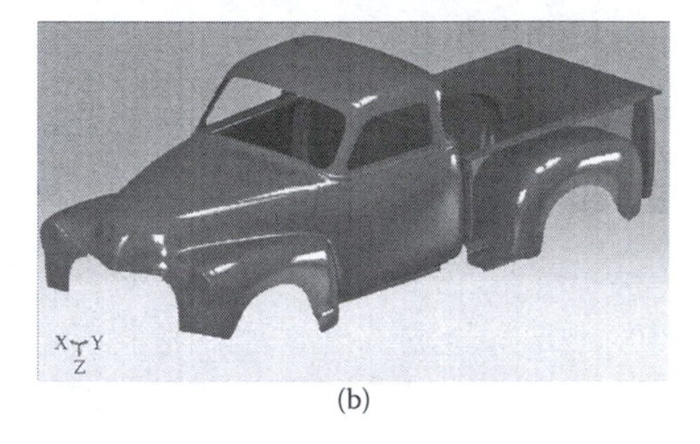

(b)

**FIGURE 2.16**
(a) STL polymesh. (b) CAD render. (Both reprinted from 3DScanCo/GKS Global Services. With permission.)

light projection for optical scanning. This optical measurement technology is based on the principle of triangulation, and the software can calculate the 3D coordinates up to 4 million object points per measurement. The complete 3D data set can be exported into standard formats for further processing.

The TRITOP is an optical coordinate measuring device. It is used to coordinate scanning and measurement. It applies the principle of photogrammetry, and uses reference markers to generate a global reference system on large or complex objects. These markers will be used for both the TRITOP and ATOS II scan processes. They are the reference grid for the individual ATOS scans needed to cover the full surface. TRITOP scanning is conducted manually with a high-resolution digital camera, which is used to take multiple pictures from varying positions around the aircraft. These images are then automatically triangulated and bundled together, producing a global reference system to be utilized later by the ATOS II scanner for scan patch placement. Figure 2.17a and b illustrates the ATOS/TRITOP scanning process of a Falcon-20 aircraft. Figure 2.17a shows the aircraft with reference marks under scanning. Figure 2.17b shows the fringe patterns that are projected onto the object's surface with a white light and are recorded by two cameras during the scanning process.

For the components where detailed features are required, multiple scans (i.e., measurements) are performed. The scanning software will align all the measurements to the same coordinate position and then generate a

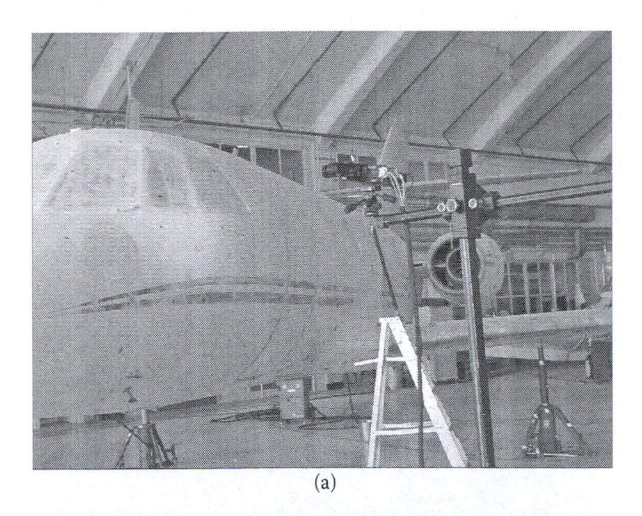

(a)

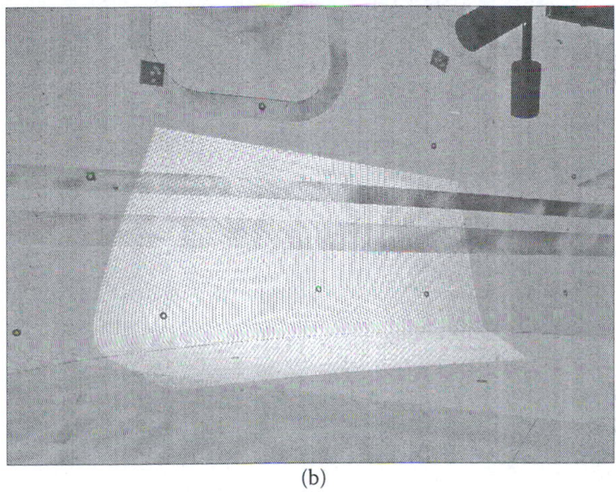

(b)

**FIGURE 2.17** **(See color insert following p. 142.)**
(a) A Falcon-20 aircraft with reference marks under scanning. (b) Fringe patterns. (Both reprinted from Capture 3D. With permission.)

normalized final data set. The ATOS system uses the TRITOP generated reference file for automatic scan patch orientation. As each scan is taken, the ATOS software responds with information on the quality of the scan and the fit of the scan patch in the global reference system. The system will then automatically merge that scan into the reference system and existing point cloud. The engineer can actually watch a real-time buildup of the point cloud on the screen as the Falcon-20 is scanned. This helps to ensure complete and effective scanning. After the aircraft has been scanned, the ATOS polygonizing module will fine-tune the alignment and generate the point cloud STL file to meet the requested density and resolution. These data can then be processed in various ways and exported out in ASCII, STL, IGES, or VDA format.

## 2.4 Part Tolerance

A part tolerance is determined by required precision and affected by variance and fluctuations in measurement. For critical reverse engineering applications, an image resemblance with high fidelity to the OEM part demands tight tolerance. For less critical applications, an image with moderate tolerance might be sufficient. Modern precision manufacturing technology has made the fluctuation of part variation smaller and smaller. As a result, the direct measurement values might show a tighter tolerance than the OEM design data allow. Without knowing the true tolerance, an engineering judgment call is often required to determine the proper tolerance to balance dimension precision and workability.

Another challenge when determining tolerance is part rigidity. The inherent flexibility of a thin-section sealer makes it too flexible to be precisely measured. In this case, corporate knowledge, the engineer's experience, and machinability all play critical roles in the final decision. Engineers also refer to the industrial standard practice as a reliable reference.

Tighter tolerance usually comes with higher manufacturing costs. The tolerances on noncritical dimensions are often allowed to be reasonably liberal to reduce manufacturing costs. In reverse engineering, the determination of noncritical dimensions is based on fit, form, and function consideration. For example, the required tolerance for a bearing depends on its grades. The Annular Bearing Engineering Committee (ABEC) of the Antifriction Bearing Manufacturers Association (AFBMA) has established four primary grades of the precision for ball bearings: ABEC grades 1, 5, 7, and 9. ABEC 1 is a standard for most normal applications. Higher grades require progressively finer tolerances. For bearing bores between 35 and 55 mm, a tolerance of 0.0000 to –0.0005 in. is sufficient for ABEC grade 1, while 0.00000 to –0.00010 in. is required for ABEC grade 9.

Whenever possible, the original design data and the OEM quality, maintenance, and repair manuals are helpful reference documents in tolerance determination. However, the allowed dimensions and tolerances listed in some OEM manuals, such as the repair manuals, are often tailored for used instead of new parts. Therefore, they might not be applicable to the new part reconstructed by reverse engineering. One of the most commonly referred to generic references for geometric tolerance is ASME Y14.5, *Dimensioning and Tolerancing*, also referred to as ANSI Y14.5. It is a language of symbols used on design drawings for geometric dimensioning and tolerancing (GD&T). This standard establishes uniform practices for stating and interpreting geometry requirements for features on parts. It is widely used in the automotive, aerospace, electronic, and manufacturing industries. The mathematical explanation of many of the principles in this standard are given in ASME Y14.5.1. ISO 1101, *Geometrical Product Specifications—Geometric Tolerancing—Tolerances of*

*Form, Orientation, Location and Run-Out*, is another internationally recognized standard. Other references and specifications are available as well.

In reverse engineering, dimensional tolerances are determined by variations in the sample measurements and accepted engineering practices. In accordance with the principle of truth in measurement, the resulting tolerances for the reverse engineered part should not exceed the minimum and maximum dimensions actually measured on the sampled OEM parts. Exceeding these limits requires justification and further substantiation.

Statistics have been used for data analysis and reliability prediction in reverse engineering. The fundamentals of engineering statistics and its applications in data analysis of dimensional measurements and property evaluation will be discussed in Chapter 6. Its application to part reliability will be discussed in Chapter 7.

## 2.5 Prototyping

Prototyping is often referred to as rapid prototyping to reflect one of the most distinctive features of this technology: much faster production of a tangible model part compared to traditional machining and other manufacturing processes. Prototyping revolutionizes the model part creation in machine design and reverse engineering, and provides designers with a tool to quickly convert a conceptual design idea into a physical model part. It helps engineers to visualize the design drawing and computer modeling. Figure 2.18a shows a simulated solid model of a sample part on the computer screen. This information is transferred to a 3D printer for prototyping, as shown in Figure 2.18b. Figure 2.18c shows the final prototype part produced by the 3D printer.

Automation with modern digital technology is the primary advantage of prototyping, which subsequently leads to other benefits, such as the aforementioned speedy production, cost savings, easy operation, and free manufacturing of complex geometric design. Therefore, rapid prototyping is from time to time referred to as rapid manufacturing or direct digital manufacturing, depending on its applications. However, the current rapid prototyping methods featured with modern high technologies are still subject to some limitations that are critical to reverse engineering, such as part accuracy, material restrictions, and surface finishing. An additive prototyping process incrementally adds layers of ceramic, wax, or plastic one atop another to create a solid part, while a subtractive prototyping process such as milling or drilling removes material to shape up the part. It is very beneficial and sometimes essential in reverse engineering to first produce a few model parts from the collected data before entering the production phase. Prototyping can quickly produce a model part at reasonable costs for form, fit, and functional test during the reverse engineering process. For instance, a model airfoil can be first

**FIGURE 2.18**
(a) Simulated solid model of a sample part on the computer screen. (b) 3D printer in process. (c) Final prototype part produced by a 3D printer.

produced by rapid prototyping and tested in a wind tunnel to measure lift and drag forces before the production. This section will focus on the advantages and limitations of the commercially available prototyping techniques, from their applications to the reverse engineering perspective (Cleveland, 2009).

### 2.5.1 Additive Prototyping Technologies

The additive prototyping process is a nonconventional fabrication technology that is supported by modern information technologies for data conversion, CAD model building and slicing, and model part fabrication. This process usually starts with the input of data from a CAD model that is an intermittent product of reverse engineering. A slicing algorithm first slices the CAD model into a number of thin layers and draws the detailed geometric information of each layer, and then transfers it to the prototyping machine to build up semi-two-dimensional sections layer by layer with skinny thickness. The final model part precision is directly related to the thickness of the slicing layers. For consistency, the STL format has been adopted as the industrial standard slicing algorithm. However, depending on the software, a CAD model can be built with various formats, such as DXF, 3DM, or IGES. The DXF, short for drawing exchange or drawing interchange, format is a CAD data file format developed by Autodesk in 1982. For many years the data exchange with the DXF file has been challenging due to lack of specifications. The 3DM is a computer graphics software format developed for free-form NURBS modeling and to accurately transfer 3D geometry between applications. The IGES, an acronym for Initial Graphics Exchange Specification, is a software format that was established in the 1980s to transfer the digital information among various CAD systems. It usually requires the CAD model presented in STL format for prototyping. For example, the NURBS CAD data have to be converted to STL format for the subsequent slicing mechanism. Additive prototyping often requires conversion from the point cloud data obtained by scanning as the first step in reverse engineering, through a surface model in NURBS, to STL format if the CAD model is so built, and then to a layer-based additive prototyping model. These data conversions and processing are often the primary sources of error in part shape and form due to the inherent discrepancies among the mathematical algorithms and software formats. A direct application of the original point cloud data to slicing algorithm and layer modeling in additive prototyping technologies, if successfully developed, will have the potential to significantly improve the model part surface finishing, precision, and tolerance.

The additive prototyping processes have the advantage to create parts with complicated internal features that are difficult to manufacture otherwise. However, most additive prototyping technologies do not provide any information on part machinability and manufacturability of the design. The additive prototyping technologies are also subject to some other restrictions, such

as part size, applicable material, and limited production. Metal prototypes are noticeably difficult to make with the additive prototyping process.

Stereolithography (SLA) is the first commercialized additive rapid prototyping process, and is still the most widely used additive prototyping technology today. In U.S. Patent 4,575,330, "Apparatus for Production of Three-Dimensional Objects by Stereolithography," issued on March 11, 1986, Hull, who invented the technology of stereolithography, defined it as a method and apparatus for making solid objects by successively "printing" thin layers of the ultraviolet curable material one on top of the other. SLA is an additive fabrication process that builds parts in a pool of resin that is a photopolymer curable by ultraviolet light. An ultraviolet light laser beam reflected from a scanner system traces out and cures a cross section of the scanned part on the surface of the liquid resin a layer at a time. Though the most common photopolymer materials used in SLA require an ultraviolet light, resins that work with visible light are also used. The solidified layer supported by a platform is then lowered just below the surface of the liquid resin, and the process is repeated with fresh material for another layer. Each newly cured layer, typically 0.05 to 0.15 mm (0.002 to 0.006 in.) thick, adheres to the layer below it. The consistent layer thickness and air entrapment prevention are often controlled with a wiper blade that clears the excess fluid resin from the top of the new layer surface. This process continues until the part is complete.

From the reverse engineering perspective, SLA provides an excellent tool to rapidly replicate a part with virtually identical geometric form and shape. It provides accurate dimensions and fine surface finishes. There is almost no limitation on part geometric complexity, but the part size is usually restricted. Most SLA machines can only produce the parts with a maximum size of about 50 × 50 × 60 cm (20 × 20 × 24 in.). The parts made of photopolymer by SLA are weaker than those made of engineering-grade resins, and therefore might not be suitable for certain functional tests.

Selective laser sintering (SLS) is another additive prototyping process that builds parts by sintering powdered materials by a laser, layer by layer, from the bottom up. This technology was developed by Carl Deckard at the University of Texas at Austin, and subsequently patented by him in 1989. Sintering is a fusion process at a temperature above one-half the material melting temperature, but below the melting temperature. Sintering welds two or more particles together and consolidates them into a solid part, usually under pressure. Compared to SLA parts, SLS parts are more accurate and durable. Nonetheless, the SLS parts are still weaker than machined or molded parts. The SLS part is usually of high porosity, and its surface finish is relatively poor, with a grainy or sandy appearance that can have an adverse effect on mechanical properties. As a result, the SLS part is generally not suitable for functional tests in reverse engineering. Another restriction is that only limited powder resins, such as nylon or polystyrene, are available for SLS applications. Direct metal laser sintering (DMLS) is a similar rapid prototyping process. It uses metal powders of steel or bronze instead of powdered

resins. The parts are built in layers as thin as 20 μm, with a typical tolerance of 0.025 mm/25 mm. This process has the potential to build production-worthy parts with targeted materials. It is best suited for small parts with complex geometries or internal passages, as shown in Figure 2.19.

Fused deposition modeling (FDM), developed by Stratasys, is also an additive process. It is a very popular rapid prototyping technology and is widely used, only second to SLA. It builds parts from the bottom up through a computer-controlled print head. In contrast to SLA, whereby a liquid resin pool is used, and SLS, whereby the compacted resin powders are used, the feedstock for FDM is a filament of extruded resin that remelts and deposits on top of the previously formed layers. The FDM process utilizes a variety of polymeric materials, including acrylonitrile butadiene styrene (ABS), polycarbonate, and polyphenylsulfone. The ceramic and metallic materials are also potential candidate materials that can be used in the FDM process in the future. The FDM parts are relatively strong with good bonding between the layers, and can be used for functional tests in reverse engineering when appropriate. However, the FDM parts are often porous, with a rough surface finish, and of relatively poor tolerance control.

The term *three-dimensional printing* is sometimes applied to all additive rapid prototyping processes because they all seem printing and building the three-dimensional part layer by layer. The following patented three-dimensional printing process discussed here was developed at Massachusetts Institute of Technology (MIT). Figure 2.20a shows a photo of the experimental three-dimensional printing prototyping machine used for studies at the Laboratory for Manufacturing and Productivity of MIT. It has a print head platform that carries the powders and can be moved upward or downward by a piston. One layer of powder is usually 100 μm thick, but it can be as thin as 20 to 25 μm. The powder geometric shape, which can be spherical or flake, often has a fair amount of effects on the final product. Several bonding stations are located on the left side of the machine. A piezoelectric inkjet installed on an air slide above the print head was designed to inject almost perfectly spherical droplets onto the powders for bonding. Figure 2.20b is a close-up look at the print head platform assembly. A spreader is located on top of the platform and used to sweep through the platform for powder

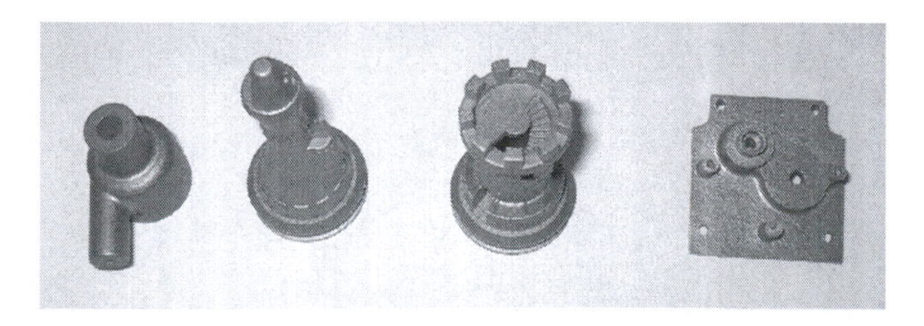

**FIGURE 2.19**
Parts produced by the direct metal laser sintering process.

(a)

(b)

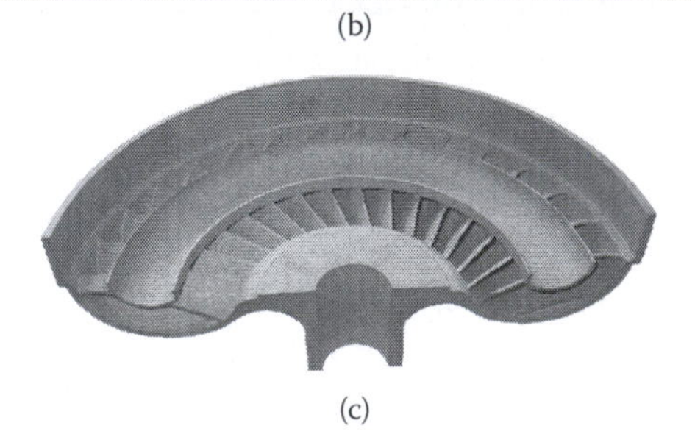

(c)

**FIGURE 2.20**
(a) An experimental 3D printing prototyping machine. (b) Close-up of the print head platform assembly. (c) A sample product of 3D printing.

feeding and layer thickness adjustment. Figure 2.20c is a sample product of this prototyping machine.

Instead of a laser device, this technology uses an inkjet head. The printer first lays down a thin layer of powder material, and then the inkjet head passes over and sprays liquid adhesive onto wherever solidification is required to build the solid part. Different colors can be easily incorporated into the finished part in this process. It is a quick and inexpensive process. However, this technology is also subject to some limitations on part surface finish, part fragility, and suitable materials for this process. The commonly used materials for this process are starch, plaster, ceramic, and metal powders. The parts produced by this process sometimes require further infiltration with another material to improve their mechanical strength. These parts are relatively weak and with rough finishing, and therefore are not usually recommended for functional testing in reverse engineering.

The three-dimensional printing technology has been integrated in many college programs. Students are doing prototyping projects with various 3D printers. The architecture students at MIT are building their architectural models with 3D printers. Figure 2.21a shows an SST (acronym for Soluble Support Technology) 1200es series three-dimensional printer manufactured by the Dimension business unit of Stratasys, Inc. It is used in the Department of Plastics Engineering at the University of Massachusetts–Lowell. Figure 2.21b is a close-up look at the compartment of this 3D printer, wherein a platform can be moved up and down.

The development of laser power-forming technology, also referred to as laser fusing in general and other names by the respective developers, was initiated at several universities and government laboratories, such as Sandia National Laboratories, who named this technology Laser Engineered Net Shaping. This technology allows the fabrication of fully dense metal parts with good metallurgical properties at reasonable speeds for reverse engineering applications. Similar to other additive prototyping technologies, this method builds the model part layer by layer using a high-power laser to melt metal powders. A variety of materials can be used, such as stainless steel, nickel-base superalloy inconel, copper and aluminum alloys, and even reactive materials such as titanium alloys. The inclusive material options of this technology allow the production of a model part with its design alloy. This is critical for part functional testing in reverse engineering.

Based on similar principles, other additive prototyping technologies, such as the polyjet process that utilizes multiple inkjets and ultraviolet light-curable material to add up very thin layers for part building, are also available for reverse engineering applications. Recently, the small desktop three-dimensional printer was introduced for model part creation in an office environment. Laminated object manufacturing is a unique, though not as widely used as other methods, additive prototyping process. It first uses a laser to profile the cross sections of a part on paper coated with polyethylene, and then cuts the model part from the consolidated stack of paper.

(a)

(b)

**FIGURE 2.21**
(a) Dimension 1200es series 3D printer. (b) 3D printer compartment.

### 2.5.2 Subtractive Prototyping Processes

In contrast to most additive prototyping technologies, which were only first commercialized in the 1980s, many subtractive prototyping processes evolved from traditional manufacturing processes updated with modern information technologies, such as computer numerically controlled (CNC) machining, and injection modeling of thermoplastic resins.

The CNC is widely used in part manufacturing, and can also be used to produce a prototype model part in reverse engineering whenever appropriate. It is an automatic machining process with programmed software in command, as opposed to manual operation. The modern CNC machine can directly input instructions from the CAD/CAM data file that is compiled during prior scanning and modeling in reverse engineering. In fact, the interaction between CNC and CAD emerged at the inception of both technologies. In the 1950s, when MIT engineers were studying the CNC technology, they also integrated the electronic systems and mechanical engineering design, and started MIT's Computer-Aided Design Project at the same time. At the discretion of the engineer, a CNC model part can be machined from a solid plastic or from the actual design alloy. A CNC machined part usually shows more homogeneous properties, stronger mechanical strength, and better surface finish than a counterpart made from any additive process. The CNC process has the ability to produce a real-life first article part with all the specified design characteristics suitable for all required fit, form, and functional tests. However, the CNC process is relatively expensive, particularly when just setting up for one or two model parts. It is also subject to the common restrictions for a typical machining process, such as limitation on geometry complexity.

### 2.5.3 Rapid Injection Molding

Rapid injection modeling satisfies all four basic requirements of prototyping in reverse engineering: quality product, limited quantity, rapid production, and reasonable costs. However, the initial costs associated with the tooling and mold fabrication often make rapid injection molding an intermittent step between the first model part produced by one of the aforementioned additive or subtractive prototyping methods and mass production. Rapid injection modeling is particularly effective for simple parts made of common thermoplastics materials such as ABS, polycarbonate, or nylon. The mold used for rapid injection molding is usually made of aluminum alloy, as opposed to steel for a production mold. Rapid prototyping molding is usually done following the standards guiding production molding, by injecting the part material into a mold. Most of the materials used for rapid injection molding in reverse engineering are thermoplastic resins. However, there are very few material restrictions for this process; a variety of engineering materials, including resins, ceramics, and metals, can be used for injection

molding. The parts modeled with rapid injection modeling are strong and, with good surface finishing, suitable for most fit, form, and functional tests. If the part is molded with the design material, the molded part is virtually a first article of the production part that not only provides a sample for performance evaluation, but also provides invaluable information on manufacturability for future production.

## 2.6 Steps of Geometric Modeling

The following exercise exemplifies a five-step process of geometric modeling practiced in reverse engineering industries:

1. Define the scope of work. The reverse engineering process begins with defining the project scope and identifying the key requirements. Once defined, appropriate methods will be utilized to obtain the relevant data of the part, such as the part geometry.
2. Obtain dimensional data. Step 2 utilizes dimensional metrology equipment to obtain all the relevant dimensional data necessary to create a design drawing or CAD model of the part. The use of digitizing or scanning may be needed. The dimensions of the part can be measured by various instruments: (a) noncontact measurement, (b) coordinate measuring machine (CMM) with contact probe, or (c) portable CMM. The 3D laser scanning is one of the most comprehensive, direct ways to reproduce complex geometries accurately. The capability of measuring hardware has been dramatically enhanced with advanced software. Though developed with different principles and often with specific strengths and shortcomings, most reverse engineering software packages are designed with comprehensive application capabilities. Table 2.1 lists some commonly used

**TABLE 2.1**

Software

| Parametric Modeling | | NURBS Modeling | | Analysis | |
|---|---|---|---|---|---|
| **Publisher** | **Software** | **Publisher** | **Software** | **Publisher** | **Software** |
| INUS Technology | RapidformXO | Innovmetric | Polyworks Modeler | Innovmetric | Polyworks Inspector™ |
| Dassault Systemes | SolidWorks | Raindrop | Geomagic Studio | Raindrop | Geomagic Qualify |
| Autodesk | Inventor | | | | |

software. RapidformXO published by INUS Technology, SolidWorks by Dassault Systemes, and Inventor by Autodesk are three widely used software packages for parametric modeling. Polyworks Modeler™ by Innovmetric, and Geomagic Studio by Raindrop are two popular software packages for NURBS modeling. Polyworks from Innovmetric and Geomagic Quality from Raindrop are used by many engineers for inspection and analysis. Nonetheless, their applications are frequently cross-referenced in both fields of modeling and analysis.

3. Analyze data. This step formulates the nominal dimensions of the part based on the measured data. It sets the CAD analytical model with the integration of industry standards and customer specifications to ensure the fit, form, and function requirements.
4. Create the CAD model. A 3D model in a suitable CAD package with the nominal dimensions is generated following a best-fit line, arc, or spline adjustment. Best practices are utilized when creating models, along with the customers' corporate standards when applicable.
5. Verify the quality. A real-life part can be scanned to verify the analytical CAD model. By comparing the point cloud data (gathered from scanning the part) with the CAD model, a comparative deviation map, usually color coded, can be generated. If any deviations are identified, the CAD model can be adjusted accordingly until the part is modeled accurately. The first article can also be effectively inspected by comparing a full scan of it to the referenced CAD geometry.

In summary, precision measurement devices, advanced software, and modern reverse engineering technologies have made the reinvention of mechanical parts feasible with tight tolerance and high fidelity.

## References

Cleveland, B. 2009. Prototyping process overview. *Adv. Mater. Processes* 167:21–23.

Chivate, P., and A. Jablokow. 1993. Solid-model generation from measured point data. *Comput. Aided Design* 25:587–600.

3DScanCo/GKS Global Services. 2009a. Case studies: ARINC in the World Wide Web. http://www.3dscanco.com/clients/case-studies/arinc.cfm (accessed October 1, 2009).

3DScanCo/GKS Global Services. 2009b. Case studies: Airbus A319 in the World Wide Web. http://www.3dscanco.com/clients/case-studies/computational-methods.cfm (accessed October 1, 2009).

3DScanCo/GKS Global Services. 2009c. Case studies: The Southern 408 in the World Wide Web. http://www.3dscanco.com/clients/case-studies/panoz-automotive.cfm (accessed October 1, 2009).

Weir, D., Milroy, M., Bradley, C., and Vickers, G. 1996. Reverse engineering physical models employing wrap-around B-spline surfaces and quadrics. *Proc. Inst. Mech. Eng. B.* 210:147–57.

# 3

# *Material Characteristics and Analysis*

Material characteristics are the cornerstone for material identification and performance evaluation of a part made using reverse engineering. One of the most frequently asked questions in reverse engineering is what material characteristics should be evaluated to ensure the equivalency of two materials. Theoretically speaking, we can claim two materials are "the same" only when all their characteristics have been compared and found equivalent. This can be prohibitively expensive, and might be technically impossible. In engineering practice, when sufficient data have demonstrated that both the materials having equivalent values of relevant characteristics will usually deem having met the requirements with acceptable risk. The determination of relevant material characteristics and their equivalency requires a comprehensive understanding of the material and the functionality of the part that was made of this material. To convincingly argue which properties, ultimate tensile strength, fatigue strength, creep resistance, or fracture toughness, are relevant material properties that need to be evaluated in a reverse engineering project, the engineer needs at least to provide the following elaboration:

1. Property criticality: Explain how critical this relevant property is to the part's design functionality.
2. Risk assessment: Explain how this relevant property will affect the part performance, and what will be the potential consequence if this material property fails to meet the design value.
3. Performance assurance: Explain what tests are required to show the equivalency to the original material.

The primary objective of this chapter is to discuss the material characteristics with a focus on mechanical metallurgy applicable in reverse engineering to help readers accomplish these tasks.

The mechanical, metallurgical, and physical properties are the most relevant material properties to reverse engineer a mechanical part. The mechanical properties are associated with the elastic and plastic reactions that occur when force is applied. The primary mechanical properties include ultimate tensile strength, yield strength, ductility, fatigue endurance, creep resistance, and stress rupture strength. They usually reflect the relationship between stress and strain. Many mechanical properties are closely related to the metallurgical and physical properties.

The metallurgical properties refer to the physical and chemical characteristics of metallic elements and alloys, such as the alloy microstructure and chemical composition. These characteristics are closely related to the thermodynamic and kinetic processes, and chemical reactions usually occur during these processes. The principles of thermodynamics determine whether a constituent phase in an alloy will ever be formulated from two elements when they are mixed together. The kinetic process determines how quickly this constituent phase can be formulated. The principles of thermodynamics are used to establish the equilibrium phase diagram that helps engineers to design new alloys and interpret many metallurgical properties and reactions. It takes a very long time to reach the equilibrium condition. Therefore, most grain morphologies and alloy structures depend on a kinetic process that determines reaction rate, such as grain growth rate.

Heat treatment is a process that is widely used to obtain the optimal mechanical properties through metallurgical reactions. It is a combination of heating and cooling operations applied to solid metallic materials to obtain proper microstructure morphology, and therefore desired properties. The most commonly applied heat treatment processes include annealing, solution heat treatment, and aging treatment. Annealing is a process consisting of heating to and holding at a specified temperature for a period of time, and then slowly cooling down at a specific rate. It is used primarily to soften the metals to improve machinability, workability, and mechanical ductility. Proper annealing will also increase the stability of part dimensions. The most frequently utilized annealing processes are full annealing, process annealing, isothermal annealing, and spheroidizing. When the only purpose of annealing is for the relief of stress, the annealing process is usually referred to as stress relieving. It reduces the internal residual stresses in a part induced by casting, quenching, normalizing, machining, cold working, or welding. Solution heat treatment only applies to alloys, but not pure metals. In this process an alloy is heated to above a specific temperature and held at this temperature for a sufficiently long period of time to allow a constituent element to dissolve into the solid solution, followed by rapid cooling to keep the constituent element in solution. Consequently, this process produces a supersaturated, thermodynamically unstable state when the alloy is cooled down to a lower temperature because the solubility of the constituent element decreases with temperature. The solution heat treatment is often followed by a subsequent age treatment for precipitation hardening. From the heat treatment perspective, aging describes a time-temperature-dependent change in the properties of certain alloys. It is a result of precipitation from a supersaturated solid solution. Age hardening is one of the most important strengthening mechanisms for precipitation-hardenable aluminum alloys and nickel-base superalloys.

Physical properties usually refer to the inherent characteristics of a material. They are independent of the chemical, metallurgical, and mechanical processes, such as the density, melting temperature, heat transfer coefficient, specific heat, and electrical conductivity. These properties are usually

measured without applying any mechanical force to the material. These properties are crucial in many engineering applications. For example, the specific tensile strength (strength per unit weight) directly depends on alloy density, and it is more important than the absolute tensile strength when engineers design the aircraft and automobile. However, most material characteristics do not stand alone. They will either affect or be affected by other properties. As a result, some material properties fall into both mechanical and physical property categories, depending on their functionality, such as Young's modulus and shear modulus. An accurate Young's modulus is usually measured by an ultrasonic technology without applying any mechanical force to the material. However, Young's modulus is also commonly referred as a ratio between the stress and strain, and they are the key elements in mechanical property evaluation. The interrelationships between metallurgical and mechanical behaviors also cause some material properties to fall into both categories, such as hardness and stress corrosion cracking resistance can be referred to as either metallurgical or mechanical properties.

## 3.1 Alloy Structure Equivalency

### 3.1.1 Structure of Engineering Alloys

Engineering alloys are metallic substances for engineering applications, and have been widely used in many industries for centuries. For example, the utilization of aluminum alloys in the aviation industry started from the beginning and continues to today; the crankcase of the Wright brothers' airplane was made of cast aluminum alloy in 1903. Alloys are composed of two or more elements that possess properties different from those of their constituents. When they are cooled from the liquid state into the solid state, most alloys will form a crystalline structure, but others will solidify without crystallization to stay amorphous, like glass. The amorphous structure of metallic glass is a random layout of alloying elements. In contrast, a crystalline structure has a repetitive pattern based on the alloying elements. For instance, the crystalline structure of an aluminum–4% copper alloy is based on the crystal structure of aluminum with copper atoms blended in. The measurable properties of an alloy such as hardness are part of its apparent character, and the underneath crystallographic structure is its distinctive generic structure. Both play their respective critical roles in alloy identification in reverse engineering.

Pure metallic elements, for example, aluminum, copper, or iron, usually have atoms that fit in a few symmetric patterns. The smallest repetitive unit of this atomic pattern is the unit cell. A single crystal is an aggregate of these unit cells that have the same orientation and no grain boundary. It is essentially a single giant grain with an orderly array of atoms. This unique

crystallographic structure gives a single crystal exceptional mechanical strength, and special applications. The single-crystal Ni-base superalloy has been developed for turbine blades and vanes in modern aircraft engines. The first single-crystal-bladed aircraft engine was the Pratt & Whitney JT9D-7R4, which received FAA certification in 1982. It powers many aircraft, such as the Boeing 767 and the Airbus A310. Compared to the counterpart with equiaxed grains, a single-crystal jet engine turbine airfoil can have multiple times better corrosion resistance, and much better creep strength and thermal fatigue resistance. Most engineering alloys, however, have a multigrain morphology. The grain size and its texture have profound effects on alloy properties. Fine grain engineering alloys usually have higher tensile strength at ambient temperature. However, for high-temperature applications, coarse grain alloys are preferred due to their better creep resistance. The effects of microstructure on the properties of engineering alloys will be discussed in detail later.

### 3.1.2 Effects of Process and Product Form on Material Equivalency

The part features, distinctive microstructure in particular, resulted from different manufacturing processes, and product forms thereby produced from raw materials are the characteristics widely used to identify material equivalency in reverse engineering. Conventional manufacturing processes used on engineering alloys to produce a specific product form include casting, forging, and rolling, as well as other hot and cold work. Power metallurgy, rapid solidification, chemical vapor deposition, and many other special processes, for example, Osprey spray forming and superplastic forming, are also used in industries for specific applications. Some near-net-shape processes directly shape the alloy into the near-final product form or complex geometry. In comparison to traditional cast and wrought products with multiple processing steps, a simpler conversion from raw material to the final product that involves fewer steps is often more desirable. For example, the Osprey spray forming process first atomizes a molten alloy, which is then sprayed onto a rotating mendrel to form a ring-shape preform hardware like an engine turbine case or seal. The near-net-shape preform is subsequently made into the final product using a hot isostatic press. An Osprey spray-formed Ni-base superalloy product is more cost effective, and typically has an average grain size of about 65 μm. It shows a similar microstructure and comparable properties to a wrought piece with the same alloy composition, and has better properties than a cast product. Recent advances in manufacturing technologies have also produced alloys with nano-microstructure.

The mechanical properties of engineering alloys are primarily determined by two factors: composition and microstructure. Though the alloy composition is intrinsic by design, the microstructure evolves during manufacturing. The microstructure and consequently the mechanical properties of an engineering alloy can be drastically different in different product forms. Figure 3.1a shows the equiaxed grain morphology of aluminum alloy casting;

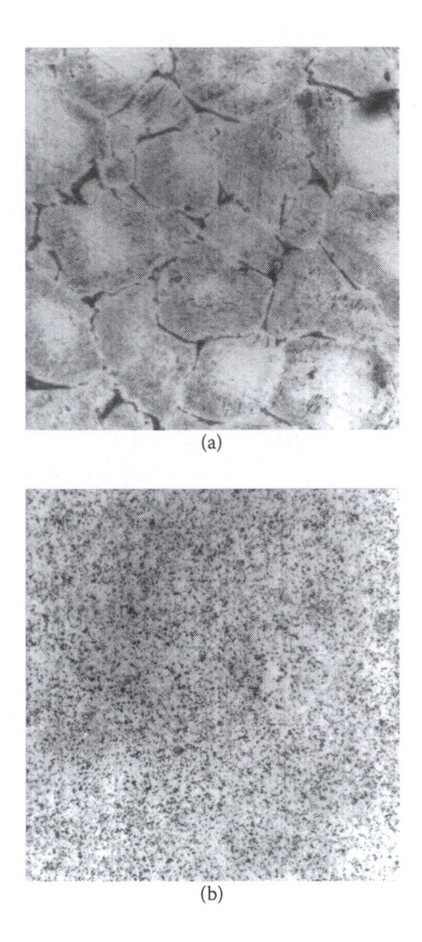

**FIGURE 3.1**
(a) Microstructures of aluminum alloy casting. (b) Microstructure of aluminum alloy extrusion.

it is vastly different from the microstructure observed in aluminum alloy extrusion, as shown in Figure 3.1b, despite that both have the identical alloy composition of Al–3.78% Cu–1.63% Li–1.40% Mg. Needless to say, cast aluminum and extruded aluminum pats have very different properties as well. In reverse engineering, the microstructure provides invaluable information to retrace the part manufacturing process.

## 3.2 Phase Formation and Identification

The phase diagram is established based on the phase transformation process. It illustrates the relationship among alloy composition, phase, and

temperature. It provides a reference guide for various manufacturing and heat treatment processes. The information that can be extracted from a phase diagram plays a key role in phase identification, and therefore is crucial for manufacturing process and heat treatment verification in reverse engineering. This section will discuss the fundamentals of phase diagrams and the related theories of thermodynamics and kinetics.

### 3.2.1 Phase Diagram

An alloy phase diagram is a metallurgical illustration that shows the melting and solidification temperatures as well as the different phases of an alloy at a specific temperature. The equilibrium phase diagram shows the equilibrium phases as a function of composition and temperature; presumptively, the kinetic reaction processes are fast enough to reach the equilibrium condition at each step. All phase diagrams hereafter in this book are referred to as equilibrium phase diagrams unless otherwise specified. Figure 3.2 is a schematic of partial iron-carbon (Fe-C) phase diagram, not complete because of the complexity of this diagram. The *y*-axis of a phase diagram is temperature, and the *x*-axis represents the alloying element composition. In the binary Fe-C phase diagram, iron is the master base element, and carbon is the alloying element. The far left *y*-axis represents pure iron, that is, 100% Fe.

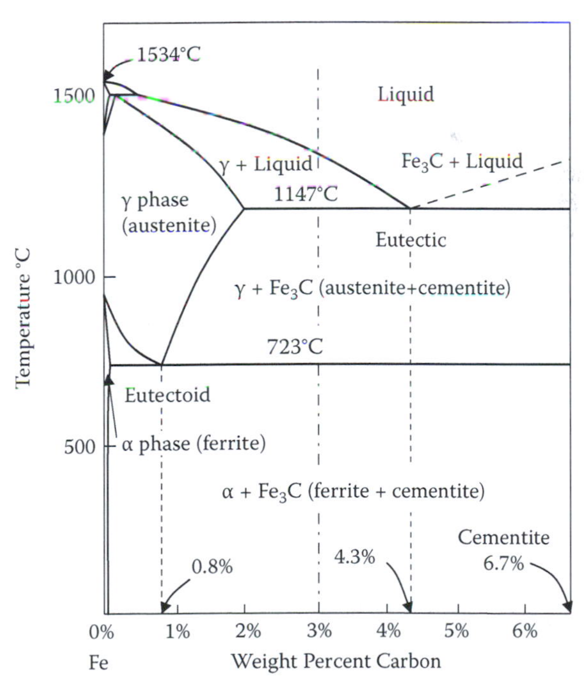


**FIGURE 3.2**
Schematic of partial Fe-C phase diagram.

The amount of carbon increases from left to right. The units for the alloying element are usually in weight percentage, but occasionally atomic percentage is used. Sometimes both are shown with one percentage scale marked at the bottom and the other on the top. In Figure 3.2 the carbon contents range from 0 to 6.7%. Fe–6.7% C is an intermetallic compound, cementite. The constituent elements in an alloy might combine into a distinct compound with a fixed or narrow composition range. These compounds are intermetallic. Most intermetallic compounds have their own identities with specific compositions and distinctive crystal structures and properties. Cementite, $Fe_3C$, is the most recognizable intermetallic compound in the iron-carbon ferrous alloys. Most Fe-C binary phase diagrams are partially presented as a Fe-$Fe_3C$ phase diagram using $Fe_3C$ instead of 100% pure carbon as another baseline.

The phase rule of J. Willard Gibbs and the laws of thermodynamics guide metallic phase transformations. The Gibbs' phase rule can be mathematically described by Equation 3.1, where $F$ is the degrees of freedom or number of independent variables, $C$ is the number of components, and $P$ is the number of phases in a thermodynamic equilibrium system:

$$F = C - P + 2 \tag{3.1}$$

The typical independent variables are temperature and pressure. Most phase diagrams assume atmospheric pressure. When an alloy melts, both solid and liquid coexist, and therefore $P = 2$. Gibbs' phase rule then only allows for one independent variable in a pure metal where $C = 1$. This explains that at atmospheric pressure, pure metal melts at a specific melting temperature and boils at a fixed boiling point. For example, the melting temperature of pure iron is marked as 1,534°C in Figure 3.2. However, for binary alloys, where $C = 2$, the Gibbs' phase rule allows one more independent variable. For a given composition, the binary alloy has the liquid–solid phase transformation extended over a range of temperatures with a coexisting liquid and solid mixture, instead of at a fixed temperature. As shown in Figure 3.2, at 1,400°C, the Fe–3% C binary alloy is in a homogeneous liquid state. It will start to solidify when the temperature decreases below the liquidus around 1,300°C. The liquidus is the temperature boundary in a phase diagram where the liquid starts to solidify. In other words, the liquidus is the locus of the starting melting temperatures of the alloys at various compositions. In Figure 3.2, it is the curve that starts at 1,534°C where pure liquid iron melts, and continues to 1,147°C, the melting temperature of Fe–4.3% C. The Fe–4.3% C is defined as eutectic composition. Despite that it is a binary alloy, it melts at a fixed temperature, 1,147°C, because two different solid phases solidified simultaneously. The eutectic temperature is the lowest melting temperature of iron-carbon alloys, as shown in Figure 3.2. The locus of the completion temperature of solidification is defined as solidus. Above the liquidus the alloy is in a homogeneous liquid state, which is commonly referred to as the liquid phase or liquid solution and labeled L in many phase diagrams. Below

solidus the alloy is in a homogeneous solid state, which is often referred to as solid phase or solid solution. The various solid phases of an alloy are usually designated with Greek letters, starting with α from the left and usually continuing as β, χ, δ, ε, φ, γ, and η phases, as one moves to the right across the phase diagram. In a Fe–3 %C alloy, the newly formed solid γ phase and the remaining liquid will coexist between the liquidus and the solidus of 1,147°C. The Fe–3% C alloy will be sequentially transformed into various solid phases, from γ to α mixed with $Fe_3C$, as the temperature continuously decreases. Thus, a molten alloy will solidify from a homogeneous liquid state into a multiphase solid state; each forms at consecutive steps during the solidification process. Quantitative analysis of each phase at a particular temperature can be conducted based on the Lever rule. The phase diagram illustrates these phase transformations, and provides invaluable "footprints" allowing engineers to retrace the process the original part experienced in reverse engineering.

The principles of thermodynamics can theoretically predict the existence of a phase in an equilibrium phase diagram. However, it might take infinite time to accomplish the phase transformation. The rate and mechanism of forming this phase are guided by the principles of kinetics, which also explain the many nonequilibrium phase transformations. A variety of nonequilibrium phase transformation diagrams are used for many engineering applications where the temperature change rate is intentionally controlled to create specific nonequilibrium phases. One example is the continuing cooling curves of ferrous alloys that are widely used in the heat treatment industry. From a reverse engineering perspective, these continuing cooling curves often provide more practical information than the equilibrium phase diagrams.

Most engineering alloys contain more than two alloying elements. If there are three constituent elements, it is called a ternary system. The ternary phase diagram is a three-dimensional space prism where the temperature axis is vertically built on top of the composition triangle base plane, with each side representing one element. It is a space phase diagram with three binary phase diagrams, one on each side.

### 3.2.2 Grain Morphology Equivalency

The three most commonly observed grain morphologies of metal microstructure are equiaxed, columnar mixed with dendritic casting structure, and single crystal. In the equiaxed microstructure, shown in Figure 3.1a, one grain has roughly equivalent dimensions in all axial directions. The columnar structure usually appears in castings when the solidification process starts from a chilly mold surface and gradually moves inward to form a coarse columnar grain morphology. The columnar structure is usually mixed with a dendritic casting structure in the end. The single crystal has no adjacent grains and no grain boundaries; the entire crystal aligns in one crystallographic direction. However, these basic grain morphologies will evolve into

**FIGURE 3.3**
Microstructure of a tungsten wire with elongated grain morphology.

more complex configurations through kinetic processes, for example, recrystallization and grain growth. Other derivative microstructures are the direct products of specific processes. For instance, cold or hot drawing can produce highly directionally textured microstructure with all the grains lined up in one direction. Figure 3.3 shows the textured microstructure with high directionality of a tungsten wire. In reverse engineering it is crucial that the replicated part have grain morphology equivalent to that of the original part for the following two reasons. First and foremost, the material properties and part functional performance heavily depend on the microstructure. Second, the grain morphology provides critical information on the manufacturing process and heat treatment schedule. Parts showing different grain morphologies are made by different manufacturing processes with different heat treatments, and have different mechanical properties.

### 3.2.3 Recrystallization, Secondary Recrystallization, and Recovery

The microstructure of deformed grains could evolve into a new morphology at above a critical temperature. It is a nucleation and growth process after the metal has been cold worked. During this kinetic process, a metal goes through a subtle microstructure evolution, which can refine coarse grains and release residual stress from prior strain. This metallurgical phenomenon is called recrystallization. A minimum critical amount of cold work is required to recrystallize metals within a reasonable time period. This required minimum cold work varies with the type of deformation, that is, tension, compression, torsion, rolling, etc. For instance, torsion can promote the recrystallization process at a relatively small amount of deformation. A metal subject to a larger amount of deformation usually recrystallizes faster than a metal that is less deformed.

The kinetic process of recrystallization can be quantitatively described by Equation 3.2. In the experimental study of recrystallization, the time is usually measured at 50% completion of recrystallization. However, the time, $\tau$, in Equation 3.2 is not restricted to represent 50% completion only; it might represent the time of full recrystallization. The recrystallization activation energy is a collective energy barrier that needs to be overcome during this process.

$$1/\tau = Ce^{-Q/RT} \tag{3.2}$$

where $\tau$ = time (usually for half recrystallization), $C$ = empirical constant, $Q$ = activation energy for recrystallization, $R$ = universal gas constant $\approx$ 2 cal/(mol K), and $T$ = absolute temperature.

The temperature at which a cold-deformed metal can be completely recrystallized in a finite period of time, usually 1 hour, is defined as the recrystallization temperature. In engineering practice it is usually acceptable to define a specific recrystallization temperature for a metal or alloy with the understanding that it is only a term for convenience. This temperature is a function of the amount of deformation, and strictly speaking, it is not an intrinsic material property. However, recrystallization is a kinetic process with a very large activation energy, $Q$. As a result, the recrystallization process is very sensitive to the annealing temperature that affects the recrystallization rate exponentially, as shown in Equation 3.2. A small change in temperature could significantly shorten or delay the recrystallization process. It therefore appears that each metal has a low limit for the recrystallization temperature, below which the recrystallization process will stall to a virtual stop. It is also worth noting that though the recrystallization temperature is practically fixed for a pure metal, it can be significantly raised up to several hundred degrees by a very small amount of impurity, as little as 0.01 atomic percent.

One of the benefits of recrystallization is grain refinement, which is dependent on the ratio of nucleus number to the grain growth rate. The higher this ratio, the finer the final grain will be because more nuclei grow slowly and compete with one another for the limited space. The smaller the grains before cold work, the greater the rate of nucleation will be and the smaller the subsequent recrystallized grain will emerge for a given degree of deformation. Occasionally, isolated coarse grains can be accidentally introduced during recrystallization. This is a result of inhomogeneous deformation throughout the alloy matrix. If a metallic object is deformed unevenly, a region containing a critical amount of cold work might exist between the worked and unworked areas. Annealing in this case can lead to a localized, very coarse grain due to recrystallization.

As discussed earlier, recrystallization is a kinetic process of nucleation and growth. It depends on alloy composition, impurity content, annealing time, prior grain size, and the complexity of deformation that initiates it. However,

the growth of newly recrystallized grains can be inhibited due to the interference of inclusions or other crystalline defects. A secondary recrystallization might occur when the annealing temperature of primary recrystallization is raised in this circumstance. In contrast to the primary recrystallization, the driving force for secondary recrystallization is surface tension, instead of strain energy. The surface-tension-induced grain boundary moves toward the curvature center. As a result, small grains with their grain boundaries concave inward will be coalesced into the neighboring large grains that have grain boundaries concave outward. This is a grain coalescence process to satisfy surface tension considerations. A grain from secondary recrystallization usually has a large grain size and multiple concave outward sides, as shown in Figure 3.4 (Reed-Hill, 1973).

Recovery is another metallurgical phenomenon observed in cold worked metals. However, it is a process very different from recrystallization. In isothermal annealing, the recovery process rate always decreases with time as the driving force, stored strain energy, gradually dissipates. On the other hand, recrystallization occurs by a nucleation and growth process; it starts very slowly and gradually builds up to the maximum reaction rate. It then finishes slowly until the entire matrix is recrystallized. The rate profile of recrystallization is like a bell curve, while the profile of recovery resembles only the second half of this curve, starting from the peak and sliding down. The analysis of recrystallization, secondary recrystallization, or recovery can provide a lot of information relative to the thermal treatment the material has experienced in reverse engineering.

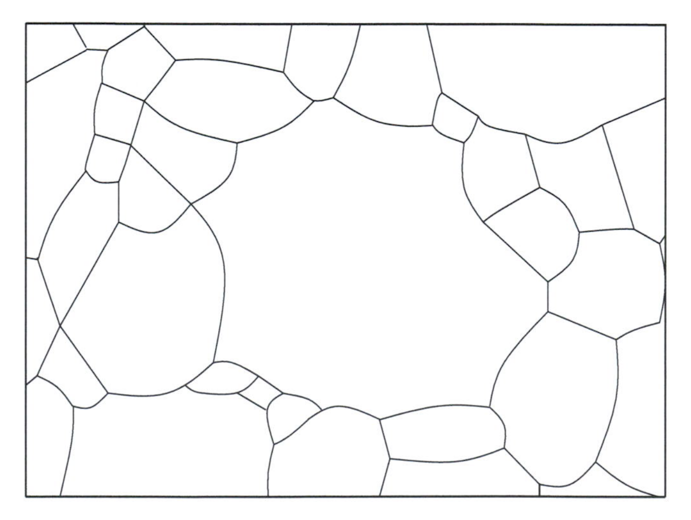

**FIGURE 3.4**
Schematic of secondary recrystallization.

### 3.2.4 Grain Size and Grain Growth

The average grain size of a polycrystalline metallic or ceramic material refers to the two-dimensional mean diameter of an aggregate of grains. The grain size measurement and comparison are indispensable for material identification in reverse engineering due to the profound effects of grain size on mechanical properties. However, the determination of three-dimensional grain size from its two-dimension sections at best only provides a statistical average measurement of the grain size. The ASTM (ASTM, 2004) defines the average grain size number in the exponential form described by Equation 3.3. A smaller grain size number represents a coarser microstructure. The microstructures with ASTM grain size numbers 1 and 2 will have the nominal grain diameters of 250 and 180 μm, respectively. ISO 643 (ISO, 2003) also provides a definition of grain size number, which is slightly smaller than the same ASTM grain size number. Since both ASTM and ISO standards are based on statistical average grain sizes, this small difference is insignificant in practice.

$$n = 2^{N-1} \tag{3.3}$$

where $n$ = the number of grains per square inch, as seen in a specimen viewed at a magnification of 100×, and $N$ = the ASTM grain size number.

The ASTM Standard E112, "Standard Test Methods for Determining Average Grain Size," provides detailed guidance on grain size measurement (ASTM, 2004). In reverse engineering, two sets of grain morphology can be visually compared to determine if they are equivalent in overall appearance. The comparison can also be quantitatively conducted by one of the following two methods. The number of grains per unit area can be actually counted by a two-dimensional planimetric procedure, also known as Jeffries' method, and then converted to the ASTM grain size number. Or, the average grain size can be estimated by a linear intercept method, known as the Heyn method. This method counts the total number of intercepts between the grain boundaries and the random test lines, and then calculates the average grain size.

Grain grows under proper metallurgical conditions. An empirical formula for isothermal grain growth is described by Equation 3.4:

$$D = kt^{n} \tag{3.4}$$

where $D$ = grain size; $k$ = material parameter, a function of temperature; $t$ = time; and $n \approx 1/2$ is the grain growth exponent.

The exponent $n$ is a function of temperature and impurity. It is usually equal to ½ or less. It increases with increasing temperature and decreases in value when impurities are present. The presence of foreign atoms or second phase inclusions could retard the movement of the grain boundary, and slow down the regular grain growth, therefore putting an upper limit on the grain size.

## 3.3 Mechanical Strength

In reverse engineering the mechanical strength and hardness number are the most verifiable characteristics to demonstrate material equivalency in comparative analysis. The broad definition of mechanical strength is the material's capability to resist mechanical failure. This ability of the material to resist mechanical load can be categorized by the maximum atomic bonding force it can sustain before separation, such as tensile strength, or the maximum plastic deformation before fracture, such as tensile elongation. It can also categorized by the energy it absorbs before fracture, which is defined as toughness. Which mechanical strength properties should be evaluated in reverse engineering are usually project specific. The following discussions will focus on their respective characteristics to provide the readers with necessary information to make educated determinations. We will first discuss the mechanical strength determined by the atomic bonding force of an engineering alloy when it is subject to externally applied stress. A component can be subject to a variety of mechanical stresses: tensile, compressive, shear, bending, or cyclic stress. The conventional ultimate tensile and yield strengths are the most widely used parameters to characterize the mechanical properties of engineering alloys that are usually assumed to be isotropic and homogeneous. However, the mechanical behavior of material is sophisticated, and there is a lot of science behind the average ultimate tensile and yield strengths. For example, the single-crystal equivalency of yield strength is the critical resolved shear stress, which is highly anisotropic, particularly for the less symmetric crystal structures, such as hexagonal close-packed (HCP). It is also worth noting that the nominal tensile strength is often less than 1% of the theoretical cohesive strength due to crystal defects, preexisting cracks, or stress concentration. Most tensile strength data are based on smooth tensile specimen test results. The existence of a sharp notch in a tensile specimen converts a simple uniaxial tensile load to a complicated triaxiality of stress, and can significantly weaken the tensile strength of brittle alloys. This section will only discuss the basic principles and mathematical equations for tensile properties and hardness for reverse engineering applications. It is essential to make sure all the data are obtained under the same test conditions prior to any direct comparison of tensile properties.

### 3.3.1 Classic Mechanics

In a typical engineering design, a mechanical component is subject to a stress below the elastic limit. Therefore, Hook's law applies in most engineering analyses and the material will linearly extend along the direction where an external load is applied. The material will elastically recover to the original dimensions after the removal of the applied load. When the applied stress is beyond the elastic limit, a permanent plastic deformation will remain.

The average strain is often referred to as an engineering or normal strain. Consider a uniform tensile test specimen that is subject to an axial static tensile load and extended to the final length $L_f$; the engineering strain, $\varepsilon_e$, is defined by Equation 3.5 as the ratio of the change in length, $\Delta L$, to the original length, $L_o$:

$$\varepsilon_e = \frac{\Delta L}{L_o} = \frac{L_f - L_o}{L_o} \tag{3.5}$$

The nominal normal stress in the axial direction, $\sigma_e$, also referred to as the engineering stress, is defined by Equation 3.6:

$$\sigma_e = \frac{P}{A_o} \tag{3.6}$$

where $A_o$ is the original cross-sectional area and $P$ is the total load. In the International System of Units (SI/Systeme International), the unit for stress is Newton per square meter, $N/m^2$, or pascal, Pa. The stress is expressed as pound per square inch, psi, or 1,000 pounds per square inch, ksi, in the U.S. customary system. One pascal only represents a very small stress, 1 Pa = 0.000145 psi. The SI stress is therefore usually expressed in MPa = $10^6$ $N/m^2$ = 145 psi. The strain is dimensionless.

To the first degree of approximation, the engineering strain is linearly proportional to the engineering stress following Hook's law when the stress is small. The proportionality constant is Young's modulus, or the modulus of elasticity, $E$, as defined by Equation 3.7. The value of Young's modulus is relatively independent of the manufacturing process, but it heavily depends on alloy composition. Both cast and extruded 2024 aluminum alloy (with a nominal composition of Al–4.5% Cu–1.5% Mg) show similar Young's modulus values. However, the addition of 1% lithium to the 2024 aluminum alloy can increase its Young's modulus by 8 to 9%. In reverse engineering, a comparison between the values of two Young's moduli can be used as a barometer to verify the equivalency of some part characteristics, such as the elastic instability of a slender column due to buckling. However, most of the mechanical properties, such as yield strength, usually are a function of the manufacturing process and independent of Young's modulus.

$$\frac{\sigma_e}{\varepsilon_e} = E \tag{3.7}$$

A typical engineering stress-strain curve in tension for ductile metals is illustrated in Figure 3.5. A proportional limit exists just below the elastic limit. The linearity between stress and strain as stated in Hook's law starts to deviate beyond the proportional limit. When the stress reaches a critical

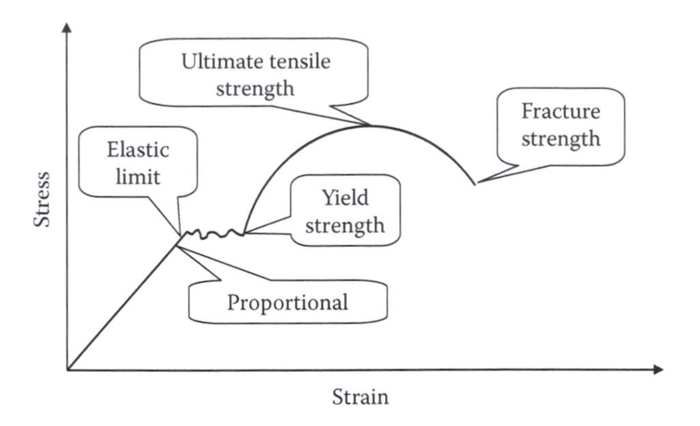


**FIGURE 3.5**
Schematic of an engineering stress-strain curve.

value, the material becomes unstable and continues to yield with permanent deformation at the same level of stress. A distinct yield point sometimes does not exist, particularly for brittle materials. In contrast to ultimate tensile strength that is well defined universally, yield strength has more than one definition. For engineering purposes, the yield strength is usually defined as the stress that will produce a small amount of permanent deformation, for example, 0.2%, the so-called 0.2% offset yield strength. In reverse engineering applications, the engineer should verify that the same definition of yield strength applies to all data before the comparison. A higher stress is required to further deform the alloy beyond the yielding point due to strain hardening, until the maximum stress is reached. The ratio of the maximum load and the original cross-sectional area is defined as ultimate tensile strength. The cross-sectional area of a ductile alloy usually begins to decrease rapidly beyond the maximum load. As a result, the total load required to further deform the specimen is decreased until the specimen fails at the fracture stress, as shown in Figure 3.5.

The yield strength depends on material composition as well as its microstructure. The grain size has a profound effect on yield strength. Equation 3.8 is the mathematical formula of the Hall-Petch equation. It is an empirical relationship between yield strength and grain size and is based on the pioneering work of Eric Hall (1951) and Norman Petch (1953). This is a functional relationship applicable to most polycrystalline alloys with grain size ranging from 1 mm to 1 μm. When the grain size is in this range, the impediment of dislocation movement is the determining factor of yield strength. The smaller the grain size is, the more grain boundaries there will be, and the more difficult dislocations can move from one grain to another—therefore, the material is stronger.

$$\sigma_y = \sigma_o + kd^{-1/2} \tag{3.8}$$

where $\sigma_y$ is the yield strength, $\sigma_o$ and $k$ are constant material parameters, and $d$ is the average grain size.

In contrast to engineering stress, the true stress actually imposed on to a tensile specimen increases continuously as the true cross-sectional area, $A$, shrinks during the test. The true stress, $\sigma_t$, is defined by Equation 3.9 as force per unit true area at that instant, where $P$ and $A$ are force and area, respectively:

$$\sigma_t = \frac{P}{A} \tag{3.9}$$

Similar to the true stress, the true or natural strain, $\varepsilon_t$, at a point is a local strain calculated against the actual length at the point of interest and at that instant. It is mathematically defined by Equation 3.10:

$$\varepsilon_t = \int_{L_o}^{L_f} \frac{dL}{L} = \ln \frac{L_f}{L_o} \tag{3.10}$$

where $L$ is specimen length at the moment, $L_f$ is final specimen length, and $L_o$ is original specimen length.

For very small elastic strains, the true and engineering strains are virtually the same value. However, the true strain more truthfully reflects the large plastic deformation. For instance, the engineering tension strain is $\varepsilon_e = 2L/L = 100\%$, while the true tension strain is $\varepsilon_t = \ln(2L/L) = 69.3\%$ when a specimen doubles its original length. A 69.3% true compressive strain $[\varepsilon_t = \ln(0.5L/L) = -69.3\%]$ implies a reduction in length by half that represents an opposite but similar deformation in tension at a 69.3% strain. In other words, in terms of true strain, 69.3% in tension means doubling the length, and 69.3% in compression means reducing the length by half. However, in terms of engineering strain, 100% in tension means doubling the length, while a 100% compressive strain ($\varepsilon_e = \Delta L/L = -L/L = -100\%$) implies a complete depression of the specimen from its original length to virtually zero length and represents a very different magnitude in deformation. Nonetheless, the simplicity in measurement of engineering stress and strain has made them overwhelmingly adopted in most engineering practices, as well as in reverse engineering applications, with only a few exceptions, such as true stress creep test.

Elastic deformation might also result in an angular or shape change, as shown in Figure 3.6. The angular change in a right angle is known as shear strain. The right angle at A was reduced by a small amount $\theta$ due to the application of a shear stress. The shear strain is the ratio between the displacement and the height, as defined by Equation 3.11, where $\tau$, $\gamma$, $a$, and $h$ are shear stress, shear strain, displacement, and height, respectively:

$$\gamma = \frac{a}{h} = \tan\theta \approx \theta \tag{3.11}$$

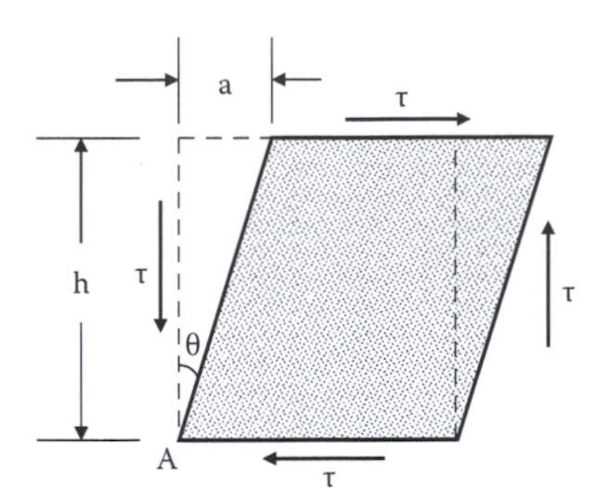


**FIGURE 3.6**
Schematic of shear strain.

For small θ values in radians, tanθ ≈ θ, and shear strains are often referred to as angles of rotation. Finally, it is worth noting that the tensile properties are functions of specimen size, loading rate, and testing environment, such as temperature. The accuracy of comparative analysis based on tensile properties in reverse engineering directly relies on careful verification of these test parameters.

### 3.3.2 Critical Resolved Shear Stress

Since the 1980s, single-crystal blades have been used in aircraft jet engines. These single-crystal blades are potential candidates for reverse engineering. In addition to the effect of grain boundary, the critical resolved shear stress theory helps engineers better understand the superior tensile properties of a single-crystal blade compared to a polycrystalline blade. In most engineering applications, polycrystalline metallic materials are assumed to be isotropic and have uniform yield strengths in all directions. However, micro slips in most alloys are often directional and follow a certain favorite orientation, as illustrated in Figure 3.7, which shows slip steps in an aluminum alloy at a very high magnification. It is also well acknowledged that different tensile loads are required to produce plastic deformation by slip in single crystals of different orientations. Atoms in a single-crystal slip in preferred crystallographic directions and planes to produce plastic deformation. These preferred directions and planes are the most closely packaged directions and planes. They form the combined slip systems. There are twelve slip systems in face-centered cubic (FCC) metals such as aluminum, forty-eight in body-centered cubic (BCC) metals such as chromium, and three in hexagonal close-packed (HCP) metals such as magnesium. For instance, slip prefers to occur in the (111) plane along the [110] direction in FCC metals, and [110](111) is one of the twelve slip systems. A single crystal starts slipping when the shearing stress reaches the critical resolved shear stress to initiate yield in the specific slip system. When a single-crystal blade is oriented with its strongest direction aligned with the stress direction, it shows better strength.

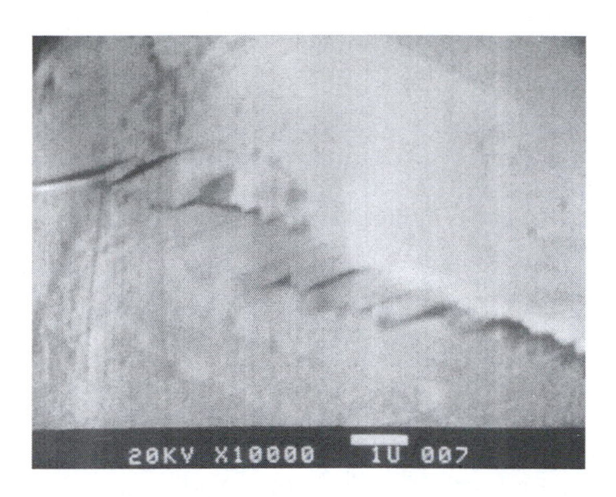


**FIGURE 3.7**
Micro slip steps in an aluminum alloy.

Consider a cylindrical single crystal with cross-sectional area $A$. It yields at a tensile load $P$. The angles between the tensile axis and the normal to the slip plane and the slip direction are $\varphi$ and $\lambda$, respectively, as illustrated in Figure 3.8. The area of the slip plane inclined at the angle $\phi$ is $A/\cos\phi$, and the stress component in the slip direction is $P\cos\lambda$. The critical resolved shear stress, $\tau_{crss}$, can therefore be calculated by Equation 3.12. The single crystal starts to yield when the shear stress reaches the critical resolved shear stress (Figure 3.8):

$$\tau_{crss} = \frac{P\cos\lambda}{A/\cos\phi} = \frac{P}{A}\cos\phi\cos\lambda \tag{3.12}$$

In the case of a single-crystal blade, the failure is not quantitatively determined by the nominal ratio between load and cross-sectional area, $P/A$, with the assumption that the material is isotropic. Instead, the strength of a single-crystal blade is directional and will not yield until the resolved shear stress on the slip plane in the slip direction has reached $\tau_{crss}$. The resolved shear stress reaches a maximum value of $(1/2)P/A$ when $\varphi = \lambda = 45°$; that is the orientation a single-crystal blade should be installed against. If the tension axis is normal to the slip plane, $\lambda = 90°$, then the stress component, $P\cos\lambda$, on the slip plane is zero. If it is parallel to the slip plane, $\varphi = 90°$, and the effective area, $A/\cos\varphi$, is infinitely large. Theoretically, the resolved shear stress is zero in both cases. Plastic deformation will not occur; instead, the single crystal tends to fracture rather than yielding by slip in these conditions.

### 3.3.3 Fracture Strength

Any reverse engineered part shall never operate beyond its fracture strength. The complexity of fracture mechanism prevents engineers from developing

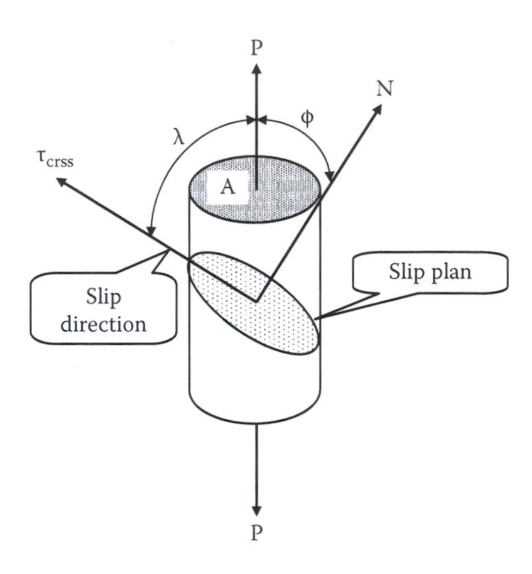


**FIGURE 3.8**
Critical resolved shear stress.

a universal fracture strength theory for all materials. However, the following theories discussed in this section will provide engineers with the fundamental knowledge on failure analysis for reverse engineering applications. The maximum shear strength theory estimates the maximum shear strength, $\tau_{max}$, of a perfect crystal by assuming that shearing results from the displacement of one whole layer of atoms over another. It approximately equals $G/2\pi$, as mathematically expressed by Equation 3.13, where $G$ is the shear modulus. However, it is 100 to 1,000 times larger than the measured value due to the line defect of dislocation. Dislocations are a linear atomic misalignment in crystalline materials. There are two basic types of dislocations: edge and screw dislocations. They also sometimes combine together to form a mixed dislocation. The required stress to move the dislocation line, one atomic distance at a time, only needs to break the atomic bond between the upper and lower atoms involved at any time. This is much smaller than the yield stress otherwise required in a perfect crystal to break all the bonds between all the atoms crossing the slip plane simultaneously. The existence of dislocations in a crystalline material has made yielding significantly easier. This explains the above-mentioned discrepancy between theoretical and nominal fracture strengths.

$$\tau_{max} = \frac{G}{2\pi} \tag{3.13}$$

Cohesive tensile strength is based on the theory that estimates fracture strength in tension. The atoms of crystalline metals are bound together by

an attractive force and simultaneously repelled apart by a repulsive force between them. These two forces balance each other to keep the atoms at equilibrium. If the crystal is subject to a tensile load, initially the repulsive force decreases more rapidly with increased atomic spacing than the attractive force. A net attractive force is therefore formed. Though the externally applied tensile load is first resisted by the net attractive force, eventually it reaches the peak point where the net attractive force starts to decrease due to increased separation between atoms. This corresponds to the maximum cohesive strength of the crystal. Beyond it an unstable state is reached. The required stress to further separate the atoms decreases, and the atoms continuously move apart at the applied stress until fracture occurs. Equation 3.14 presents a good approximation for the theoretical cohesive tensile strength, $\sigma_{max}$, where $E$ is the Young's modulus. The theoretical cohesive strength in tension can only be observed in tiny, defect-free metallic whiskers and very fine diameter silica fibers. The measured fracture strength for most engineering alloys is only 1/100 to 1/1,000 of the theoretical strength. This leads to the conclusion that existing flaws or cracks are responsible for the nominal fracture stress of engineering alloys.

$$\sigma_{max} = \frac{E}{2\pi} \tag{3.14}$$

Fracture mechanics was first introduced to the engineering community in the 1930s by A. A. Griffith, an English aeronautical engineer, to explain the discrepancy between the actual and theoretical strengths of brittle materials. Later it was further developed by G. R. Irwin at the U.S. Naval Research Laboratory (NRL) in the 1940s. It explains that a part failure is dependent on not only a material's inherent strength, but also the preexisting cracks in the subject part. Fracture mechanics is a theory that analyzes the mechanical strength with the acknowledgment of existing cracks. This is in contrast to the classic mechanics that calculates the mechanical strength with the assumption that the part is defect-free. When fracture occurs in a brittle solid, all the work consumed goes to the creation of two new surfaces. This theory leads to the fracture strength described by Equation 3.15, where $\gamma_s$ is surface energy and $a_o$ is the atomic distance:

$$\sigma_{max} = \left(\frac{E\gamma_s}{a_o}\right)^{1/2} \tag{3.15}$$

Consider an infinitely wide plate subject to an average tensile stress $\sigma$, with a thin elliptical crack of length $2c$ and a radius of curvature at its tip of $\rho_t$,

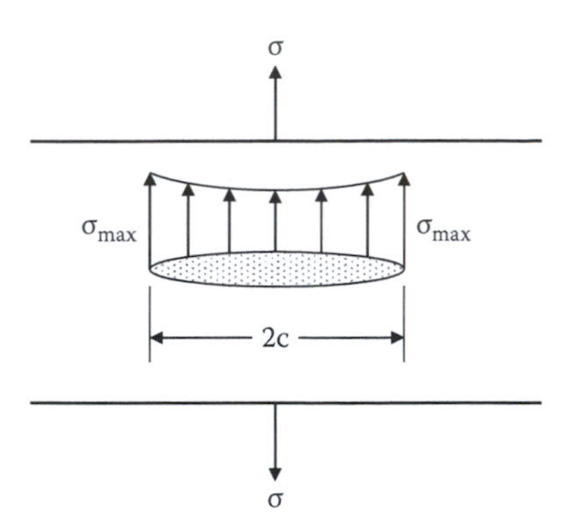


**FIGURE 3.9**
Elliptical crack in an infinitely wide plate.

as depicted in Figure 3.9. The maximum stress at the tip of the crack due to stress concentration is given by Equation 3.16 (Inglis, 1913):

$$\sigma_{max} = \sigma\left[1+2\left(\frac{c}{\rho_t}\right)^{1/2}\right] \approx 2\sigma\left(\frac{c}{\rho_t}\right)^{1/2} \tag{3.16}$$

Assume that the theoretical cohesive strength can be reached at the crack tip, while the average tensile stress represents the nominal fracture strength, $\sigma_f$. Set Equations 3.15 and 3.16 equal to each other; then the nominal fracture strength can be calculated by Equation 3.17:

$$\sigma_f \approx \sigma = \left(\frac{E\gamma_s\rho_t}{4a_oc}\right)^{1/2} \tag{3.17}$$

The sharpest possible crack has a radius of curvature at the tip equal to the atomic distance, $\rho_t = a_o$, and the fracture strength can be approximated by Equation 3.18:

$$\sigma_f \approx \left(\frac{E\gamma_s}{4c}\right)^{1/2} \tag{3.18}$$

If a crack of length $2c = 10$ μm exists in a brittle material having $E = 100$ GPa, $\gamma_s = 2\ J/m^2$, the nominal fracture strength can be numerically calculated by Equation 3.19:

$$\sigma_f = \left(\frac{E\gamma_s}{4c}\right)^{1/2} = \left(\frac{100 \times 10^9 \times 2}{4 \times 5 \times 10^{-6}}\right)^{1/2} = (10^{16})^{1/2} = 0.1 \text{ GPa} \tag{3.19}$$

This example demonstrates that the existence of a very small crack can significantly reduce the fracture strength from the theoretical cohesive strength by a factor of 100. In some cases, it can even be reduced by a factor of 1,000. The level of preexisting cracks in a part is primarily determined by the manufacturing process and the quality control system. From a reverse engineering perspective, the reproduced part should be manufactured under such a quality control system that only introduces the same level of or less preexisting cracks than what the original part is allowed.

### 3.3.4 Material Toughness

The toughness of a material depends on strain rate, temperature, and crack size. As a result, different tests and measurements for material toughness are established to reflect the profound effects of these factors. The Charpy or Izod test is designed to measure material toughness under dynamic loads. It measures the energy absorbed by the specimen at fracture when exposed to a heavy impact. The study of crack size effects on material toughness has led the development of fracture mechanics. It introduces several quantitative parameters to measure material toughness in terms of stress and preexisting crack size.

The effects of strain rate and temperature on material toughness are best exemplified by the sudden brittle fracture reported in many steel ship hulls in the 1940s and 1950s. These failures usually occurred at low ambient temperature, and were accompanied by the high stress imposed by heavy waves. While the tensile strength of these steels suggested they should have sufficient strength and ductility at the normal service temperature, they failed with brittle fracture appearance. The Charpy impact tests demystified these failures. The ductile–brittle transition temperature for impact load could be 100°C higher than the tensile elongation transition temperature. Figure 3.10 is a schematic illustration of the ductile–brittle transitions of ferric steels measured by tensile elongation, Charpy impact toughness, and angular torsion displacement, respectively. The transitions of tensile elongation and Charpy impact toughness occur in rather narrow temperature ranges, while the transition of angular torsion displacement is milder and smoother. This has dramatically changed the design criterion in material selections for engineering structures to include material toughness in many applications. The complexity of material toughness is beyond external loading condition and temperature; the material's own crystallographic structure also plays a critical role. Metals with a BCC crystal structure, for example, ferritic steels, usually show a transition in toughness from ductile high-energy fracture to brittle low-energy fracture in a narrow transition temperature range, as discussed above. In contrast, austenitic stainless steels, with an FCC crystal

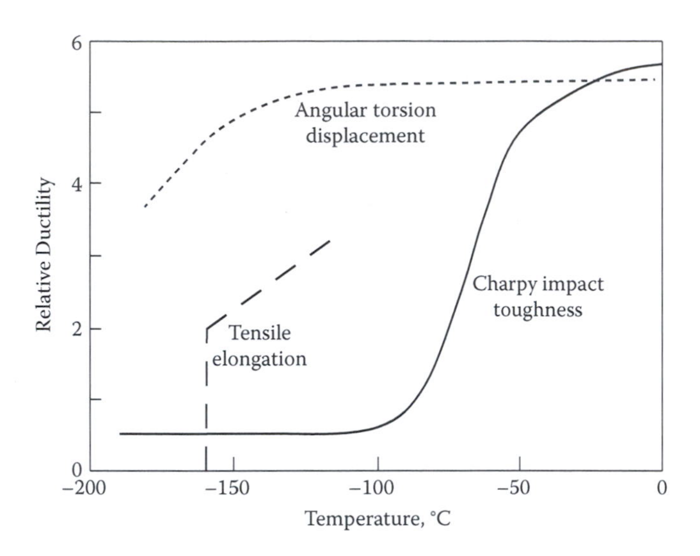


**FIGURE 3.10**
Ductile–brittle transition of ferric steels.

structure, have excellent toughness down to −273°C, with no steep ductile-to-brittle transition. Therefore, the identification of alloy phase and crystallographic structure might be required in some reverse engineering applications to ensure proper ductile–brittle transition behavior.

The modern fracture mechanics goes beyond simple stress and strain measurements. Fracture toughness has become a primary criterion in failure analysis based on fracture mechanics. Its calculation quantitatively integrates stress with existing crack size. The fracture toughness is a material property obtained by a valid test that first introduces a parameter defined as a stress intensity factor. Equation 3.20a is the mathematical formula of stress intensity factor for a thin plate of infinite width with an existing crack of length 2c in the center and subject to a tensile stress, $\sigma$, as illustrated in Figure 3.9. The same equation also applies to a wide thin plate under a tensile stress $\sigma$ with an edge crack of length c. This formula is virtually identical to the mathematical formula for fracture toughness expressed by Equation 3.20b. Equation 3.20a and b is based on linear elastic fracture mechanics. It assumes a linear relationship between stress and strain and small elastic deformation. It works best for brittle fracture. The fracture toughness, $K_{IC}$, is a validated material property that satisfies a set of defined conditions, such as meeting the test specimen thickness requirement, while the stress intensity factor is just a numerical quantity describing the loading condition. Any set of stress and crack length will generate a numerical value of stress intensity factor calculated by Equation 3.20a, but will not always produce a valid fracture toughness value as expressed by Equation 3.20b. In contrast to the material toughness measured by the Charpy impact test, which has a unit of energy

per unit volume, the fracture toughness has a unit of MPa$\sqrt{m}$. The fracture toughness symbol itself, $K_{IC}$, reflects the complexity of this material parameter. The Roman numeral subscript *I* refers to the externally applied load categorized as mode I, that is, tension; *C* refers to a critical value. It implies that the fracture toughness is a critical reference parameter, and when the combined effect of stress and crack length is beyond this value, the material will fail. Different mathematical equations with various stress and geometric parameters are integrated together to quantitatively describe fracture toughness for different load modes and crack configurations. When the loading mode is a shear or torsion, the symbols for fracture toughness are $K_{IIC}$ and $K_{IIIC}$, respectively. It is worth noting that fracture mechanics predicts a part to fail at different stress levels depending on the size, shape, and form of the existing crack.

$$K = \sigma\sqrt{\pi c} \tag{3.20a}$$

$$K_{IC} = \sigma\sqrt{\pi c} \tag{3.20b}$$

The following example demonstrates the combined effects of stress and crack size on fracture criterion from a fracture mechanics perspective. Figure 3.11 shows a wide thin plate with an edge crack of $c = 0.002$ m long and extending through the full thickness. The width, $w$, is 0.2 m, and the thickness, $t$, is 0.001 m. The plate is made of aluminum alloy with a yield strength of $\sigma_y = 350$ MPa, and a fracture toughness value of $K_{IC} = 40$ MPa$\sqrt{m}$. Determine the maximal load this plate can sustain under tension. What will be the maximal tensile load this plate can sustain when the edge crack grows to 0.02 m long?

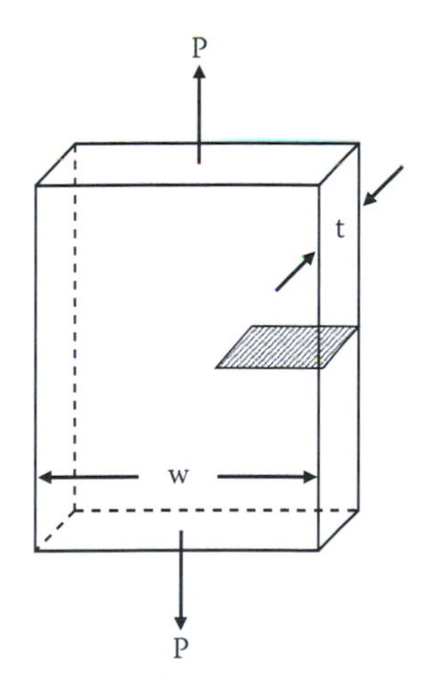


**FIGURE 3.11**
Edge crack in a wide thin plate.

The maximal load, $P_{max}$, this plate can sustain without yielding can be approximately calculated by the following equation:

$$\sigma_y = \frac{P_{max}}{t(w-c)} = \frac{P_{max}}{(0.001)(0.2-0.002)} = \frac{P_{max}}{0.000198} = 350 \text{ MPa}$$

$$P_{max} = 350 \text{ MPa} \times 0.000198 \text{ m}^2 = 69{,}300 \text{ Newtons}$$

For a wide thin plate under tension with an edge crack, the fracture toughness can be mathematically described by Equation 3.20b:

$$K_{IC} = \sigma\sqrt{c\pi}$$

In fracture mechanics, the maximal load this plate can sustain without fracture can therefore be calculated as

$$\sigma = \frac{K_{IC}}{\sqrt{c\pi}} = \frac{40}{\sqrt{(0.002)\pi}} = 504.6 \text{ MPa}$$

$$P_{max} = \sigma \times [t \times (w - c)] = 504.6 \times [0.001 \times (0.2 - 0.002)] = 99{,}991 \text{ Newtons}$$

With a 0.002 m edge crack, the plate will yield at 69,300 Newtons first, before it fractures. The determining factor is yield strength, and the maximum load this plate can sustain under tension is 69,300 Newtons.

When the crack grows to 0.02 m long, the plate will yield at

$$P_{max} = 350 \text{ MPa} \times [(0.001) \text{ m} \times (0.2 - 0.02) \text{ m} = 63{,}000 \text{ Newtons}$$

However, the allowed maximal stress from a fracture mechanics perspective before fracturing will be

$$\sigma = \frac{K_{IC}}{\sqrt{c\pi}} = \frac{40}{\sqrt{0.02\pi}} = 159.6 \text{ MPa}$$

The plate will fracture at $P_{max} = \sigma[t \times (w - c)] = 159.6 \text{ MPa} \times [0.001 \text{ m} \times (0.2 - 0.02) \text{ m}] = 28{,}729$ Newtons. Therefore, based on the fracture mechanics calculation, the maximal load this plate can sustain is only 28,729 Newtons when the crack grows to 0.02 m. In other words, the load-carrying capability of this plate decreases more rapidly according to fracture mechanics as the crack length increases, and the determining factor shifts from yield strength when the crack length is 0.002 m to fracture toughness when the crack length grows to 0.02 m.

The complexity of the fracture toughness test requires specialized engineering expertise to obtain a valid value, and it can be costly. Nonetheless, the test of fracture toughness and life calculation based on fracture mechanics are

warranted for the reverse engineered part that is a critical structure element serving in an environment across the ductile–brittle transition temperature. In many reverse engineering projects, the determination of fracture toughness is yet to become a mandatory requirement despite its being a critical parameter that determines if a part will fail. However, when more and more reverse engineered parts are life-limited or critical parts, and the original part was designed based on fracture toughness, the fracture toughness test should be conducted to demonstrate the equivalency whenever feasible.

### 3.3.5 Notch Effects

A working knowledge of notch effects on mechanical properties is critical for reverse engineering practices. The existence of a notch on a test specimen would have significant effects on the test results. The properties obtained from smooth test specimens are not comparable with those obtained from notched specimens in a comparative analysis. The presence of a sharp notch could strengthen ductile metals, but will usually weaken brittle materials. Whether a test on a notched specimen should be conducted is one of the frequently asked questions in reverse engineering. The most significant impact of a sharp notch is the introduction of a triaxial stress state and a local stress concentration at the notch root. Other effects include the ductile–brittle transition temperature increase (of mild steel), higher local strain rate, the enhancement of local strain hardening, etc. This section will first explain why the triaxial stress state is formed at the notch. It will then discuss how the resultant stress profile can affect the mechanical properties.

Metal deforms elastically at the notch if the applied normal stress (normal to the notch), $\sigma_{norm}$, in the longitudinal $y$ direction is relatively low. The longitudinal stress, $\sigma_y$, distribution in a thin plate with a sharp notch is illustrated in Figure 3.12. A transverse elastic stress, $\sigma_x$, is induced by the notch. The introduction of $\sigma_x$ can be understood by imaging a series of small tensile

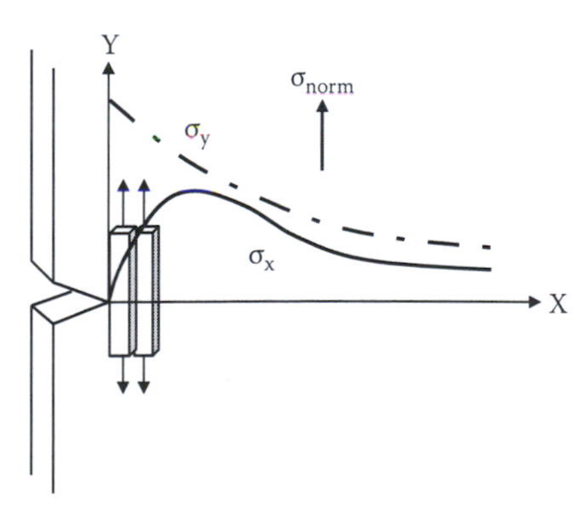


**FIGURE 3.12**
Elastic stress profiles at a sharp notch in a thin plate.

specimens at the tip of the notch. If each tensile specimen is free to deform, a lateral strain $\varepsilon_x$, resulting from contraction, will be produced due to Poisson's ratio, ν. To maintain the material continuity, a tensile stress $\sigma_x$ must exist across the tensile specimen interface. At the free surface of the notch ($x = 0$) the tensile specimen can be laterally contracted freely without any restriction, and $\sigma_x = 0$. The $\sigma_x$ rises steeply near the tip, and then falls slowly as the $\sigma_y$ distribution flattens out, $\sigma_x = E\varepsilon_x = -E(\nu\varepsilon_y) = -E[\nu(\sigma_y/E)]$. In the plane stress condition of a thin plate, the stress in the thickness direction $z$ is negligibly small, that is, $\sigma_z = \tau_{xz} = \tau_{yz} = 0$, and can be ignored.

As the thickness, $B$, increases, it becomes a plane strain condition. The strain in the thickness direction is approximately zero, that is, $\varepsilon_z = 0$. It is assumed that all deformation occurs in one plane, and the stress in the $z$ direction, $\sigma_z = \upsilon(\sigma_x + \sigma_y)$, becomes more significant and cannot be deemed as zero anymore. The principal stresses and strains of plane stress and plane strain conditions are summarized in Table 3.1. The stress distributions for a thick-notched plate loaded uniaxially in the $y$ direction are illustrated in Figure 3.13a, showing a high degree of triaxial stress configuration with stress components in all three, $x$, $y$, and $z$, directions. The value of $\sigma_z$ falls to zero at the notch root where $x = 0$ on both surfaces of the plate ($z = \pm B/2$), but rises rapidly with distance from the free surfaces. The distribution of $\sigma_z$ with $z$ at the notch root is shown in Figure 3.13b. As the thickness $B$ decreases, the values of $\sigma_x$ and $\sigma_y$ only fall by less than 10%; however, the value of $\sigma_z$ decreases to 0 as the thickness approaches zero to assume a plane stress configuration (Dieter, 1986).

For a ductile metal, as the applied stress increases to yield strength, it starts yielding plastically, and a plastic zone will be established at the notch tip. According to the maximum shear stress yielding theory, the existence of transverse stresses $\sigma_x$ and $\sigma_z$ will raise the yielding stress in the longitudinal $y$ direction, where the external stress is applied. The maximum shear stress yielding theory predicts yielding when the maximum shear stress reaches the value of the shear strength in the uniaxial-tension test. This criterion is mathematically expressed as $\sigma_{yield} = \sigma_1 - \sigma_3$, where $\sigma_{yield}$ is the yield strength, and $\sigma_1$ and $\sigma_3$ are the algebraically largest and smallest principal stresses, respectively. In an unnotched tension specimen subject to a uniaxial stress, the material yields at $\sigma_{yield} = \sigma_1 - 0$, and therefore $\sigma_{yield} = \sigma_1 = \sigma_y$. In a thick plane strain plate, $\sigma_1 = \sigma_y$ and $\sigma_3 = \sigma_x$, as illustrated in Figure 3.13a. The yielding first starts at the notch root, which requires the smallest stress for yielding because $\sigma_3 = \sigma_x = 0$ at the free root surface. The required externally

**TABLE 3.1**

Principal Stresses and Strains for Plane Stress and Plane Strain Conditions

| Condition | Stress | Strain |
|---|---|---|
| Plane stress (thin plate) | $\sigma_x \neq 0, \sigma_y \neq 0, \sigma_z = 0$ | $\varepsilon_x \neq 0, \varepsilon_y \neq 0, \varepsilon_z \neq 0$ |
| Plane strain (thick plate) | $\sigma_x \neq 0, \sigma_y \neq 0, \sigma_z = \nu(\sigma_x + \sigma_y) \neq 0$ | $\varepsilon_x \neq 0, \varepsilon_y \neq 0, \varepsilon_z = 0$ |

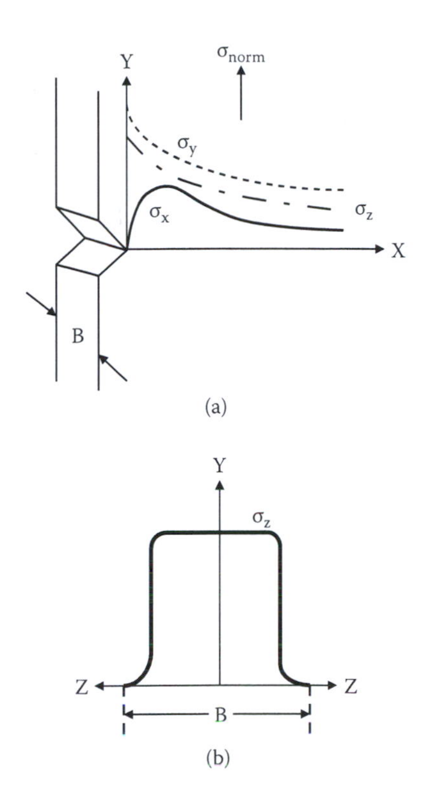


**FIGURE 3.13**
(a) Elastic stress profiles at a sharp notch in a thick plate. (b) Stress distribution of $\sigma_z$ at the notch tip $x = 0$.

applied stress in the longitudinal $y$ direction ($\sigma_y$) to yield increases with the distance from the notch root following the criterion $\sigma_{yield} = \sigma_y - \sigma_x$, because $\sigma_{yield}$ is a constant material parameter, while $\sigma_x$ increases with the distance from the notch root near the tip.

As explained above, the existence of a sharp notch can strengthen the ductile metal due to the triaxiality of stress. The ratio of notched-to-unnotched yield stress is referred to as the plastic constraint factor, $q$. In contrast to the elastic stress concentration factor that can reach values in excess of 10, the value of $q$ does not exceed 2.57 (Orowan, 1945). However, brittle metals could prematurely fail due to stress increase at the notch before plastic yielding occurs.

When plastic deformation occurs at the notch root, $\sigma_y$ drops from its high elastic value to $\sigma_{yield}$. Once the first imaginary tensile element at the notch root starts yielding, it deforms plastically at a constant volume that requires Poisson's value to be $\nu = 0.5$ instead of about 0.3 during elastic deformation. Therefore, a higher transverse stress, $\sigma_x = E\varepsilon_x = -E(\nu\varepsilon_y) = -E[\nu(\sigma_y/E)]$, will be developed to maintain the material continuity. The stress $\sigma_x$ will also increase with the distance from the notch root more quickly than in the

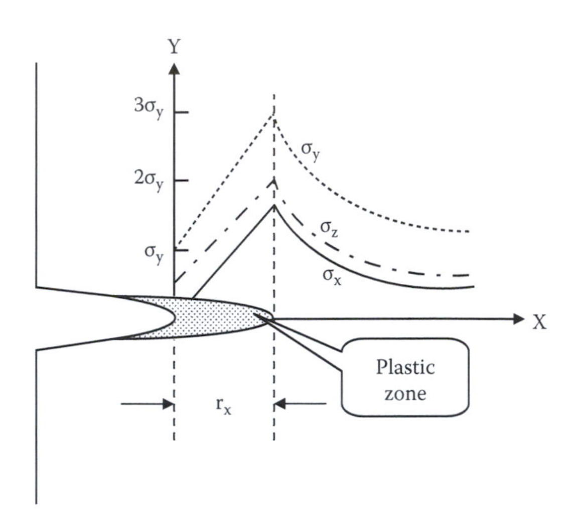


**FIGURE 3.14**
Stress profile at the sharp notch with a plastic zone.

elastic case. Within the plastic zone, the stresses $\sigma_y$ and $\sigma_z$ increase according to $\sigma_y = \sigma_{yield} + \sigma_x$ and $\sigma_z = 0.5(\sigma_y + \sigma_x)$ until they reach the plastic-elastic boundary. The three principal stress, $\sigma_x$, $\sigma_y$, and $\sigma_z$, profiles at the notch with a plastic zone are illustrated in Figure 3.14, where $r_x$ is the length of the plastic zone in the $x$ direction (Dieter, 1986).

The stress profiles for elastic and plastic deformation in front of a notch, illustrated in Figures 3.12 to 3.14, have demonstrated the complex notch effects that could drastically affect a comparative analysis of mechanical strength in reverse engineering. Many questions related to notch effects frequently come up in reverse engineering when a component test is not conducted. Can the tensile strength obtained from a smooth test specimen be used to estimate the strength of a notched specimen? How much debit should be factored in if an analysis is based on smooth tensile strength against a part with a notch? Why and when should a test with a notched specimen be mandated? The answers to these questions are often based on the specific part configurations and criticality.

### 3.3.6 Bending, Torsion, and Hoop Stress

Bending is a combination of tension and compression. Reverse engineering a structural beam or a transmission gear requires an engineer to know the bending stress imposed on the beam or the gear tooth. Consider a beam bending upwards; the top portion is under compression, and the bottom portion under tension. The boundary between tension and compression in a cross section is referred to as the neutral axis. The bending stress, $\sigma_b$, is expressed in terms of bending moment, $M$; moment of inertia of cross-sectional area, $I$; and the distance from the neutral axis, $c$, as expressed in Equation 3.21:

$$\sigma_b = \frac{Mc}{I} \tag{3.21}$$

Gathering data on torsion is essential when reverse engineering a shaft. Torsion stress is a variation of shear stress that results from a force applied parallel to the surface. Torsion stress twists the part in response to a torque, as can be seen in an automobile driveshaft. Considering a cylindrical shaft subject to a torque, $T$, the maximum torsion stress, $\tau_t$, occurs on the shaft surface and is expressed by Equation 3.22, where $r$ is the shaft radius and $J$ is the polar moment of inertia of cross-sectional area:

$$\tau_t = \frac{Tr}{J} \tag{3.22}$$

Hoop stress is a circumferential tensile stress in a container wall like the pressure vessel wall, or in a round part such as a turbine disk. Figure 3.15 is the top view of the right-side cross section of a cylindrical gas tank under an internal pressure, $p$. The tank has an inside radius of $r$, and a thin wall of thickness $t$. The force tending to separate the two sides is $p(2r)$ per unit cylinder length. This force is resisted by the tangential stress, also referred to as the hoop stress, $\sigma_h$, which acts uniformly and is pointed leftward, as shown in Figure 3.15. In equilibrium, we have $p(2r) = 2t\sigma_h$, and the hoop stress, which functions like a hoop to bind the two halves tighter, can be calculated by Equation 3.23. The static components of a pressurized container are only subject to static loads; however, their failure might result in high hazardous effects to cause fire or explosion. These parts are deemed as critical, such as a turbine jet engine compressor case, which is required for stringent tests to ensure its durability when exposed to sporadic overload. Following a similar analysis, the hoop stress for a spherical container can be

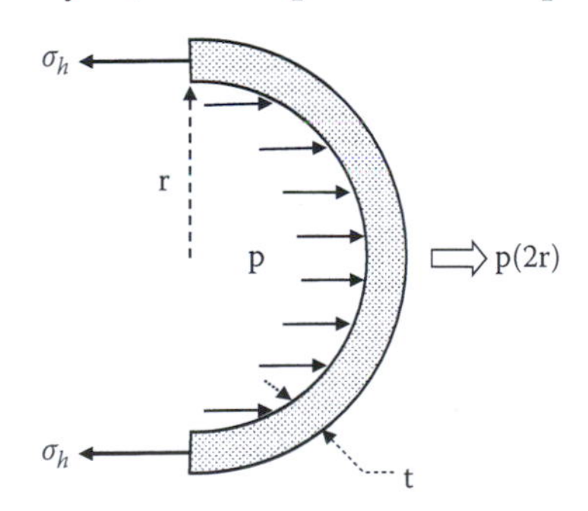


**FIGURE 3.15**
Hoop stress.

calculated as $\sigma_h = pr/2t$, where $r$ and $t$ are the radius and wall thickness of the spherical container, respectively.

$$\sigma_h = \frac{pr}{t} \tag{3.23}$$

## 3.4 Hardness

Hardness is a measurement of material resistance to plastic deformation in most cases. It is a simple nondestructive technique to test material indentation resistance, scratch resistance, wear resistance, or machinability. Hardness testing can be conducted by various methods, and it has long been used in analyzing part mechanical properties. In reverse engineering, this test is also widely used to check the material heat treatment condition and strength, particularly for a noncritical part, to save costs. The hardness of a material is usually quantitatively represented by a hardness number in various scales. The most utilized scales are Brinell, Rockwell, and Vickers for bulk hardness measurements. Knoop, Vickers microhardness, and other microhardness scales are used for very small area hardness measurements. Rockwell superficial and Shore scleroscope tests are used for surface hardness measurements. Surface hardness can also be measured on a nanoscale today.

### 3.4.1 Hardness Measurement

The Brinell hardness number (BHN) was first introduced by J. A. Brinell of Sweden in 1900. It is calculated by Equation 3.24 based on the stress per unit surface area of indentation, as illustrated in Figure 3.16. Brinell hardness numbers were usually tabulated in reference charts before the test machine was computerized. The Brinell hardness test is not suitable for very thin specimens due to the depth of indentation impressed onto the part, or very hard materials because of the induced deformation of the tester itself.

$$BHN = \frac{F}{\frac{\pi}{2} D(D - \sqrt{D^2 - D_i^2})} \tag{3.24}$$

where $F$ = indenting force, $D$ = diameter of indenter ball, and $D_i$ = diameter of indention.

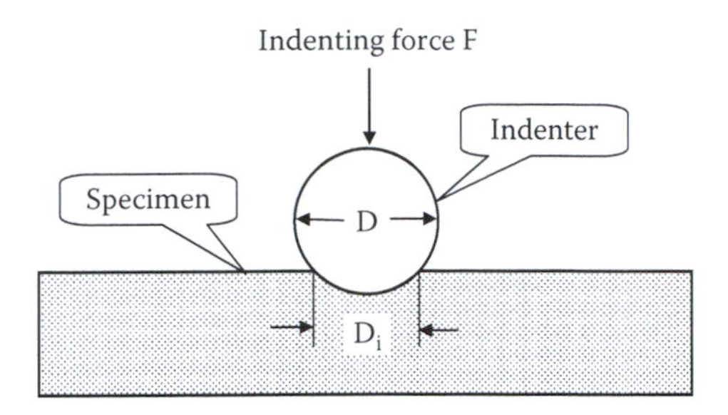


**FIGURE 3.16**
Brinell hardness measurement.

The Rockwell hardness test was first introduced to check if the bearing race of a ball bearing was properly heat treated. It is the most widely utilized hardness testing method in the United States today. Figure 3.17a shows a stand-alone Rockwell hardness tester; similar testers, though possibly of different models and configurations, can be found in many laboratories. It has a control panel on the top and a load selection dial on the right side. The tester is equipped with an intuitive liquid crystal display (LCD) atop the control panel to show the test result and a Universal Serial Bus (USB) port at the upper right corner of the control panel for data transfer. Figure 3.17b illustrates the further integration of modern computer and image technologies with the hardness measurement. It shows the indention on the screen and the data process of hardness measurements in the computer. The installed software allows automatic hardness measurement and digitalized image processing. It also makes the statistical analysis on hardness measurement, such as standard deviation, easy. The software enhances the data filing and exporting capacities as well.

Figure 3.18 illustrates the Rockwell hardness measurement process, where the indenter is first applied to the specimen surface with a minor load $F_1$ of 10 kg to introduce an initial indention, *e*, and to establish a zero reference position. The major load, $F_2$, which may be either 60, 100, or 150 kg, is then applied for a specified dwell time on the surface. The major load is then released, leaving only the minor load on during the hardness reading. The Rockwell hardness number is a measurement of the indention depth, *h*, on the test specimen. It is worth noting that it is a linear measurement of indentation resistance with a different unit from stress. There are many scales in the Rockwell hardness readings, ranging from A through F, and continuing on. The two most commonly used are the B and C scales. The B scale is used for relatively soft materials such as aluminum alloys, and the C scale is for hard materials such as stainless steel. Each scale has 100 divisions and the hardness numbers are designated as $R_B$ or $R_C$, respectively. If a material has a hardness number close to or above $R_B = 100$, the C scale should be used. Most materials have their Rockwell hardness numbers below $R_C = 70$. The detailed test methods are explained in the following standards: ASTM E18

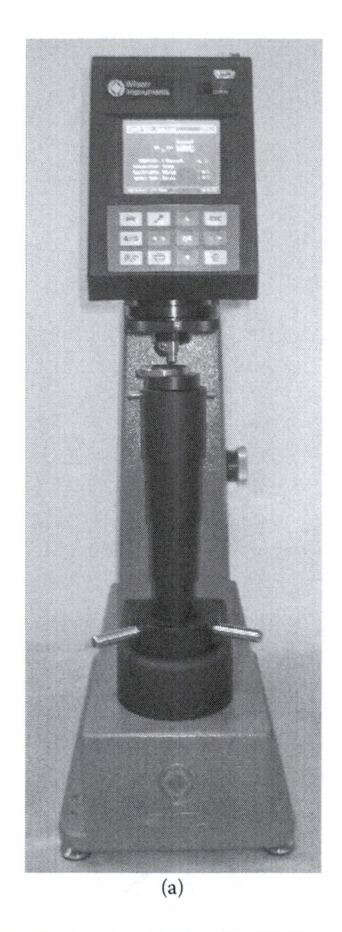

(a)

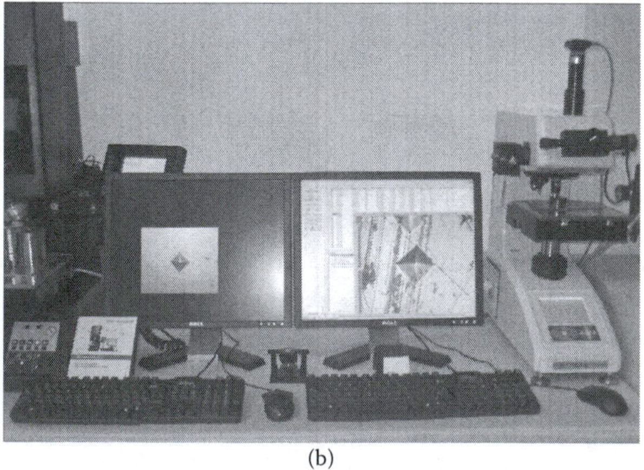

(b)

**FIGURE 3.17**
(a) Rockwell hardness test machine. (b) Computer and image technologies utilized in hardness measurement.

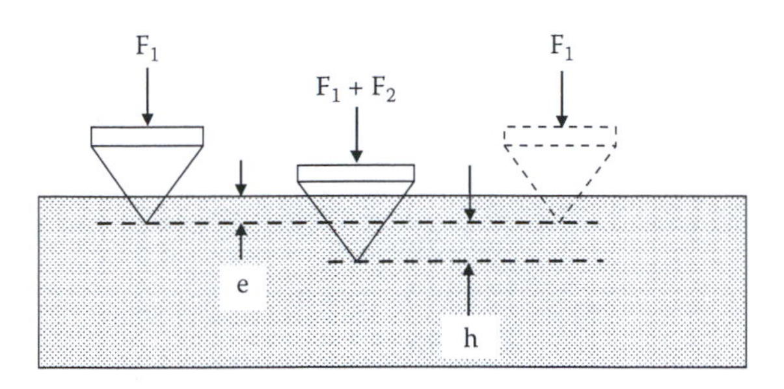


**FIGURE 3.18**
Rockwell hardness measurement.

(ASTM, 2008) and ISO 6508 (ISO, 2005) for metallic materials, and ASTM D787 (ASTM, 2009) for plastic materials.

The Vickers pyramid hardness (VPH) test uses various diamond pyramid indenters. Therefore, Vickers hardness is also referred to as diamond pyramid hardness (DPH). Vickers and Brinell hardness numbers are both calculated based on the applied load per unit area of indentation. Their values are very close to each other at the low hardness range. The Vickers hardness number retains its accuracy at higher values up to 1,300 (about BHN 850), while the BHN shows noticeable deviation from the VPH number at hardness numbers higher than 500. This deviation is due to the induced deformation on the steel ball indenter used for the Brinell hardness test, as demonstrated in Figure 3.19, which shows the conversion among the five hardness scales: Rockwell A, Rockwell B, Rockwell C, Brinell, and Vickers for nickel alloys. Considering that the Rockwell hardness number is based on the indentation depth, while the Brinell and Vickers hardness numbers are based on the load per unit area, the conversion between them is indirect and just an approximation. This conversion should be avoided in reverse engineering analysis if possible. It is also worth noting that the complexity of elastic and plastic sample deformations during a hardness test and the deformation of the indenter itself make the reproduction of the same hardness test results virtually impossible, even using the same hardness scale, as shown in Figure 3.20 (Low, 2001). It plots the hardness data in consecutive tests for 10 days, all on the same sample and in the Rockwell C scale. Five consecutive tests were conducted each day; each individual test usually produced a slightly different result every day. Also shown is a line plot of the daily average, which varies from day to day as well. For a representative average hardness number, four or more tests on a sample are usually conducted for most hardness tests.

Vickers microhardness is most widely used for microhardness tests of thin coatings. The Shore scleroscope hardness test is a dynamic test that measures rebound height/energy as an indicator of surface hardness by dropping a test hammer onto the surface. The rebound height/energy is heavily dependent on the material elasticity; therefore, the Shore hardness should only be

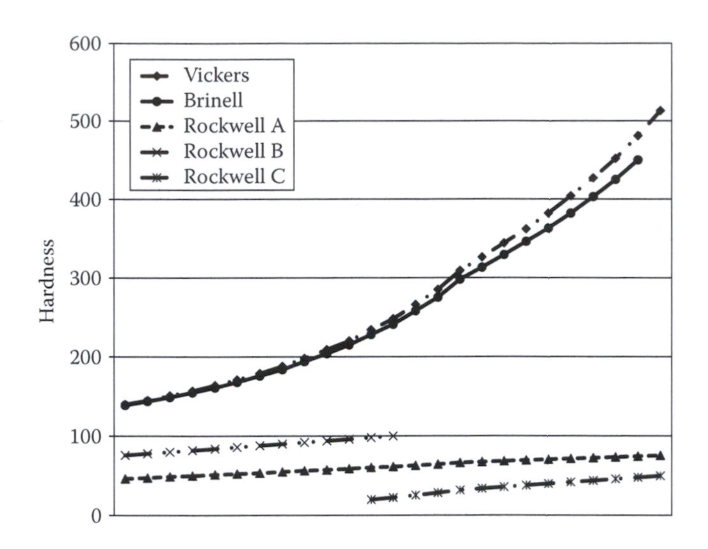


**FIGURE 3.19**
Hardness conversion.

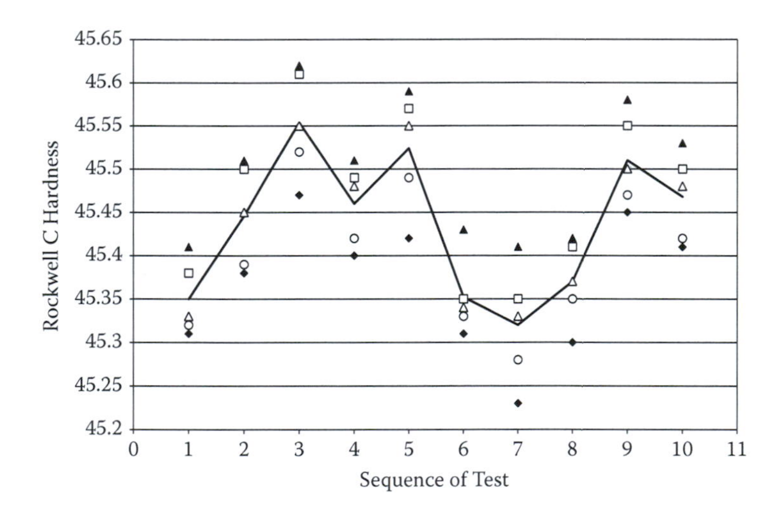


**FIGURE 3.20**
Scattering hardness test data.

used to compare materials with similar elasticity. It has the potential to be used for hardness comparison between the surfaces that are hardened by a thin coating, like a hard-coating wear resistance surface, or case-hardened gear surfaces in reverse engineering analysis.

In light of the complexity and variance of hardness measurement, it is essential that the hardness numbers are measured on the same scale as specified in the reference material specification for direct comparison. If not

feasible, only hardness numbers measured under comparable conditions can be converted to a same scale for comparison.

### 3.4.2 Hardness and Tensile Strength Relationship

For mild steels, the Brinell hardness number shows a simple empirical relation with the ultimate tensile strength (UTS), as described by Equation 3.25a and b (Budynas and Nisbett, 2008). For certain cast irons, the empirical relationship between ultimate tensile strength and Brinell hardness number is described by Equation 3.25c and d (Krause, 1969). However, the relationships between hardness numbers and tensile strengths are neither universal nor precise. Great caution should be exercised when applying these relationships in reverse engineering analyses.

$$\text{UTS (MPa)} \approx 3.4\ \text{BHN} \tag{3.25a}$$

$$\text{UTS (ksi)} \approx 0.5\ \text{BHN} \tag{3.25b}$$

$$\text{UTS (MPa)} \approx 1.58\ \text{BHN} - 86\ \text{MPa} \tag{3.25c}$$

$$\text{UTS (ksi)} \approx 0.23\ \text{BHN} - 12.5\ \text{ksi} \tag{3.25d}$$

Similar to Brinell hardness, Vickers hardness has also been the subject of study to search for a possible relationship between hardness and other mechanical properties. Some semiempirical relationships were reported. Based on a study of a magnesium alloy AZ19 with a nominal composition of Mg–8% Al–0.7% Zn–0.2% Mn–0.002% Fe–0.002% Cu, the flow stress can be approximately calculated by Equation 3.26, where the Vickers hardness number has a nominal unit of $kg/mm^2$, and the flow stress is measured in MPa (Cáceres, 2002).

$$\text{Vickers hardness} \approx 0.3\ \text{flow stress} \tag{3.26}$$

The flow stress is the instantaneous stress for continuous material flow, and is defined as the stress required to sustain plastic deformation, usually at a specific strain. The flow stress is closely related to yield strength, and its value is affected by alloy composition, phase constituent, microstructure, and grain morphology. Equation 3.26 is a semiempirical correlation for a specific alloy under specific plastic strain, 2.3%. It is very tempting to just measure the simple nondestructive hardness instead of conducting the expensive tensile or fatigue test to decode the OEM part. Indeed, hardness measurement is widely used for reverse engineering noncritical parts. However, the inherent alloy-specific restrictions and the complexity of mechanical behavior of material have significantly limited the applications of hardness-strength relationships in reverse engineering for critical parts.

The relationship between hardness and other mechanical properties will be further discussed in Chapter 6 from the perspective of statistical regression.

## References

ASTM. 2004. *Standard test methods for determining average grain size*. ASTM E112-96(2004)e2. West Conshohocken, PA: ASTM International.

ASTM. 2008. *Standard test methods for Rockwell hardness of metallic materials*. ASTM E18-08b. West Conshohocken, PA: ASTM International.

ASTM. 2009. *Standard specification for ethyl cellulose molding and extrusion compounds*. ASTM D787-09. West Conshohocken, PA: ASTM International.

Budynas, R. G., and Nisbett, J. K. 2008. *Mechanical engineering design*. New York: McGraw-Hill.

Cáceres, C. H. 2002. Hardness and yield strength in cast Mg-Al alloys. *AFS Transaction* 110:1163–1169.

Dieter, G. E. 1986. *Mechanical metallurgy*. New York: McGraw-Hill.

Hall, E. O. 1951. The deformation and aging of mild steel: III discussion of results. *Proc. Phys. Soc.* B64:747.

Inglis, C. E. 1913. Stresses in a plate due to the presence of cracks and sharp corners. *Trans. Inst. Nav. Archit.* 55 pt. I:219–230.

ISO. 2003. *Steels—Micrographic determination of the apparent grain size*. Geneva, Switzerland: ISO.

ISO. 2005. *Metallic materials—Rockwell hardness test*. ISO 6508-1. Geneva, Switzerland: ISO.

Krause, D. E. 1969. *Gray iron—A unique engineering material*. ASTM Special Publication 455. West Conshohocken, PA: ASTM.

Low, S. R. 2001. *Rockwell hardness measurement of metallic materials*. NIST SP 960-5. Washington, DC: NTSB.

Orowan, E. 1945. Notch brittleness and the strength of metals. *Trans. Inst. Eng. Shipbuild. Scot.* 89:165.

Petch, N. J. 1953. Cleavage strength of polycrystals. *JISI* 174:25.

Reed-Hill, R. E. 1973. *Physical metallurgy principles*. Monterey, CA: Brooks/Cole, Wadsworth.

# 4

# *Part Durability and Life Limitation*

Many mechanical components have life limits in their service due to the deterioration of their durability over time. These limitations are either explicitly defined by mandating the replacement of the part within a defined amount of time, or implicitly defined by the need for periodic inspections in accordance with the maintenance manual. Although it is more technically challenging to reverse engineer a life-limited part, market demands and higher profit margins provide strong incentives for their reproduction using reverse engineering. The life cycle of a part is determined either by the total load cycles the part has experienced or by the total time period the part has been placed in service. For example, a jet engine turbine disk usually needs to be replaced after a certain number of cycles or hours. This life limitation is because certain material properties, and therefore the performance of the part made of this material, are time dependent. This chapter will focus on the following three material properties that can affect part performance throughout time and impose life limitations on a part: fatigue, creep, and corrosion.

## 4.1 Part Failure Analysis

Advances in technology and improvements in material quality have made part failures less frequent in modern machinery. In the 2000s, approximately 80% of all aviation accidents were related to human factors, 15% were related to material deficiencies, and 5% were due to machine malfunction. The reverse engineered parts are expected to maintain the same level of safety attributable to the integrity of materials and machine functionality. This expectation requires reverse engineered parts to match the same level of perfection in material and part production. The failure of mechanical components can be categorized into two primary categories: instantaneous failure due to overload, such as a surge of tension or compression; and progressive failure in service under stress, such as fatigue, creep, or stress corrosion cracking. Instantaneous failures are rare because most components are designed with sufficient strength to sustain the expected loading condition. Proper material identification and manufacturing process verification in reverse engineering ensure equivalent performance to the OEM parts. However, prolonged

service under stress might cause a part to fail if proper maintenance is not conducted in a timely manner. This type of failure is relatively unpredictable based on material characterization and theoretical calculation, and therefore poses a tough challenge in reverse engineering. Metallurgical failure analysis focuses on the relationship between material characteristics and service conditions, alloy microstructure and applied load in particular.

In the most delicate materials, such as semiconductors, a comprehensive metallurgical failure analysis starts with the "sub-micro" crystal structure at the atomic scale. Transmission electron microscope (TEM), Auger electron spectroscopy, and X-ray spectroscopy are used to check the crystallographic lattice layouts and directions, as well as crystallographic defects like dislocation of atomic misalignments. For a typical mechanical component, TEM, scanning electron microscope (SEM), X-ray spectroscopy, and light microscope are used to examine the alloy microstructural characteristics like metallurgical phase and grain size to elaborate the interrelationship between them and the mechanical properties. Metallurgical failure analysis also checks the "macro" material features, such as surface cracks, for stress concentration induced by these cracks and their effects on material strength.

The failures resulting from overloading tension, compression, or torsion are static in nature, and fractures will rapidly occur at the application of load. While the deformations of creep and failure of fatigue are time dependent, immediate fracture does not occur at the time of loading. The mechanical components subject to creep or cyclical fatigue can sustain the load for a long time, up to several years. Instead of instant failure, creep and fatigue impose a limit of life expectancy on these parts. Proving that a reverse engineered part has an equivalent or better life limit than the OEM counterpart is a challenge, partially because it is usually a theoretical prediction based on accelerated test results. Furthermore, many failures result from the combined effects of externally applied loads and the operating environment, e.g., stress corrosion. The prevention of metallurgical failure plays a critical role in reverse engineering, and an understanding of the typical causes of metallurgical failure will be very beneficial when practicing reverse engineering.

A mechanical component usually fails due to one of the following four reasons, or a combination of them:

1. Excessive elastic deformation. The elastic deformation, e.g., deflection of shaft or buckling of column, is controlled by various elastic moduli such as Young's modulus. The values of these elastic moduli are primarily determined by alloy composition. Heat treatment does not change the value of Young's modulus. The elastic modulus values listed in most engineering handbooks are associated with the alloy compositions regardless of the manufacturing processes. The most effective way to increase the stiffness of a component with a

given composition is usually by changing its geometrical shape or increasing the cross-sectional dimension.

2. Excessive plastic deformation. Yielding and creep are the two most commonly observed plastic deformations in engineering alloys. Several empirical yielding criteria were established to predict the stress conditions under which yielding starts. The Von Mises criterion suggests that yielding occurs when the distortion energy reaches a critical value. The maximum shear stress or Tresca criterion suggests that yielding occurs when the maximum shear stress reaches the value of the shear stress in the uniaxial tension test. At elevated temperatures, alloys will develop permanent deformation at a stress lower than the yield strength due to creep. The effect of creep is particularly critical to high-temperature engine or power plant applications.
3. Fracture. Most mechanical parts will endure a certain amount of plastic deformation before fracture. However, a sudden fracture can occur in brittle materials without warning. Fracture mechanics has been widely used to analyze brittle fracture problems. Fatigue is a progressive fracture caused by a cyclical load. It is the most concerning fracture mode in machine design and for operational safety. Stress rupture is a delayed failure by fracture, which occurs when a metal is subject to a static load, usually much lower than the yield strength, at elevated temperatures.
4. Environmental effect. The fourth common root cause of part failure results from environmental effects, such as corrosion and hydrogen embrittlement. This is a prolonged failure mode that progresses slowly and is difficult to monitor. When a part is exposed to a corrosive environment during service, various protective measures, such as coating and plating, are often taken to minimize the environmental effects.

The prevention of part failure requires full knowledge of material characteristics, loading condition, and service environment to examine the microstructure, analyze the mechanical and physical properties, compare the failure modes, and understand the effects of material processing. A thorough understanding of the part design functionality and operation is critical for reproducing an equivalent mechanical component using reverse engineering.

## 4.2 Fatigue

Fatigue is a dynamic and time-dependent phenomenon. When a component is subject to alternating stresses repeatedly, it fails at a much lower stress than the material yield strength due to fatigue. Most mechanical failures are related

to dynamic loading; therefore, the safety assessment in fatigue life plays a critical role in reverse engineering. Two basic types of fatigue are low-cycle fatigue (LCF) and high-cycle fatigue (HCF). There are many types of derivative fatigue, such as thermal mechanical fatigue, a variation of LCF; and fretting fatigue, a variation of HCF. One fatigue cycle is defined as a complete application of stress or strain deformation from minimum to maximum and back to minimum. An LCF cycle for a jet engine is defined as a complete flight cycle from taking off to cruise to final landing. A minimum stress is applied to the engine part when the engine first starts. The stress reaches a maximum at takeoff, and then it turns back to a minimum at landing. However, the complete cycle for HCF of a jet engine part is different from a flight cycle. One HCF cycle is completed by a single revolution of a low-pressure compressor because this part experiences the same cyclic stress pattern at every revolution, which is typically related to vibratory or thermal stress.

There are two conditions that must be met for fatigue failure to occur: repetitive alternating stress or strain and a sufficient number of cycles. For many engineering materials there is a third criterion that must also be met: the maximum alternating stress or strain has to be beyond a certain value. The alternating stress cycles do not have to follow any specific profile. Figure 4.1a to c shows three potential fatigue stress cycles. Figure 4.1a represents a simple sinusoidal tension-compression stress cycle with the same stress amplitude, $\sigma_a$, for the maximum tensile stress and the minimum compression stress. The mean stress, $\sigma_m$, is zero. When the mean stress moves up or down, the cyclically

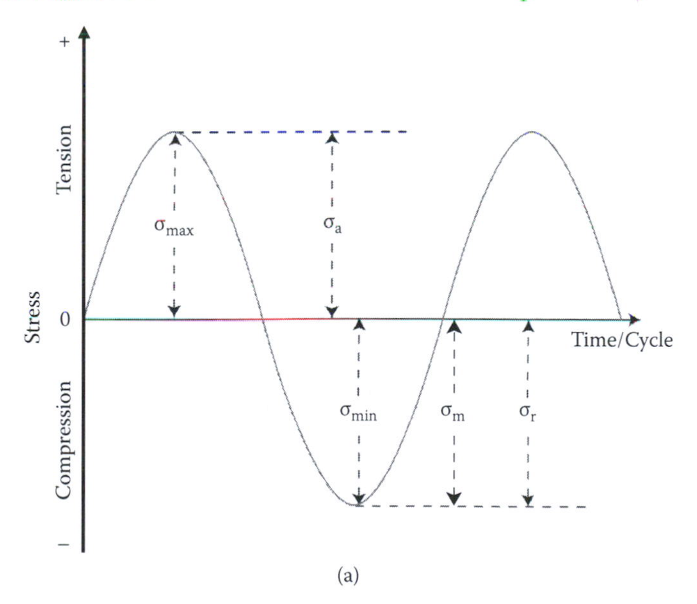


**FIGURE 4.1**
Fatigue stress cycles: (a) tension-compression stress cycle.

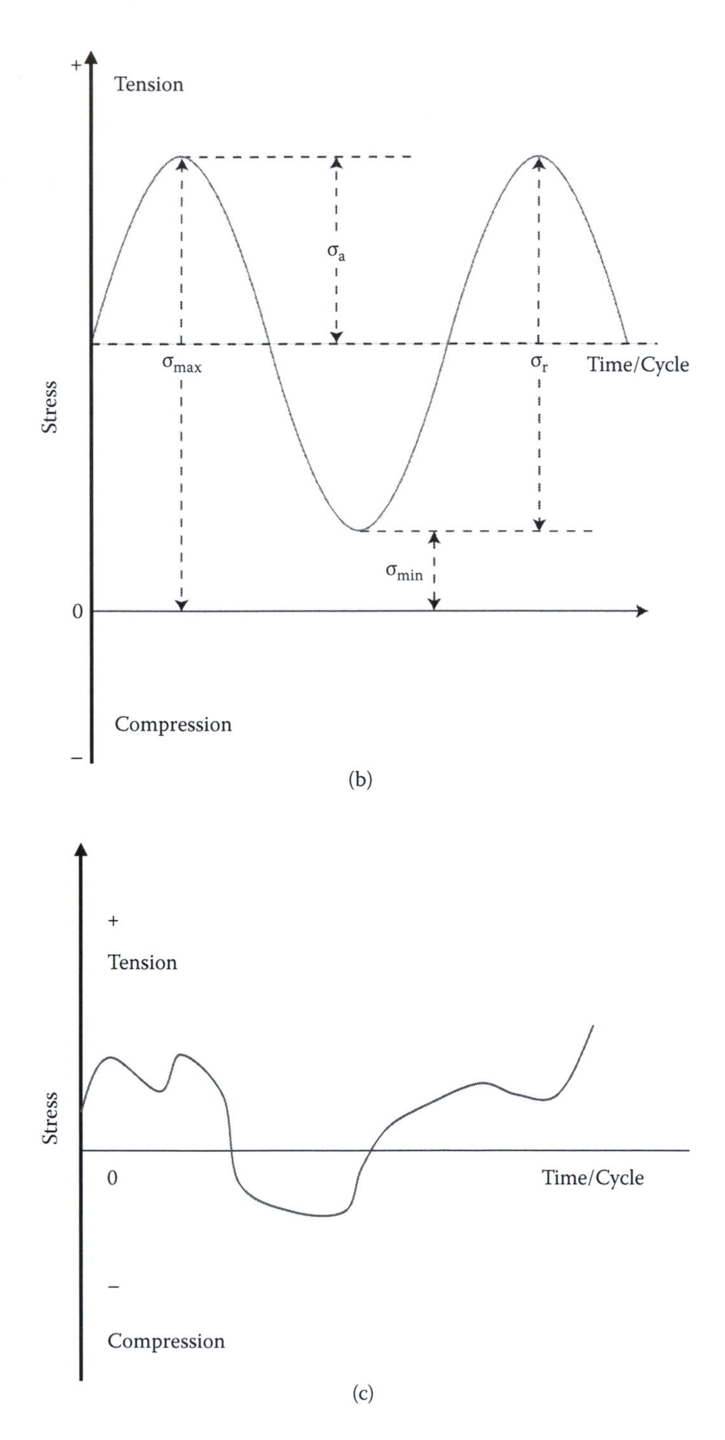


**FIGURE 4.1**
Fatigue stress cycles: (b) tension-stress cycle, and (c) irregular stress cycle.

applied stress can be either all in tension or all in compression, as shown in Figure 4.1b, where all the stresses are in tension. However, in many real-life conditions, the stress profile is rather irregular, as shown in Figure 4.1c.

The maximum and minimum stresses are designated as $\sigma_{max}$ and $\sigma_{min}$, respectively. The amplitude of the alternating stress, $\sigma_a$, is one-half of the stress range, $\sigma_r$, which is the difference between $\sigma_{max}$ and $\sigma_{min}$, and defined by Equation 4.1:

$$\sigma_a = \frac{\sigma_r}{2} = \frac{\sigma_{max} - \sigma_{min}}{2} \tag{4.1}$$

The mean stress, $\sigma_m$, is the algebraic average of $\sigma_{max}$ and $\sigma_{min}$, as defined by Equation 4.2:

$$\sigma_m = \frac{\sigma_{max} + \sigma_{min}}{2} \tag{4.2}$$

Many fatigue test data are reported along with two ratios, $R$ and $A$, to identify the cyclic stress profile. The $R$ ratio, as defined by Equation 4.3, is the stress ratio between $\sigma_{min}$ and $\sigma_{max}$:

$$R = \frac{\sigma_{min}}{\sigma_{max}} \tag{4.3}$$

The $A$ ratio, as defined by Equation 4.4, is the ratio between the amplitude of the alternating stress, $\sigma_a$, and mean stress, $\sigma_m$:

$$A = \frac{\sigma_a}{\sigma_m} \tag{4.4}$$

When $R = -1$ or $A = \infty$, the stress cycle is completely reversed and the mean stress equals zero. One simple example is the sinusoidal tension-compression stress cycle as shown in Figure 4.1a. The correlation between the $R$ and $A$ ratios is described by Equation 4.5:

$$A = \frac{1-R}{1+R} \tag{4.5}$$

Figure 4.2 is a schematic of three fatigue stress profiles with different $R$ and $A$ ratios. The stress profile of $R = 0$ or $A = 1$ represents a cyclic loading condition starting from zero, reaching the maximum stress, then falling down to zero again. Most reference fatigue data reported in the engineering handbooks or other databases are under this loading condition. For comparative analysis in reverse engineering, the actual alternating stress experienced by a component in service is often normalized so that the component life can be determined by comparing it to the reference data based on the

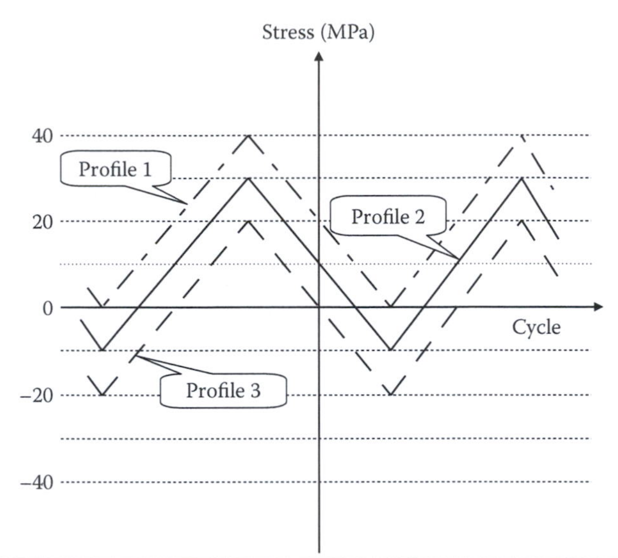


| Profile | $\sigma_{max}$ (MPa) | $\sigma_{min}$ (MPa) | $\sigma_m$ (MPa) | $\sigma_a$ (MPa) | $R = \sigma_{min}/\sigma_{max}$ | $A = \sigma_a/\sigma_m$ |
|---|---|---|---|---|---|---|
| Profile 1 | 40 | 0 | 20 | 20 | 0/40 = 0 | 20/20 = 1 |
| Profile 2 | 30 | -10 | 10 | 20 | -10/30 = -1/3 | 20/10 = 2 |
| Profile 3 | 20 | -20 | 0 | 20 | -20/20 = -1 | 20/0 = ∞ |

**FIGURE 4.2**
Fatigue stress profiles.

$R = 0$ or $A = 1$ loading condition. This normalized (zero-to-maximum) tensile stress is referred to as the Walker equivalent stress, which is defined by the Walker relationship as expressed by Equation 4.6 (Walker, 1970). It provides an equivalent algebraic maximum tensile stress for fatigue life prediction by comparing data under $R = 0$ (or $A = 1$).

$$\sigma_{\text{Walker}} = \sigma_{R,\max}(1 - R)^m \tag{4.6}$$

where $\sigma_{Walker}$ = Walker equivalent stress = equivalent algebraic maximum stress at $R = 0$ (or $A = 1$), $\sigma_{R,\max}$ = algebraic maximum stress at a specific $R$ ratio, and $m$ = Walker exponent.

For a component subject to multiaxial cyclic stress, the effective alternating and mean stresses can be calculated following the Von Mises theory. In a special zero-max-zero biaxial cyclic stress profile in both $x$ and $y$ directions, the effective alternating stress can be calculated using Equation 4.7:

$$\sigma_{eff,\max} = \sqrt{\sigma^2_{x,\max} + \sigma^2_{y,\max}(\sigma_{x,\max})(\sigma_{y,\max})} \tag{4.7}$$

where $\sigma_{eff,\max}$ = effective maximum stress, $\sigma_{eff,\max}/2$ = effective alternating stress, $\sigma_{x,\max}$ = maximum stress in the $x$ direction, and $\sigma_{y,\max}$ = maximum stress in the $y$ direction.

An $A$ or $R$ ratio is first calculated with this effective alternating stress, $\sigma_{eff,\max}/2$, and then an effective uniaxial Walker equivalent stress is calculated with Equation 4.6 for fatigue life comparison.

### 4.2.1 The *S-N* Curve and High-Cycle Fatigue

High-cycle fatigue is characterized by a relatively high number of cycles before failure, typically beyond $10^4$ or $10^5$ cycles, with a relatively low stress only causing elastic strain. In other words, the HCF is primarily controlled by the material's elastic behavior. A frequently encountered HCF problem in the aviation industry is the fatigue induced by thermal cycles. For instance, during engine operation, the temperature rises and cools, generating a cyclic thermal stress imposed on the engine parts. Other HCF failures may result from vibration. The HCF data of a material are often presented by an *S-N* curve, where *S* is the applied stress and *N* is the number of cycles to failure. The stress plotted in the *S-N* curve is usually $\sigma_a$. It can also be $\sigma_{\max}$ or $\sigma_{\min}$, where $\sigma_a$, $\sigma_{\max}$, and $\sigma_{\min}$ are the alternating, maximum, and minimum stresses, respectively. The cycles are usually plotted on a logarithmic scale due to the large *N* values, which can go up to $10^9$ cycles. There are many factors that affect fatigue strength, including alloy microstructure, specimen surface condition, temperature, and frequency of stress cycle. Due to the inherent complexity of fatigue behavior and the difficulty to duplicate exactly the same test conditions, fatigue test data are more widely scattered than tensile test data. From time to time a band instead of a curve is used to reflect the data scattering in the *S-N* curves.

For mild steel and many other engineering alloys, for example, nickel alloys, the *S-N* curve levels off when the applied stress is below a certain value. The critical minimum stress, below which fatigue failure will not occur, is defined as the fatigue endurance limit. For ferrous alloys the ratio between ultimate tensile strength and the fatigue endurance limit, which is also known as the endurance ratio, usually ranges from 0.4 to 0.5. However, there is no precise quantitative relationship between fatigue endurance limit and ultimate tensile strength. Great caution should be exercised when inferring fatigue properties based on tensile properties in reverse engineering applications. Some engineering alloys, for example, aluminum alloys, have an *S-N* curve that never completely levels off. The cycles to failure will continuously increase with ever decreasing stress. In this case, the fatigue endurance limit or strength is defined as the stress at which the alloys will not fail at a reasonable cycle, for example, $10^7$ or $10^8$ cycles. Figure 4.3 shows the plots of *S-N* curves of aluminum alloys at room temperature. The alloys tested for curves 1 and 2 are modified 2024 aluminum alloys with the addition of lithium that increases tensile strength. Curves 3 and 4 are the *S-N* curves of 2024 aluminum alloys. The alloy for curve 3 is manufactured through a rapidly solidified process, and therefore has a finer grain size than the alloy for curve 4, which is an ingot alloy.

Stress-controlled fatigue tests and the properties generated by these tests are useful for HCF analysis where elastic strains are dominant. One

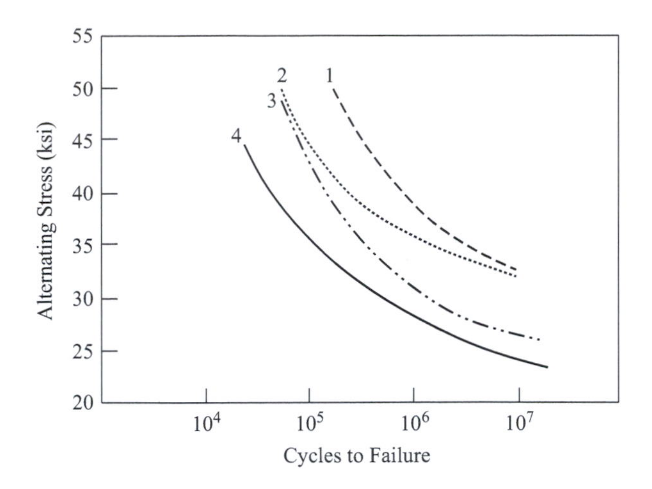


**FIGURE 4.3**
Stress vs. strain fatigue curves.

simple and commonly used stress-controlled fatigue test is the rotating-beam fatigue test that generates *S-N* curves. The Basquin equation, as described by Equation 4.8, is derived from the *S-N* curve. It shows lower HCF life cycles with increasing alternating stresses:

$$N\sigma_a^p = C \tag{4.8}$$

where $\sigma_a$ = alternating stress, $N$ = number of cycles to failure, $p$ = empirical constant, usually having a value of ≈1/10, and $C$ = empirical constant.

The Basquin equation can be reformulated as Equation 4.9 to show the mathematical relationship between HCF and elastic strain:

$$\sigma_a = \frac{\Delta\varepsilon_e}{2}E = \sigma'_f(2N)^b \tag{4.9}$$

where $\Delta\varepsilon_e/2$ = elastic strain amplitude; $E$ = Young's modulus; $\sigma'_f$ = fatigue strength coefficient approximately equal to the monotonic true fracture stress, $\sigma_f$; $2N$ = number of load reversals to failure ($N$ = number of cycles to failure); and $b$ = fatigue strength exponent, which varies between −0.05 and −0.12 for most metals.

The Basquin equation can also be remodified as Equation 4.10 to include the effect of mean stress.

$$\sigma_a = (\sigma'_f - \sigma_m)(2N_f)^b \tag{4.10}$$

where $\sigma_m$ = mean stress. Equation 4.10 explains the well-acknowledged observation that a positive tensile mean stress can shorten fatigue life.

For example, shot peening introduces residual compressive stress on part surface, lowers the mean tensile stress, and therefore improves part's fatigue life.

### 4.2.2 Low-Cycle Fatigue

Low-cycle fatigue is usually related to high local stresses that result in failure after less than about $10^4$ or $10^5$ cycles. Plastic strain is the primary deformation mode. Figure 4.4 is a photo of a fatigue test machine equipped with modern computer technology for operation control and data processing. The installed software often offers engineers the capacities to design the customized stress profiles with high precision and automization of the test accordingly. The software usually also provides graphic and calculation tools to make posttest data analysis, reporting, and filing easy. The machine in Figure 4.4 is installed with a furnace in the center as well, which allows it to conduct fatigue tests at both room and elevated temperatures.

The fatigue fractures observed in turbine engine disks are usually LCF failures. In the LCF regimes, where large plastic strains determine the rate

**FIGURE 4.4**
A fatigue test machine.

of damage accumulation, stress-based analyses become inaccurate, and the Basquin equation no longer holds. Strain-controlled fatigue behavior dominates in LCF. The most common method to present the LCF data is to plot the plastic strain range, $\Delta\varepsilon_p$, or the total strain range, $\Delta\varepsilon$, vs. the cycles. The data points converge to a straight line and show a linear relationship between $\Delta\varepsilon_p$ and $N$ in a log-log coordinate system; that is, $\log(\Delta\varepsilon_p)$ is linear, inversely proportional to $\log(N)$. Mathematically this relationship is expressed by the Coffin–Manson law, as defined by Equation 4.11, based on empirical observations:

$$\frac{D\varepsilon_p}{2} = \varepsilon_f'(2N)^c \tag{4.11}$$

where $\Delta\varepsilon_p/2$ = plastic strain amplitude; $\varepsilon_f'$ = fatigue ductility coefficient, approximately equal to the true fracture strain, $\varepsilon_f$, for many metals; $2N$ = number of strain reversals to failure (one cycle, two reversals); and $c$ = fatigue ductility exponent, which varies between −0.5 and −0.7 for many metals.

Table 4.1 compares LCF and HCF in terms of strain, controlling factor, equations that prevail, and testing methods.

### 4.2.3 Component Low-Cycle Fatigue Life Prediction

The primary application of LCF theory in reverse engineering is for LCF life prediction to demonstrate that the duplicated counterpart is equivalent to or better than the original OEM part. Many mechanical components, for example, rotating shafts and jet engine turbine disks, are subject to LCF life limits in service. The methodology applied for LCF life prediction of a critical component is usually part specific, depending on its criticality, and applied under the most severe conditions to ensure maximum safety. The reliability of material data is critical to the theoretical calculation of a component LCF life. The test results directly from a laboratory report and the field data collected in service are the most reliable data and should take precedence in reverse engineering practice. Design experience is another important factor that is commonly referred to as corporate knowledge in designing the subject

**TABLE 4.1**

Comparison between Low-Cycle Fatigue and High-Cycle Fatigue

| Parameter | Low-Cycle Fatigue | High-Cycle Fatigue |
|---|---|---|
| Strain | Plastic strain (related to material ductility) | Elastic strain (related to material strength) |
| Controlling parameter | Elastic and plastic strain | Stress |
| Equation | Coffin-Manson equation | Basquin equation |
| Cycle | $<10^4$ or $10^5$ | $>10^4$ or $10^5$ |
| Testing | Strain-controlled fatigue test | Stress-controlled fatigue test |

part. The selection of a computer model in a reverse engineering application may rely on previous experience with this software system. The experience of specific features, for example, effect of a bolt hole in an engine turbine disk on its vulnerability to crack initiation, also plays a significant role in LCF life prediction.

The LCF life prediction is profoundly affected by the design philosophy applied to the part, such as fail-safe, damage tolerance, and statistical minimum LCF life. The concept of fail-safe is widely adopted in the design of critical components. The philosophy is that despite precautions that have been taken to avoid failure, if failure does occur, the structure will still be safe. In many designs the concept of damage tolerance is also integrated with basic requirements to ensure that the component can tolerate unexpected damage during its operation. A part might be designed with acceptable tolerance of existing cracks up to 1/10 in. In reverse engineering the parts designed with this level of conservatism require more information than just meeting the typical strength requirements based on service loads. What level of sporadic overload can a structure sustain while still functioning safely? What is the acceptable damage a part can tolerate? The answers to these questions vary for different industries. For example, crack initiation is the underlying base used to establish LCF life limits in the aviation industry. In other words, the LCF life of a new part is calculated based on the assumption of no preexisting cracks. The question is then whether a part has to be immediately replaced when a crack is observed. If not, when should it be replaced? In the aviation industry, a part is usually considered unsafe when it has a crack of 1/32 in. in length, and when this happens a replacement is required. The minimum LCF life can be defined as the B.1 statistical life for finding a 1/32-in. crack. It means if 1 out of 1,000 parts (e.g., B.1 statistically) has been observed showing a 1/32-in.-long crack, the corresponding number of fatigue cycles is the minimum LCF life cycle for this part.

The estimated component LCF life varies depending on the methodology used in life prediction. It can be predicted based on the material LCF properties, theoretical calculations with various assumptions, or component testing results. The local LCF life cycles are also significantly different at different locations within the same component due to geometrical variations. In a conservative approach, the approved component LCF life adopted in a design is the lowest life cycle value at issue. For example, the predicted LCF life at the bolt hole area of a disk is usually less than that in the hub area of the same disk, and the LCF life at the bolt hole is used as the LCF life for the disk in design. A safety factor is usually integrated into the calculation of an "approved life" to ensure the safety margin. For instance, in the aviation industry, the approved life published in the engine manual is usually the product of the theoretically allowed component LCF life and a safety factor, ranging from 1/3 up to 1.

The material properties listed in the *Metallic Materials Properties Development and Standardization* (MMPDS) *Handbook* are generally acceptable for most

component LCF life cycle analyses. This handbook is an engineering data source available to the U.S. public through the National Technical Information Service, and is a replacement for the obsolete Military Handbook 5 (MIL-HDBK-5), *Metallic Materials Properties Development and Standardization*. The component LCF life used for engineering design can be established by using the MMPDS median material LCF life along with a scatter factor. Neither the industries nor government safety regulatory authorities have mandated a universal scatter factor value in component LCF life calculation. However, the use of a proper scatter factor is highly advisable. A scatter factor of 3 is generally acceptable in the aviation industry to estimate the component LCF life that has a damage tolerance capability, and 5 for a component that is not fully damage tolerant. For a damage-tolerant structure the component LCF test would run to three times the LCF life goal. If the LCF life goal is 40,000 flights, then the test article would be run to 120,000 cycles to justify the 40,000 LCF flight cycles when damage tolerance is factored in and a scatter factor of 3 is used. For a structure that is not fully damage tolerant, a factor of 5 would be used, with the test run to 200,000 cycles to justify a 40,000 LCF flight life.

The component LCF life also depends on the manufacturing process. For example, despite that the typical component LCF life is calculated based on grain facet, the LCF life of a component made from power metallurgy needs to be assessed for the effects of both grain facet and inclusion, because fatigue cracks might initiate at either the grain facet or inclusion sites. The inclusion is usually heterogeneous in nature. Its effect on LCF life depends on the inclusion size, location, and distribution, and is estimated by probabilistic fatigue analysis that incorporates fracture mechanics principles and crack incubation time.

Usually the theoretical LCF lifing calculation is accomplished using finite element modeling. A two-dimensional model can be used for an axisymmetric part, for example, engine fan disk, while a three-dimensional model is more accurate for parts with nonaxisymmetric geometry, for example, discontinuity caused by bolt holes, or attachment slots for fan blades on the engine fan disk. One common concern is that the finite element model often fails to accurately predict the stress concentration factor, and can overestimate the fatigue life cycles. An accurate three-dimensional modeling can predict a stress concentration factor five to ten times larger than that estimated by a similar two-dimensional modeling. In reverse engineering, the selection of a proper finite element model is essential for accurate LCF life prediction.

Fatigue cracks are usually initiated at a free surface, for example, part surface. In those rare instances where fatigue cracks initiate in the interior, there is always an interface involved, such as the interface of a nonmetallic inclusion and the base metal. Surface hardening by carburizing or nitriding treatment, or shot peening often enhances fatigue strength by preventing fatigue crack initiation on the surface.

The integrity of a component surface or subsurface plays a key role in determining the component fatigue life. Today's ever-improved material process control has minimized material anomalies for some high-quality alloys made by conventional casting or forging processes. For example, in a high-quality aerospace-grade titanium alloy, the occurrence of the detrimental hard alpha phase is less than twice in every 1 million kilograms of material; and in a nickel-base superalloy, an oxide white spot can only be found a few times in every 1 million kilograms. However, statistically the occurrence of material anomalies in powder metallurgy alloys is usually much higher. It is not unusual to find thousands of small micro anomalies in a single turbine disk made of powder metallurgy alloy. This has led the aerospace industry to adopt different methodologies to predict the fatigue life for components made of powder metallurgy alloys.

Many parts reproduced by reverse engineering are subject to cyclical stresses of fluctuating magnitudes in service. Their fatigue lives can be estimated based on the linear cumulative fatigue damage described by Equation 4.12, the Palmgren-Miner's rule:

$$\frac{n_1}{N_1}+\frac{n_2}{N_2}+\frac{n_3}{N_3}+\ldots=\sum\frac{n_i}{N_i}=1 \tag{4.12}$$

where $N_1$ is the number of cycles to failure under stress $\sigma_1$, $n_1$ is the number of cycles the component is exposed to while under stress $\sigma_1$, $N_i$ is the number of cycles to failure under stress $\sigma_i$, and $n_i$ is the number of cycles the component is exposed to while under stress $\sigma_i$. The total fatigue life is $\sum n_i$. The Palmgren-Miner's rule states that the total fatigue life can be estimated by adding up the percentage of the life that is consumed at each stress level to which the component has been exposed. There are many exceptions to this simple linear damage summation rule; however, it does provide a first order of engineering approximation to estimate the fatigue life when the component is subject to irregular alternating stresses.

It is worth noting that in service, a component is often subject to multiaxial loads, such as axial and radial stresses, at the same time, and multiple fatigue modes, such as tension, torsion, and bending cyclic stresses, simultaneously. The most reliable life perdition of a component is based on a direct component test with a real-life simulated loading condition.

### 4.2.4 Effect of Mean Stress on Fatigue

In engineering service a component is often subject to both dynamic alternating stresses and static steady-state stresses. A static steady-state stress resembles a mean stress upon which the cyclical stress is imposed. The alternating stress range, $\sigma_{max} - \sigma_{min}$, that can be imposed onto a mean stress without fatigue failure decreases when the value of the tensile mean stress

increases, where $\sigma_{max}$ and $\sigma_{min}$ are the maximum and minimum stresses, respectively. The curves that show the dependence of the alternating stress range on mean stress are generally referred to as Goodman diagrams. The Goodman diagrams are presented in various formats. One is schematically illustrated in Figure 4.5, showing the basic principle of a Goodman diagram. The mean stress is plotted along the *x*-axis as the abscissa, and the total stress is plotted along the *y*-axis as the ordinate, where $\sigma_u$ and $\sigma_y$ are the ultimate tensile and yield strengths, respectively; $\sigma_e$ is the fatigue endurance limit; $\sigma_m$ and $\sigma_a$ are the mean and alternating stresses, respectively; and $\sigma_r$ is the alternating stress range. Also plotted in the diagram is a supplementary line with a 45° inclination showing the middle mean stress between the maximum and minimum alternating stresses. It shows the allowed stress boundary for a fatigue life, with the maximum stress on top and the minimum stress at the bottom. If the yield strength is the design criterion for failure, the maximum and minimum stress boundaries converge to yield strength with decreasing stress amplitude when the mean stress increases.

Several modified Goodman diagrams are schematically illustrated in Figure 4.6, where the ordinate *y*-axis is the alternating stress and the abscissa *x*-axis is the mean stress. The diagram shows two additional stress boundaries, the Gerber parabolic curve and the Soderberg line, where $\sigma_{yt}$, $\sigma_{ut}$, and $\sigma_{yc}$ are yield strength in tension, ultimate tensile strength, and yield strength in compression, respectively. It is assumed that both tensile and compressive yield

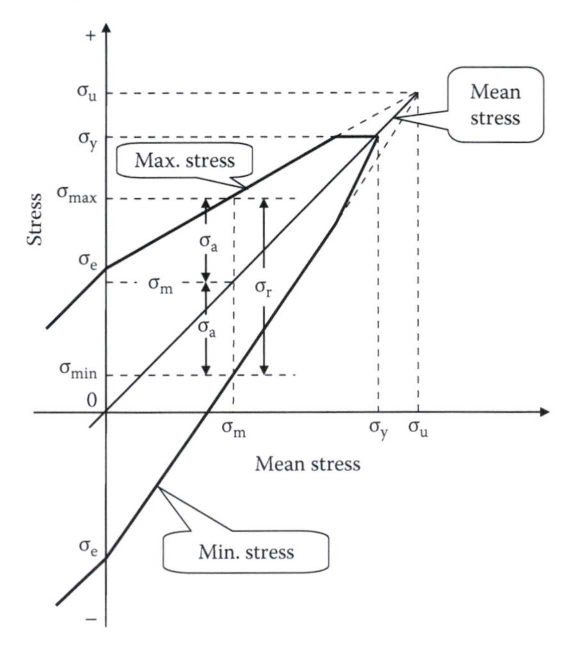


**FIGURE 4.5**
Goodman diagram.

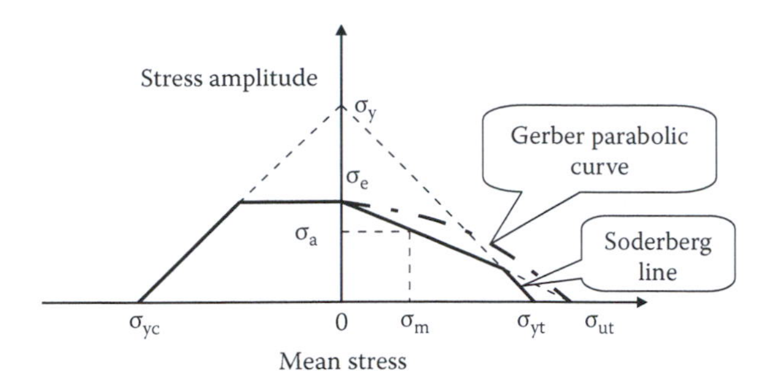


**FIGURE 4.6**
Revised Goodman diagrams.

strengths are the same and simplified as $\sigma_y$ in the ordinate axis. Following the classic Goodman theory, the straight line connecting the fatigue endurance limit, $\sigma_e$, and the ultimate tensile strength in tension, $\sigma_{ut}$, presents a boundary of fatigue limit. Any combination of mean stress and alternating stress that falls below this boundary meets the fatigue safety criterion. Gerber took a more liberal approach to reflect a better match with the experiment data. The Gerber parabolic curve connects $\sigma_e$ and $\sigma_{ut}$ with a parabolic curve instead of a straight line. Soderberg suggested a more conservative approach. He replaced the ultimate tensile strength with the yield strength in tension, $\sigma_{yt}$; therefore, the Soderberg line bends the $\sigma_e$–$\sigma_{ut}$ connection to $\sigma_{yt}$. The alternating stress is reduced as the mean stress in tension increases, and it eventually reduces to zero when the mean stress reaches the ultimate tensile strength in the Gerber parabolic curve, or yield strength in the Soderberg line. However, the mean stress in compression shows little effect on fatigue strength. The alternating stress essentially remains the same when the compressive mean stress increases within a boundary. Therefore, a straight line usually applies in the compressive mean stress region to reflect the marginal effect of compressive mean stress on fatigue strength until it reaches the $\sigma_y$–$\sigma_{yc}$ boundary line, when the alternating stress is subject to the limitation of yielding. Mathematically these diagrams can be expressed as Equation 4.13, where $\sigma_a$, $\sigma_e$, $\sigma_m$, and $\sigma_u$ are alternating stress, fatigue endurance limit, mean stress, ad ultimate tensile strength, respectively, and $x$ is an exponent constant. When $x = 1$, Equation 4.13 represents the Goodman linear diagram, and when $x = 2$, the Gerber parabolic diagram. Equation 4.13 represents the Soderberg diagram when $\sigma_u$ is replaced by $\sigma_y$. Currently, there is no established methodology to decode which theory was used by an OEM in fatigue life analysis. This is still a dilemma that reverse engineering faces today. The best solution is to make an educated judgment call based on industrial standards, corporate knowledge, and tests if necessary.

$$\sigma_a = \sigma_e \left[ 1 - \left( \frac{\sigma_m}{\sigma_u} \right)^x \right] \tag{4.13}$$

In engineering analysis the Goodman diagram is often plotted with constant fatigue life curves and referred to as a constant fatigue life diagram. Figure 4.7 is a simulated Goodman diagram depicted based on the data extracted from Military Handbook K-5 of an engineering alloy. It is for general discussion purposes only. It is a master Goodman diagram showing various $R$ and $A$ ratios for smooth unnotched and notched specimens. This master diagram summarizes the relationship between fatigue life and the following parameters: maximum stress, minimum stress, alternating stress, mean stress, stress and strain ratios, symbolized as $\sigma_{max}$, $\sigma_{min}$, $\sigma_a$, $\sigma_m$, $R$ and $A$, respectively. There are two sets of coordinate systems in this diagram. The inside one is established by turning the referenced Goodman diagram 45° counterclockwise. The $y$-axis of the internal coordinate system represents the alternating stress. It is coincident with the $R = -1$ ($A = \infty$) line where $\sigma_{min} = -\sigma_{max}$, $\sigma_a = \sigma_{max}$, and $\sigma_m = 0$. The $x$-axis of the internal coordinate system reflects the mean stress. It is coincident with the $R = 1$ ($A = 0$) line, which represents a simple tensile test condition, $\sigma_{min} = \sigma_{max} = \sigma_m$, and $\sigma_a = 0$. The $y$-axis of the external coordinate system is marked with the maximum stress. The $x$-axis of the external coordinate system is marked with the minimum stress; tension is on the right as a positive value, and compression is to the left as a negative value. The stress condition $R = 0$ and $A = 1$ is represented by the vertical line perpendicular to the minimum stress $x$-axis at $\sigma_{min} = 0$, and to the right the dashed line represents another stress condition, $R = 0.2$, $A = 0.67$. There are also two sets of constant fatigue life curves, one in solid and another in dashed lines. The solid curves are boundaries confining the safe combined stress conditions for smooth unnotched specimens. The dashed curves apply to the notched specimens. The lower fatigue lives for the notched specimens are due to the effects of stress concentration at the notches.

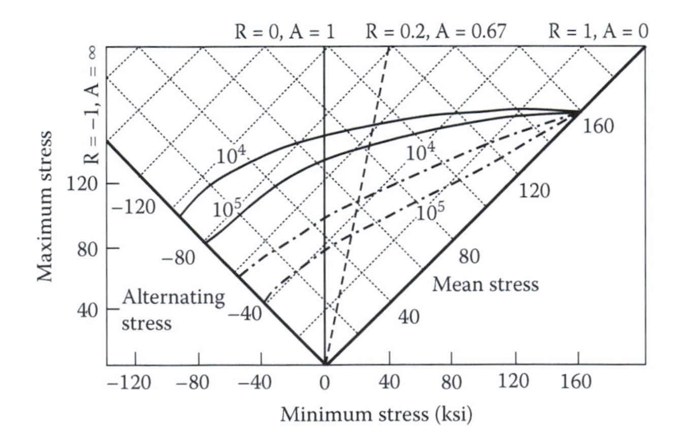


**FIGURE 4.7**
Master Goodman diagram of AISI 4340 steel. Data from Military Handbook 5, U.S. Department of Defense (Dieter, G. F., *Mechanical Metallurgy*, McGraw-Hill, New York, 1986, p. 386).

This master Goodman diagram is useful for predicting fatigue life with any two independent parameters. If the mean stress is 80 ksi and $R = 0$ (or $A = 1$), the estimated fatigue life for a smooth unnotched specimen can be predicted at the intersection of the two lines of constant $\sigma_m = 80$ ksi and $R = 0$ (or $A = 1$). In this case, the estimated fatigue life will be less than $10^4$ cycles.

The superimposed curves and scales make Figure 4.7 a very busy diagram. However, the complexity also allows predictions of fatigue life to be made using alternate data sets. The maximum and minimum stresses, $\sigma_{max}$ and $\sigma_{min}$, are the two most commonly cited stress data in fatigue analysis. Their inclusion as the ordinate and abscissa coordinates in Figure 4.7 makes the fatigue life analysis easier, particularly for those who are not familiar with the terminologies used in fatigue analysis, such as $R$ or $A$ ratio. If a notched specimen is subject to a minimum stress, $\sigma_{min} = 20$ ksi in a fatigue loading condition with $R = 0.2$, the corresponding maximum, mean, and alternating stresses will be $\sigma_{max} = 100$ ksi, $\sigma_m = 60$ ksi, and $\sigma_a = 40$ ksi, respectively. The fatigue life can be estimated as approximately $3 \times 10^4$ cycles by locating the intersection of the two lines of constant $\sigma_{min} = 20$ ksi and $R = 0.2$ in Figure 4.7. Alternatively, the same fatigue life can also be found by locating the intersection of the two lines $\sigma_{min} = 20$ ksi and $\sigma_{max} = 100$ ksi, or $\sigma_m = 60$ ksi and $\sigma_a = 40$ ksi, etc., because all these lines intersect at the same point.

### 4.2.5 Fatigue Crack Propagation

A fatigue failure usually starts with a localized minute crack, like a scratch, tool mark, or corrosion pit, and progressively deteriorates to eventual failure. A fatigue also occurs at discontinuous areas like a weld-repaired area or an area adjacent to a bolt hole. The stresses at these localized defects or discontinuities can be significantly higher due to stress concentration. At the fatigue crack tip, the material is subject to plastic deformation. However, it will convert to elastic deformation at a short distance. A typical fatigue failure often initiates before it can be detected, and progresses in three stages. First, a small crack initiates at a location associated with either material irregularities (e.g., inclusion or void) or stress concentration due to sharp geometrical variation (e.g., small fillet radius or keyway on the shaft). The discovery of this original crack initiation site often helps explain the root cause of the fatigue failure. Then the crack will grow and propagate, reducing the effective load-carrying cross-sectional area and weakening the component. Fatigue crack propagation is ordinarily transgranular. Corresponding to the cyclical stress, synchronized striations are generated in ductile alloys during the crack propagation process. Typically, one striation is generated during each fatigue cycle. Measurements of these striations provide a method of estimating crack growth rate. Finally, the component fails due to overload when the remaining cross section can no longer sustain the load. Fatigue

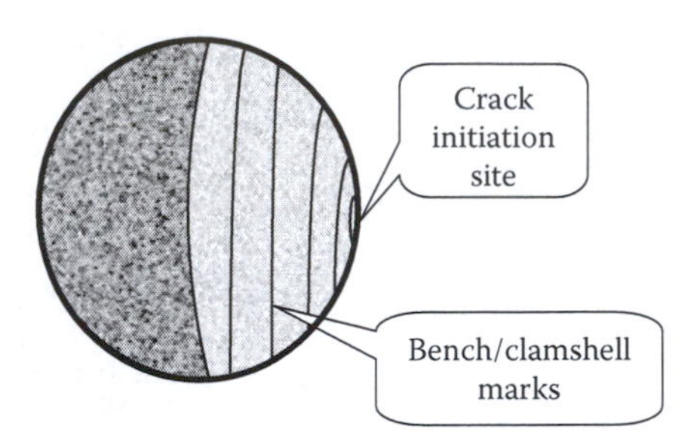


**FIGURE 4.8**
Schematic of fatigue benchmarks.

striations can only be observed under a high-magnification microscope. Fatigue striations should not be confused with benchmarks, which appear as irregular ellipses or semiellipses and can be observed without a microscope. Benchmarks are also referred to as clamshell marks or arrest marks, as schematically illustrated in Figure 4.8. These marks are created by drastic stress changes during fatigue that cause severe deformation and alter crack growth rate. Benchmarks usually converge to the origin of the fracture, which helps determine the location of the crack initiation site. Figure 4.9a and b (NTSB, 2005) shows the fractography of a mechanical part failed due to fatigue in the adjacent area of two drilled holes, as marked by the arrows. The benchmarks emerging from the crack origin are clearly visible on the fracture surface in Figure 4.9a. Figure 4.9b is a close-up view of the fatigue initiation site. The two brackets indicate fatigue origin areas at the surfaces of the fastener hole, and the dashed lines indicate the extent of the fatigue region.

After initiation, a fatigue crack propagates slowly in the order of angstroms in the early stage, and shows featureless fracture surface. The propagation rate increases to a few microns per cycle after reaching the steady state. For ductile metals such as beta-annealed Ti–6% Al–4% V alloy, the fracture surface generated in this stage typically shows distinctive fatigue striations. However, the presence of striations is not the defining condition for fatigue crack propagation. Many brittle alloys fail by fatigue showing no striations at all, and others show striations only in certain areas, as illustrated in Figure 4.10. It is the fatigue fractography of an aluminum alloy failed after $2.8 \times 10^6$ cycles. The crack propagation rate rapidly increases in the final stage, quickly becoming unstable and resulting in a final total fracture.

Figure 4.11 is a schematic representation of fatigue crack propagation rate. The fatigue crack propagation or growth rate, $da/dn$, is most often plotted against the range of stress intensity factors $\Delta K$. The stress intensity factor, $K$, is a measurement of fracture toughness. The maximum, the minimum, and the range of stress intensity factors involved in fatigue crack growth

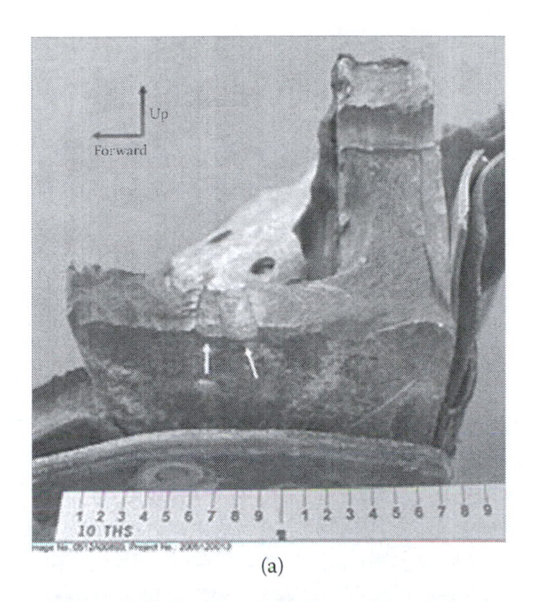


(a)

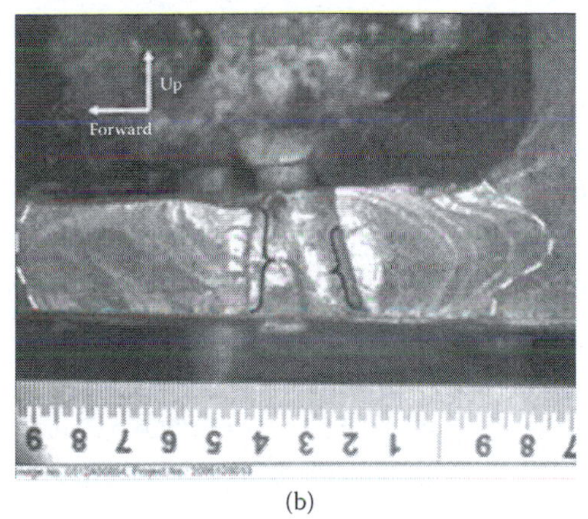


(b)

**FIGURE 4.9**
(a) Fractography of a fatigue failure. (b) Close-up view of fatigue initiation site. (Both photos courtesy of NTSB.)

are defined by Equations 4.14 to 4.16, respectively, for thin plates with edge cracks under tension.

$$K_{\max} = \sigma_{\max}\sqrt{\pi a} \tag{4.14}$$

$$K_{\min} = \sigma_{\min}\sqrt{\pi a} \tag{4.15}$$

$$\Delta K = K_{\max} - K_{\min} \tag{4.16}$$

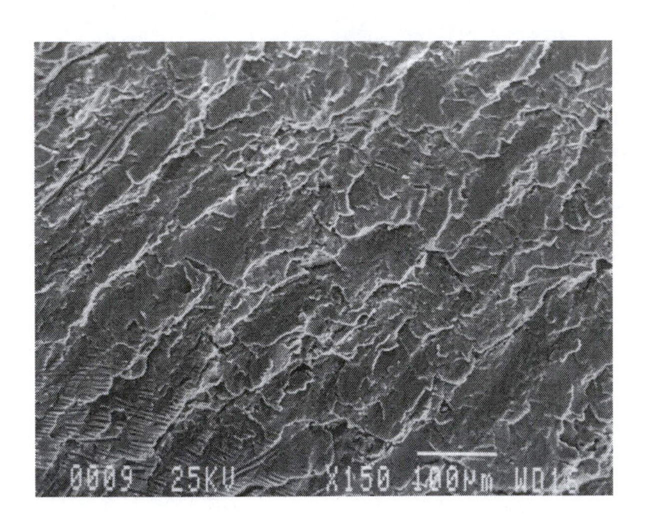


**FIGURE 4.10**
Fatigue fractography of an aluminum alloy.

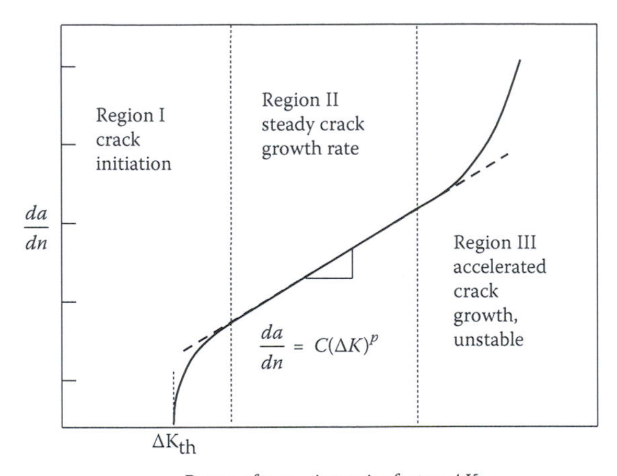


**FIGURE 4.11**
Schematic of fatigue crack growth rate.

Figure 4.11 shows three distinctive regions of fatigue crack propagation behavior. In region I, the fatigue crack does not propagate when $\Delta K$ is below a critical threshold value, $\Delta K_{th}$. In region II, a linear empirical relationship as expressed by Paris' law in Equation 4.17 exists between $da/dn$ and $\Delta K$ in the logarithm scale.

$$\frac{da}{dn} = C(\Delta K)^p \tag{4.17}$$

where $n$ is the number of cycles, and $C$ and $p$ are empirical constants. The value of $p$ is approximately 3 for steels, and 3 to 4 for aluminum alloys. Paris'

law offers an important linkage between fatigue phenomena and fracture mechanics through fatigue crack propagation rate, *da/dn*, and the range of stress intensity factor, $\Delta K$. Region III is a highly unstable region where the crack propagates at an accelerated rate.

The effects of grain size on fatigue life depend on the deformation mode. Grain size has its greatest effect on fatigue life in the low-stress, high-cycle regime, in which slip band crack propagation predominates. In high stacking-fault energy materials, such as aluminum, cell structures develop readily and they control the slip band cracking propagation. As a result, the dislocation cell structure masks the influence of grain size, and the grain size has less effect on fatigue life. However, in the absence of cell structure because of planar slip in low stacking-fault energy materials, such as $\alpha$ brass, grain boundaries will control the rate of fatigue cracking. In this case, the fatigue life, $N_f$, is inversely proportional to the square root of the grain size, as shown in Equation 4.18.

$$N_f \propto \frac{1}{\sqrt{\text{grain size}}} \tag{4.18}$$

In general, the fatigue strength of metals decreases with increasing temperature with only a few exceptions, for example, mild steel. Fine grain size often results in better fatigue properties at low temperatures. As the temperature increases, the difference in fatigue properties between coarse and fine grain materials decreases. When the temperature reaches a value that is about half the melting point, creep becomes the predominant mechanism in determining material strength. Coarse grain materials have higher creep resistance and become stronger. At elevated temperatures, the fracture mechanism will also shift from transgranular, which is typical for fatigue failure, to intergranular, which is typical for creep failure.

### 4.2.6 Thermal Mechanical Fatigue and Fatigue Initiated from Wear Cracking

Thermal mechanical fatigue is a derivative of LCF. It occurs due to the combined effects of thermal and mechanical stresses. The engine turbine blade is exposed to both very hot and very cold environmental conditions during engine operation. The blade expands to various degrees due to these temperature variations and is subject to thermal stress if these expansions are constrained due to physical restraint. At the same time, the blade is also subject to mechanical stresses, such as the bending stress from gas flowing through the core section of the engine during its operation or the centrifugal stress from the rotation of the disk. This repetitive combination of thermal and mechanical stresses can cause thermal mechanical fatigue. Therefore, when reverse engineering a jet engine turbine blade, thermal fatigue is one of the properties that needs to be evaluated.

Several other types of fatigue resistance also require proper consideration in reverse engineering, as discussed below. Wear occurs between two contact surfaces due to friction. Fractures that occur due to wear are referred to as fretting, erosion, galling, or spalling, depending on the configurations of the contact surface. Fretting occurs between two tightly contacted surfaces that make oscillating movements of extremely small amplitudes, typically 5 to 50 μm. This repetitive movement under pressure can be a hidden cause of fatigue failure because the fretting process may cause local material break-off and initiate fatigue propagation. For example, high-cycle fretting fatigue cracking occurs at the dovetail joint where blades are attached to the rotor disks in aircraft engines. It also occurs in bearing housing assemblies and mechanically fastened joints, such as bolted or riveted joints. A catastrophic failure due to fretting fatigue was seen in the in-flight disintegration of a portion of an Aloha Airlines 737 fuselage section in 1988. The interaction and subsequent rapid linkage of small and often undetectable cracks emanating from, at, and around the fastener-sheet interfaces in aircraft joints was widely believed to be the cause of this accident. The occurrence of fretting fatigue can be evidenced by part surface conditions, such as roughened metallic surfaces. Additionally, surfaces made of steel are usually decorated with reddish brown deposits from fretting fatigue.

Erosion is a type of wear caused by either an abrasive moving fluid or small particles striking on a surface. The leading edge of an engine compressor blade is often eroded by dirt or sand. The damage caused by erosion can be the origin of an HCF failure. Most of the surface damage from erosion and its subsequent effect on fatigue life occurs in the first 20 to 100 hours. After these first 100 hours, the part surface will stabilize with an "eroded" layer. The continuous damage in fatigue life primarily results from the wear of the part itself, such as reduction of dimension, rather than from the surface defects due to erosion.

Galling occurs when two surfaces rub together with friction, as can be seen between the two contact surfaces of seals. Between these two surfaces there are isolated protrusion spots where excessive friction might cause localized welding or "smearing." Therefore, galling between steels is sometimes referred to as cold welding. The rubbing surfaces of the mating parts might fall off and appear rough. Grease or surface coatings are frequently applied to avoid galling; for instance, turbine engine blades are coated with anti-galling materials. The design tolerances, surface finish, hardness, and microstructure of the metals in contact are the key factors affecting the tendency for galling. These part details are a challenge in reverse engineering because they are difficult to match; the exact design tolerances of a part cannot easily be duplicated. Galling can occur even if the parts move slowly. However, using different materials that are individually susceptible to galling sometimes can reduce the risk of galling between them. For example, galling is not a concern when fastening a bolt made of 18-8 stainless steel with a nut made of 17-4 stainless steel.

A spall is a chip or a flake broken off from a solid body, such as a brick, stone, or mechanical component. Spalling is the process of surface failure in which a spall is shed. Mechanical spalling occurs at high-stress contact points, such as the contact in a ball bearing. The cracking and flaking of chips out of a surface on an inner ring, outer ring, or balls of a ball bearing assembly are often resulted from spalling. This type of failure is progressive and, once initiated, will continue to spread. The HCF fractures observed in bearings and gears are often attributable to cracks initiated by spalling due to repeated concentrated stress at the contact surfaces. It is a common failure mode in bearings and is also referred to as rolling contact fatigue.

### 4.2.7 Fatigue and Tensile Strengths

Whether any inference on fatigue strength between two parts can be drawn by comparing their respective tensile strengths is a frequently asked question in reverse engineering. Some material-specific relationships between fatigue and tensile strengths were reported. A U.S. patent was even issued for a method to create a steel with high fatigue strength based on its high tensile strength (Sawai et al., 2003). However, the controlling factors of fatigue and tensile strength are different, and so are their respective failure mechanisms. The tensile stress concentration factor is also different from the fatigue stress concentration factor due to their different surface sensitivities. Fatigue usually initiates on surface or interior irregularities, and shows transgranular fracture at room temperature. These observations do not apply to the failure resulting from overload tensile stress. The tensile fractography of the same alloy can be vastly different from fatigue fractography of the same alloy due to different fracture mechanisms. Figure 4.12 shows ductile

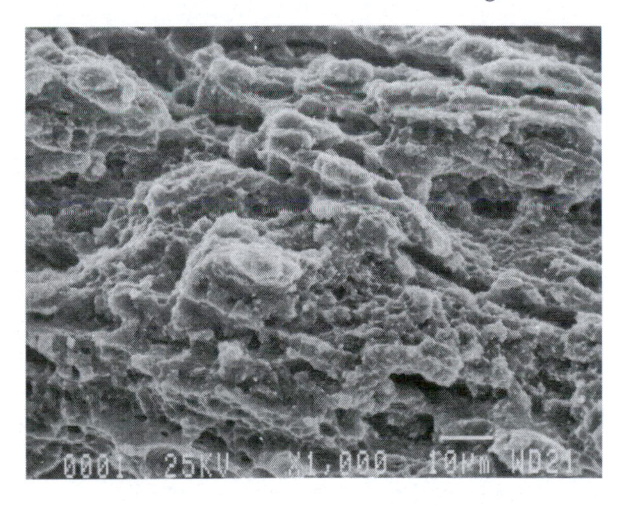


**FIGURE 4.12**
Tensile fractography of an aluminum alloy.

dimple tensile fractography of the same alloy that shows a brittle cleavage fracture in fatigue, as illustrated in Figure 4.10.

The effects of residual stress on fatigue and tensile strengths are different as well. Compressive residual stress is beneficial to fatigue strength, while tensile residual stress is detrimental. Most noticeably, the compressive residual stress induced by shot peening is very beneficial to fatigue strength. On the other hand, if machining or grinding left a tensile residual stress on the part surface, it could be a convenient fatigue crack initiation site later. Empirical testing data might show some relationships between fatigue and tensile strengths, but great caution is urged in applying these relationships in reverse engineering because of the underlying differences between these two failure modes. The relationship between fatigue endurance limit and tensile strength will be further discussed from the perspective of statistical regression in Chapter 6.

## 4.3 Creep and Stress Rupture

### 4.3.1 High-Temperature Failure

The mechanical strength of a metal at elevated temperatures is usually limited by creep rather than by yield strength or other mechanical properties. Creep is one of the primary concerns that could cause the failure of engine turbine blades, which might operate at temperatures above 1,000°C. The primary metallurgical factor affecting metal rupture behavior at elevated temperatures is the transition from transgranular to intergranular fracture. Figure 4.13 is a schematic diagram of grain and grain boundary cohesive strengths as a function of temperature. The grain cohesive strength is lower

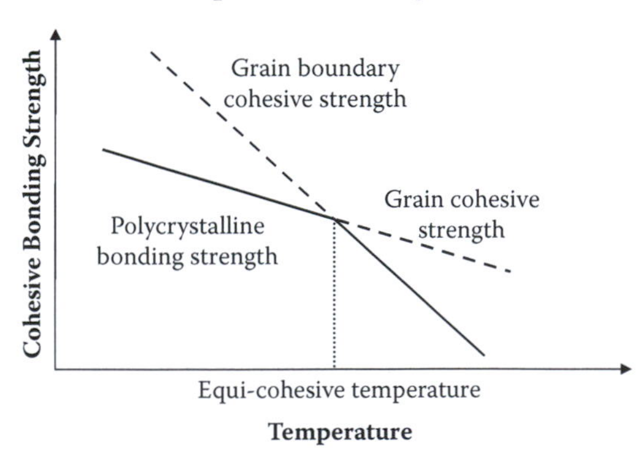


**FIGURE 4.13**
Cohesive bonding strength of polycrystalline metals.

than the grain boundary cohesive strength at lower temperatures. When the temperature increases, the grain boundary cohesive strength decreases more quickly than the grain cohesive strength does. At the equi-cohesive temperature, the controlling strength of a polycrystalline metal shifts from grain cohesive strength to grain boundary cohesive strength. This explains why most fracture modes at elevated temperature are intergranular.

Creep is a time-dependent progressive deformation that occurs under stress at elevated temperatures. In general, creep occurs at a temperature slightly above the recrystallization temperature of the metal involved. The atoms become sufficiently mobile to allow gradual rearrangement of positions at this temperature. A creep test explores the creep mechanism and studies the relationship between stress, strain, and time. Figure 4.14 is a schematic of a typical engineering creep curve tested under constant load. It is a record of strain or elongation against time. The microstructure has a profound effect on creep behavior. For example, the Ti–6% Al–2% Sn–4% Zr–2% Mo alloy shows distinctive creep curves with different microstructures, and the presence of β or pseudo-β microstructure will give the highest creep strength.

A typical creep curve has three stages: primary, secondary, and tertiary. The test specimen has an instant extension as soon as the load is applied. It is marked as the initial strain, $\varepsilon_o$, in Figure 4.14. The deformation rate will gradually slow down in the primary creep stage, and reaches a constant creep rate in the secondary creep stage. This constant creep rate is also the minimum creep rate and is usually referred to as the steady-state creep rate, or simply the creep rate. The slope of the curve can be calculated using Equation 4.19, where $\dot{\varepsilon}$ is the creep rate and $\varepsilon$ and $t$ are creep deformation and time, respectively.

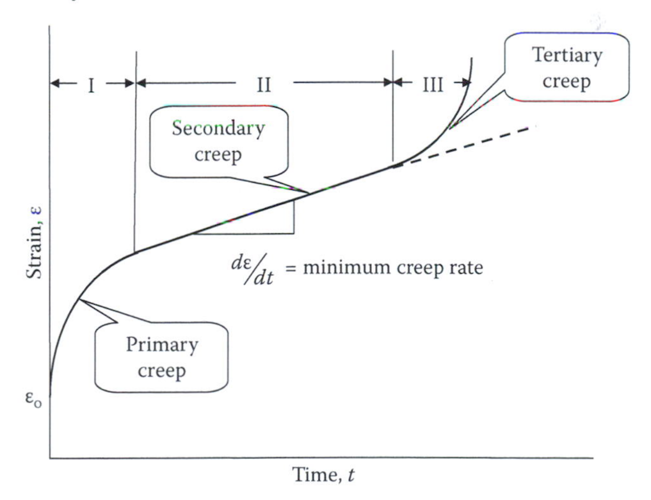


**FIGURE 4.14**
Schematic of creep curve under constant load.

$$\dot{\varepsilon} = \frac{d\varepsilon}{dt} \tag{4.19}$$

The creep rate increases very rapidly in the tertiary creep stage until the specimen finally fractures. The acceleration of creep rate in the final stage can be attributed to many factors, such as the reduction of the load-carrying cross-sectional area due to specimen necking, void formation, or metallurgical changes such as recrystallization, grain or precipitate coarsening, etc.

The stress rupture test is very similar to the creep test except that it is tested at a higher load to cause fracture in a shorter period of time. In contrast to the creep test, the primary focus of the stress rupture test is to study the relationship between stress and rupture time, but not creep mechanism. The stress rupture test fills in the gap between the tensile and creep tests. It provides a set of short-time test data to predict long-time performance by extrapolation. The stress rupture test data are usually presented with a plot of stress against rupture time at a specific temperature on a logarithmic scale, as illustrated in Figure 4.15. Curve 1 is based on a naturally aged aluminum alloy with a composition of Al–3.78% Cu–1.40% Mg–1.63% Li tested at 200°C. Curves 2 and 3 are based on an aluminum alloy with a composition of Al–4.16% Cu–1.80% Mg–0.96% Li–0.50% Mn tested at 200°C as well. Specimens for curve 2 are solution treated at 510°C and naturally aged, while specimens for curve 3 are solution heat treated at 510°C and artificially aged. The stress rupture data might be composed of sections of linear straight lines on the logarithmical scale, as illustrated by curve 3 with different slopes due to metallurgical evolutions, such as the transition from transgranular to intergranular fracture, recrystallization or grain growth, etc. The stress rupture data do not report deformation rate, and can only be used to determine the amount of deformation after fracture or average deformation rate indirectly. The creep deformation rate reflects the combined effects of elastic and plastic deformation. However, the deformation measured after failure in a stress rupture test only shows plastic deformation. Creep failure is often initiated by

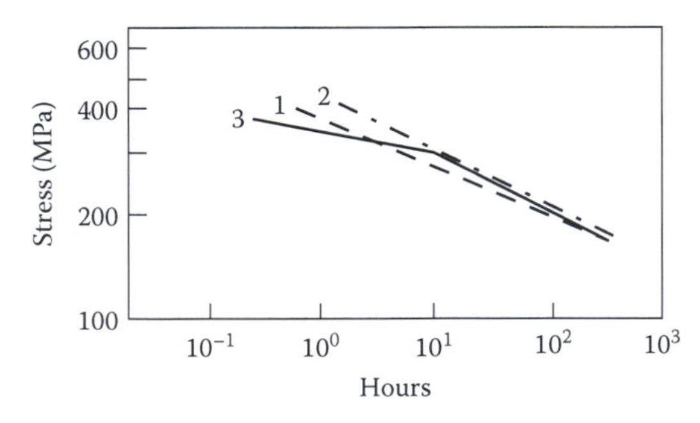


**FIGURE 4.15**
Stress rupture curves.

a distinctive primary crack, and it subsequently grows to a point when the specimen ultimately fails. In contrast, multiple cracks are usually observed in a stress rupture specimen. The adjacent cracks sometimes grow and link together. Figure 4.16a shows multiple cracks observed at the surface of a titanium specimen subject to 379 MPa (55 ksi) at 648.9°C (1,200°F). Figure 4.16b shows the linkage between two cracks. The linkage between separate cracks can form a continuous crack that eventually fails the specimen. Figure 4.16c is an SEM fractography of this alloy showing a mixed intergranular ductile dimple and brittle stress-ruptured surface. Nonetheless, the stress rupture data are still of great engineering value in machine design, and therefore in reverse engineering. With a given operating temperature and required service life (rupture time), the design engineer can easily determine the allowed stress from the stress rupture curve. It can also demonstrate that the reverse engineered part has an equivalent or better stress rupture (or creep) resistance than the original OEM counterpart.

### 4.3.2 Larson–Miller Parameter (Prediction of Long-Term Creep Properties)

Engineering design often requires engineers to predict material properties at high temperatures where no experimental data are available. The creep deformation rate can be so slow that it might require 10 years test time to reach 1% deformation. Reliable predictions based on accelerated test data obtained over a shorter period of time are essential. Several theoretical parameters were proposed to predict long-term metal creep or stress rupture life based on short-term test data. One of the most utilized parameters is the Larson–Miller parameter, as defined by Equation 4.20:

$$P = T(\log_{10} t + C) \tag{4.20}$$

where $T$ = the test temperature in Rankin, °R = °F + 460; $t$ = the time to rupture (or reach a certain strain), in hours; $C$ = the Larson–Miller constant, approximately 20; and $P$ = Larson–Miller parameter.

The Larson–Miller parameter is also often converted to and expressed as Equation 4.21:

$$P = (T + 460)(\log_{10} t + 20) \tag{4.21}$$

where the temperature is in Fahrenheit, °F.

Under the same stress, a higher test temperature results in a shorter stress rupture life, and vice versa. Assuming there are no structural changes, the Larson–Miller parameter is used to predict the long-term rupture behavior by extrapolating information from the short-term experimental data

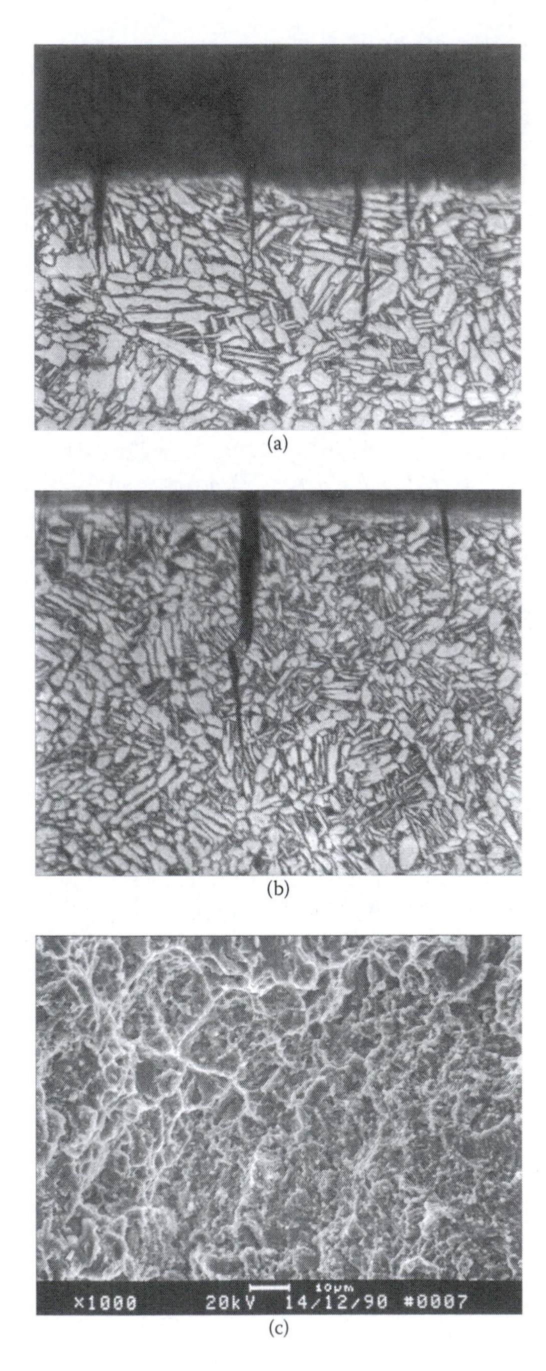


**FIGURE 4.16**
(a) Multiple cracks observed at the surface of a stress-ruptured titanium specimen. (b) Linkage between two cracks. (c) Fractography of a stress-ruptured titanium alloy.

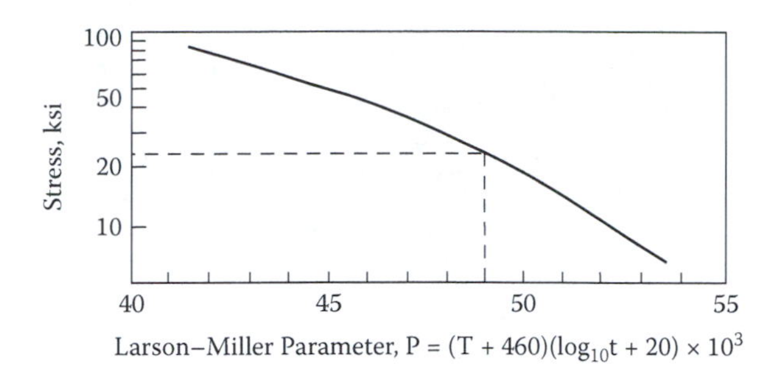


**FIGURE 4.17**
Larson–Miller master curve.

obtained at higher temperatures under the same stress. The value of Larson–Miller parameter is a function of stress. A Larson–Miller master curve can be established for a specific material with experimental data obtained over a range of temperature, *T*, and time, *t*. To explain the application of Larson–Miller parameter, a Larson–Miller master curve is simulated based on the data from 760°C (1,400°F) to 982.2°C (1,800°F) in the literature (Dieter, 1986), as shown in Figure 4.17.

To predict the allowed maximum stress for a component made of this alloy operating at 815.6°C (1,500°F) for a minimal 100,000-hour service time, the Larson–Miller parameter can first be obtained by Equation 4.21a:

$$P_{1,500°\text{F; }100,000\text{ hr}} = (1,500 + 460)\ (\log_{10} 10^5 + 20) = 1,960 \times 25 = 49 \times 10^3 \quad (4.21\text{a})$$

Applying this Larsen-Miller parameter to the master curve, Figure 4.17, the allowed maximum stress is estimated to be 165 MPa (24 ksi).

Whenever possible, an experimental proof of the allowed maximum stress is recommended because not all engineering alloys obey the Larson–Miller prediction or any other predictions due to metallurgical changes during prolonged exposure at elevated temperatures. In reverse engineering, the lack of the alloy-specific master Larson–Miller parameter curve, such as Figure 4.17, also often presents a challenge for engineers.

### 4.3.3 Creep Mechanisms

There are many factors that contribute to creep deformation. At relatively high stresses, dislocation glide is the predominant creep mechanism. The controlling mechanism gradually shifts to diffusional creep when the stress decreases and temperature increases. The diffusional creep is a self-yielding process in solid grain or along the grain boundary by atom movement. Atoms diffuse and relocate themselves in response to externally applied stress within the grain; the resultant creep deformation rate is proportional to the applied stress and inversely proportional to the square of the grain

size. When grain size decreases, atomic diffusion along the grain boundary becomes more significant, and the controlling mechanism once again shifts from lattice diffusion in the grain to grain boundary diffusion. The creep deformation rate is then inversely proportional to the third power of grain size for fine grain polycrystalline materials. The creep deformation rate resulting from combined diffusion flow from within the grain lattice and along the grain boundary is mathematically described by Equation 4.22:

$$\dot{\varepsilon} = 14 \frac{\sigma \, \Omega}{k \, T \, d^2} D_l \left\{ 1 + \frac{\pi \, \delta \, D_b}{d \, D_l} \right\} \tag{4.22}$$

where $\dot{\varepsilon}$ = creep rate, $\Omega$ = atomic volume, $k$ = Boltzmann's constant ($13.8 \times 10^{-24}$ J/K), $T$ = absolute temperature, $\sigma$ = normal stress, $D_l$ = lattice diffusivity, $d$ = grain size, $\delta$ = grain boundary width, and $D_b$ = grain boundary diffusivity.

Equation 4.22 shows that grain size has adverse effects on the diffusional creep rate. The smaller the grain size, the higher the creep deformation rate, and the weaker the material becomes at elevated temperatures. This phenomenon reflects both the significance and complexity of the effects of grain size on mechanical strength in reverse engineering. A finer grain size in the reverse engineered part than that observed in the OEM counterpart is usually beneficial to tensile strength at room temperature; however, it is detrimental to creep resistance at elevated temperatures. In reverse engineering, the same grain size should be sought to demonstrate the equivalency between the performance of the duplicated part and the OEM part. When different grain sizes are observed, a detailed analysis is required before any conclusion can be drawn as to whether it is beneficial, detrimental, or only has negligible effects on part performance.

## 4.4 Environmentally Induced Failure

Environmentally induced failures cost industries billions of dollars every year. A 1995 study reported that the cost impact of corrosion to the U.S. economy totaled nearly $300 billion annually, about 4% of the gross domestic product (Kuruvilla, 1999). Corrosion failures are usually caused by electrochemical reactions on the surfaces between the components and the environments. Typically, these corrosion failures occur way into their life cycles, otherwise defined by the loading conditions. They often occur unexpectedly in service. It is paramount to manage corrosion control and assess part corrosion resistance for a critical structural component of a machine. In automotive industries, corrosion management is part of the automotive

design, and corrosion protection is often integrated into its warranty. In aviation industries, corrosion inspection plays a critical role in aging aircraft management. The Italian Air Force introduced a Corrosion Control Register Program for corrosion management of the Italian Air Force fleet. It is a flexible and integrated program for making decisions on both prevention and operational measures (Colavita and De Paolis, 2001). Unfortunately, the prolonged corrosion test is often prohibitively time-consuming and impractical in reverse engineering. Most estimated corrosion resistance of a reverse engineered part is based on comparative analysis.

Environmentally induced failures can also be caused by cracking resulting from embrittlement. Several types of embrittlement are induced by the presence of certain chemicals or other environmental effects, such as hydrogen embrittlement, cadmium embrittlement, and cryogenic embrittlement. Some high-strength steels and body-centered cubic (BCC) metals fail without warning or yielding when they are statically loaded in the presence of hydrogen. Cadmium embrittlement is often associated with high-temperature protective coatings. In contrast, cryogenic embrittlement occurs at cryogenic temperatures. Some alloys, such as carbon steels, lose their ductility and fail abruptly at very low temperatures.

### 4.4.1 Classification of Corrosion

Corrosion is the most commonly used generic terminology for all environmentally induced degradation. Strictly speaking, corrosion is a chemical and electrochemical reaction between a material and its surrounding environment that results in a deterioration of the mechanical and physical properties of the material. The electrochemical nature of a corrosion process is best demonstrated in many automobile batteries, as shown in Figure 4.18. Severe corrosion appears in the positive post and other areas with direct contact between the battery and the frame. However, the term *corrosion* is also loosely applied to mechanically assisted corrosive attack, such as fretting corrosion and erosion corrosion. In many cases, metal embrittlement, such as cadmium embrittlement, or hydrogen embrittlement, and oxidization are also referred to as corrosion. There is no unified terminology used to describe the forms of corrosion. The following discussions are based on the terminology acceptable to most engineering communities.

The overlapping characteristics of various corrosion forms and mechanisms made it very challenging to completely separate one mechanism from another. The most frequently observed corrosion forms are categorized into the following seven classes based on how the corrosion process manifests itself: uniform or general, galvanic, crevice, pitting, intergranular, erosion, and stress corrosion cracking. Figure 4.19 is a photo of a mining cart exhibited at the Bingham Canyon Mine Visitors Center of Kennecott Utah Copper in Bingham Canyon, Utah, that shows general corrosion in the form of rust, resulting from exposure to the atmosphere. This type of uniform

**FIGURE 4.18** **(See color insert following p. 142.)**
Corrosion due to electrochemical reaction.

**FIGURE 4.19** **(See color insert following p. 142.)**
General corrosion observed on a mining cart.

environmental degradation is observed in many outdoor exhibits and structures. Any reverse engineered part is expected to demonstrate sufficient rust resistance when it is used outdoors. The corrosion processes are typical electrochemical processes. Intergranular corrosion is heavily influenced by alloy metallurgical properties. Erosion corrosion is only observed in the presence of moving corrosive fluid. Stress corrosion cracking is a combined effect of corrosive environment and applied stress. Further subclassification

is also used to define the unique corrosive attack under these primary corrosion forms. For example, the term *exfoliation corrosion* is widely used to identify a unique corrosion class in aluminum alloys caused by intergranular corrosion.

The uniform or general corrosion is characterized by corrosive attack proceeding evenly over the entire or most of the surface area. Compared to most other corrosion mechanisms, the uniform corrosion is more predicable. The measurement of weight loss is commonly used to quantitatively calculate the corrosion rate of uniform corrosion.

Galvanic corrosion is an electrochemical process between two dissimilar metals in which one corrodes preferentially. There are three necessary conditions for galvanic corrosion to occur. First, two electrochemically different metals are present: one functions as an anode and the other as a cathode. Second, an electrically conductive path exists between these two metals. Third, a conductive path of metal ions is available between the anodic and cathodic metals. The corrosive interactions between two metals are often referenced in a galvanic series table or chart. This table ranks the metals in the order of their relative nobility in a corrosive environment such as seawater. This table begins the list with the most active anodic metal and proceeds down to the least active cathodic metal. In a galvanic couple that consists of two dissimilar metals, the metal higher in the series, representing an anode, will corrode preferentially. The galvanic series table provides very useful guidance to galvanic corrosion protection in joint metals. The closer two metals are in the series; the more electrochemically compatible they are, and therefore less a chance the galvanic corrosion will occur when they are in contact. Conversely, the farther apart the two metals are, the worse the galvanic corrosion that occurs will be. A galvanic series applies only to a specific electrolyte solution. Different galvanic series tables are used for different environments and different temperatures.

Crevice corrosion is a localized corrosion occurring in narrow openings such as crevices. There are many of these crevices in the part joint areas or in a machine itself, such as the areas under gaskets or seals, or inside cracks and seams. These crevices are often filled with muddy deposits, solid sediments, or slushy precipitates. These sludge piles can develop a local chemistry of the electrolyte that is very different from that of the surroundings. The diffusion of oxygen into the crevice is usually restricted. As a result, a differential aeration cell can establish between the crevice and the external surface. An electrochemical potential drop in the crevice might also occur because of deoxygenation of the crevice and a separation of electroactive areas, with net anodic reactions occurring within the crevice and net cathodic reactions occurring exterior to the crevice. Unfortunately, this local corrosive environment stagnates because of lack of electrolyte flow, and induces crevice corrosion. In contrast to galvanic corrosion where corrosion occurs between two dissimilar metals immersed in one electrolyte, crevice is a corrosive

action that occurs between two metal parts made of the same alloy while surrounded with two different electrolytic environments.

Figure 4.20 shows a localized corrosion. However, it is not a pitting corrosion. Carbon steel typically does not pit; the observed localized corrosion is breakthrough of a galvanized coating that allows red-rust formation from the underlying steel substrate. Pitting is a corrosion confined to a small area, penetrating deep into the metal surface. It appears as small and irregular pit holes on the surface. Pitting is most likely to occur in the presence of chloride ions, combined with such depolarizers as oxygen or oxidizing salts. The distinct features typifying pitting corrosion have long classified it as a unique form of corrosion. However, the driving force of pitting corrosion is very similar to galvanic corrosion. In pitting corrosion, the lack of oxygen around a small area makes this area anodic, while the surrounding area with an excess of oxygen becomes cathodic. This leads to a localized galvanic corrosion that corrodes into the part and forms pit holes. These tiny pit holes limit the diffusion of ions and further pronounce the localized lack of oxygen. The formation of pit holes makes the mechanism of pitting corrosion very similar to that of crevice corrosion.

Intergranular corrrosion is also referred to as intercrystalline or interdendritic corrosion. The detailed microstructure characteristics of intergranular corrosion can only be examined under a microscope. However, the accumulated damage, such as part exfoliation, is readily visible when the intergranular corrosion just underneath the surface expands and blists the part surface. Exfoliation corrosion is most often observed on extruded or rolled aluminum products where the grain thickness is relatively shallow. It may also occur on parts made of carbon steel. Without proper microstructure analysis the actual grain morphology of the plates used for the box shown in Figure 4.21 can not be absolutely confirmed. Nonetheless, the subject box does show distinct macro characteristics of exfoliation corrosion.

**FIGURE 4.20 (See color insert following p. 142.)**
Localized corrosion.

**FIGURE 4.21** **(See color insert following p. 142.)**
A box with distinct macro characteristics of exfoliation corrosion.

Intergranular corrosion occurs along the grain boundaries or immediately adjacent to the grain boundaries, which usually have a different crystallgographic structure and chemical composition than the interior grain matrix. Figure 4.22 shows the grain morphology of an aluminum alloy with various second phases attached to the grain boundaries. Heat-treated stainless steels and aluminum alloys are noticeably susceptible to grain boundary corrosion attack, partially due to the segregation and precipitation induced by heat treatment. Such segregation or precipitation can form a zone in the immediate vicinity of grain boundary, leading to preferential corrosive attack. The intergranular precipitation of chromium carbides ($Cr_{23}C_6$) during a sensitizing heat treatment or thermal cycle often causes the intergranular corrosion of austenitic stainless steels. Intergranular corrosion occurs in many aluminum alloys either due to the presence of some chemical elements or second phases anodic to aluminum or due to copper depletion adjacent to grain boundaries in copper-containing alloys. Small quantities of iron segregation to the grain boundaries in aluminum alloys can induce intergranular corrosion. Precipitation of some second phases, such as $Mg_5Al_8$, $Mg_2Si$, $MgZn_2$, or $MnAl_6$, in the grain boundaries will also cause or enhance intergranular attack of high-strength aluminum alloys, particularly in chloride-rich media.

### 4.4.2 Environmental Effects and Protection

Most environmentally induced material degradations result from corrosion, oxidation, stress corrosion, and hot corrosion. Corrosion is a universal

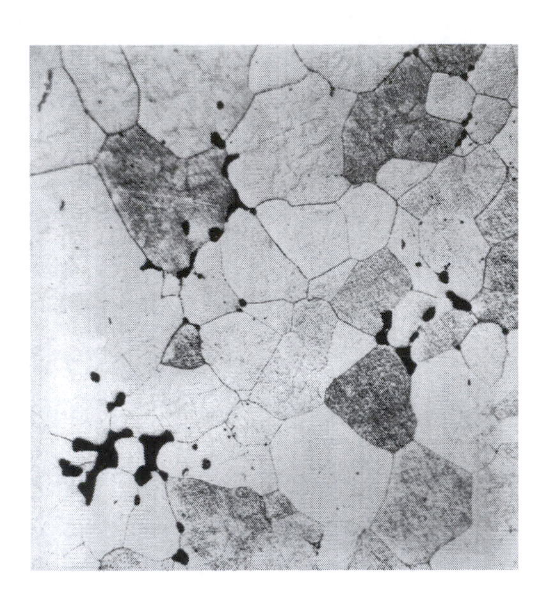

**FIGURE 4.22**
Grain boundary morphology of an aluminum alloy.

phenomenon that has been observed in many engineering structures, from automobiles to bridges. Oxidation only occurs in an oxidizing environment. Stress corrosion is a combined result of mechanical stress and corrosive attack. Hot corrosion is caused by a few specific contaminating elements, like sodium, potassium, vanadium, lead, and carbon, when they react with sulfur or oxygen. The resultant effect of weakening the strength of mechanical components due to environmental degradation is a primary concern for safety in many industries, from transportation to construction. It is also a concern of machine design and reverse engineering. A reverse engineered part must demonstrate equivalent or better resistance to environmental degradation compared to the OEM counterpart. However, a comparative analysis on environmental effects is probably one of the most time-consuming and expensive tasks in reverse engineering. An accelerated test is often proposed to predict the long-term effect. This requires a comprehensive understanding of the underlying principles. The following sections provide an introductory glance at these subjects. Interested readers are urged to read the books and literature specializing in these fields.

### 4.4.3 Aqueous Corrosion

Metals corrode in aqueous environments by an electrochemical mechanism in which an anodic and a cathodic reaction occur simultaneously. The anodic reaction is an oxidation process. The metal loses electrons and dissolves into the solution, $Fe \rightarrow Fe^{2+} + 2e^{-}$. The excess electrons generated in the electrolyte are usually consumed in two ways at a cathodic site where a

reduction process occurs. In acid solutions, they reduce hydrogen ions and that hydrogen gas is liberated from the metal, according to Equation 4.23:

$$2H^+ + 2e^- \rightarrow H_2 \tag{4.23}$$

Or, they might create hydroxyl ions by the reduction of dissolved oxygen, according to Equation 4.24:

$$O_2 + 4e^- + 2H_2O \rightarrow 4OH^- \tag{4.24}$$

The corrosion rate is therefore associated with the flow of electrons or an electrical current. Two reactions, oxidation and reduction, simultaneously occur at anodic and cathodic sites, respectively, on the metal surface. If the metal is partially immersed in water, there is often a distinct separation of the anodic and cathodic areas, with the latter near the waterline where oxygen is readily dissolved. Figure 4.23 illustrates the formation of such a differential aeration cell where $Fe^{2+}$ ions dissolve into solution from the "bottom" anode, $OH^-$ ions from the "top" cathode, and they meet to form hydroxide $Fe(OH)_2$, $Fe(OH)_3$, $Fe_2O_3 \cdot H_2O$, or $Fe_3O_4$. In this case, the corrosion rate is controlled by the supply of oxygen to the cathodic areas. If the cathodic area is large, intense local attacks on small anode areas, such as pits, scratches, and crevices, can occur.

### 4.4.4 Stress Corrosion

Stress corrosion occurs when a part is under mechanical stress and at the same time is being exposed to a corrosive environment, for instance, a steel tie rod or bolt connecting the two flanges of a tank that is immersed in corrosive fluid. Stress corrosion failure is brittle in nature, and its fracture surface

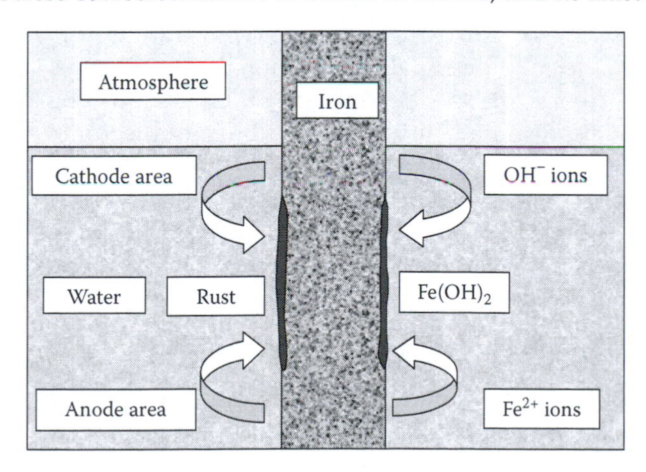


**FIGURE 4.23**
Differential aeration cell of iron corrosion.

is usually discolored and appears rough. The cracking mechanism of stress corrosion crack (SCC) is rather complicated. It can be either intergranular or transgranular. Figure 4.24a shows chloride stress corrosion cracking near the surface of a part made of austenitic stainless steel. Figure 4.24b shows multi-branched transgranular stress corrosion cracking in a cold-drawn 316 stainless steel connector pin from a marine vessel (Metallurgical Technologies).

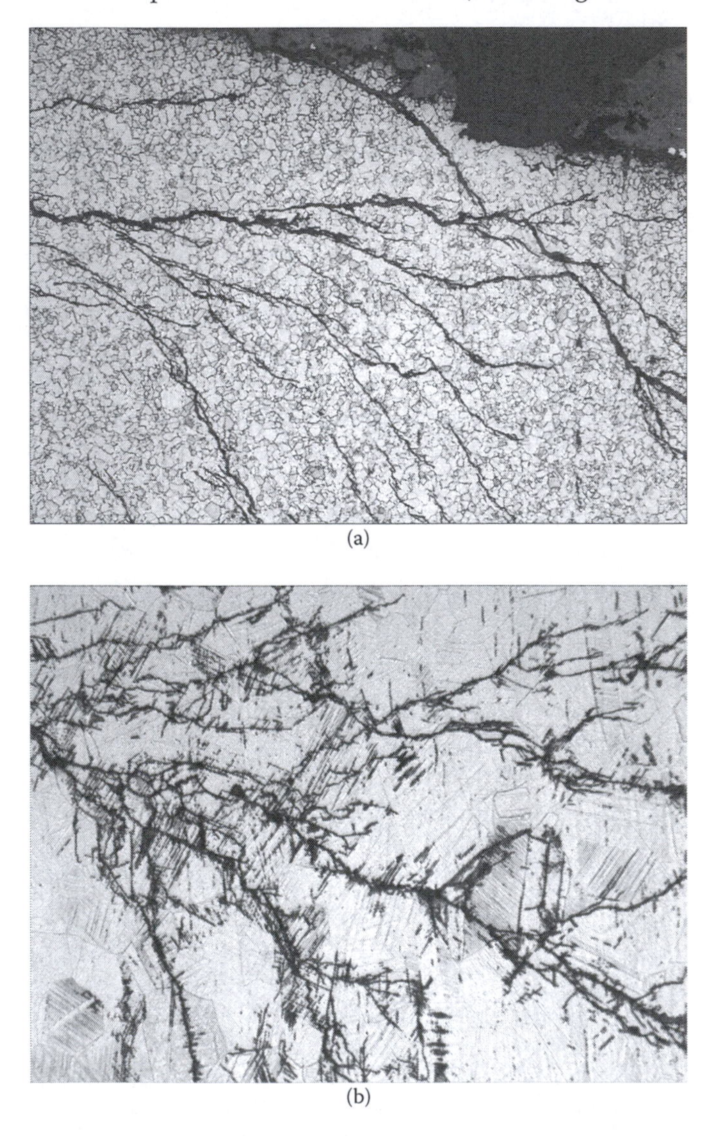

(a)

(b)

**FIGURE 4.24**
(a) Transgranular stress corrosion cracking. (Reprinted from Metallurgical Technologies. With permission.) (b) Multibranched chloride stress corrosion cracking. (Reprinted from Metallurgical Technologies. With permission.)

The fine crack induced by stress corrosion often penetrates into the part and is difficult to detect from the outside surface. However, the resulting damage can be catastrophic. A disastrous failure may occur unexpectedly with minimal warning. The experimentally tested SCC data are notoriously scattering. To demonstrate an equivalent SCC resistance is both technically challenging and financially costly. Nonetheless, it is a necessary test for many load-bearing parts used in a corrosive environment.

### 4.4.5 Oxidation and Protective Coating

Many alloys react with oxygen and alter their surface microstructures and properties. Figure 4.25 shows the oxide scale and second phase observed in a titanium alloy when it is exposed to oxygen. Corrosive oxidation is an electrochemical reaction where a metal loses its electrons and becomes a cation. Metallurgically, it is a reaction between metal and oxygen to form an oxide. For instance, Equation 4.25 shows the oxidation of aluminum forming aluminum oxide, that is, alumina $Al_2O_3$:

$$4Al + 3O_2 = 2Al_2O_3 \tag{4.25}$$

Engineering alloys are commonly developed with corrosion protection elements in their compositions. Both nickel- and cobalt-base superalloys contain one or more reactive elements, for example, chromium, aluminum, or silicon. These reactive alloying elements can form a protective oxide film on the part surface in an oxidizing environment wherein these superalloy parts are exposed. The specific weight change is an index showing oxidation degradation. For example, the effect of chromium content on the oxidation of Ni–Cr–Al alloys can be quantitatively measured in terms of weight gain.

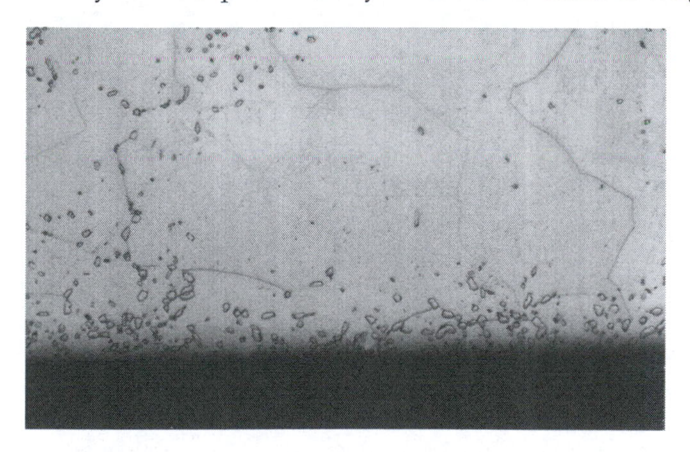

**FIGURE 4.25**
Surface morphology of a titanium alloy.

(Kvernes and Kofstad, 1972). The alloys with higher chromium content show less oxidation.

When being exposed to an oxidizing environment without protective coating, superalloys can lose their strength at a temperature below their capacity due to oxidation. The oxygen attacks the grain boundaries and weakens them to allow rapid fatigue crack initiation at temperatures above 700°C. The oxidation attack accelerates at temperatures above 950°C. The oxides found on engine turbine blade surface are dark and form relatively rough surfaces compared to the unoxidized surface. When the OEM material specification is substituted by another industrial material specification in reverse engineering, a deviation in alloy composition can have significant effects on oxidation resistance. Proper evaluations should be conducted.

Coating is one of the most widely used corrosion protection methods, particularly for high-temperature applications, where other protective methods are usually less effective. For instance, engine turbine blades are coated with a thermal barrier coating for their protection. Chrome carbide, platinum aluminide, and CoCrAlY are some examples of widely used coating materials. These coatings can be applied to the substrate using various methods, including plasma or thermal spraying. In a thermal spray coating process, the coating alloy powder is injected into a mixture of high-pressure oxygen and fuel (i.e., hydrogen or propylene) that is ignited to produce a heated gas jet propelling the coating alloys onto the surface. As a result, most of the coating is via a mechanical bonding instead of a metallurgical bonding. Other commonly applied coating processes in industries include galvanizing, electroplating, and deposition. Galvanizing is also referred to as hot dipping. Numerous metals are used for electroplating processes, such as chromium, zinc, titanium, nickel, copper, and cadmium. Therefore, these processes are referred to as chromium plating, zinc plating, or titanium plating, etc. Many fasteners, such as bolts, are Zn plated. Three primary deposition processes are physical vapor deposition, chemical vapor deposition, and ion implantation. Reverse engineering the coating process is a challenging, yet critical step in material process verification for many mechanical components.

### 4.4.6 Hot Corrosion

Hot corrosion results from the reaction between a metal and sulfur and is therefore sometimes referred to as sulfidation. Most gas turbine engines are susceptible to sulfidation. It is usually detected in the blade root and shroud areas. Unfortunately, these areas are also vulnerable to fatigue cracking. Sulfidation requires constant monitoring to avoid potential catastrophic failures. An engine can be injected with seawater and contaminated. The sodium from the seawater reacts with the sulfur from the fuel to form $Na_2SO_4$ (sodium sulfate) in the turbine engine gas stream. When $Na_2SO_4$ precipitates on the hot surface, around 820 to 950°C, of the downstream components, like turbine blades, it reacts with the protective surface oxide, for example, $Al_2O_3$,

and dissolves it. Some elements in the base metal are particularly detrimental, for example, titanium, which competes with aluminum to form its own oxide, and dissolves the base metal. Hot corrosion is evidenced by irregular greenish voids and crackings on the metal surface. The continuation of sulfidation decreases the wall thickness of a turbine blade, and makes it susceptible to fatigue cracking. Hot corrosion is observed in the components of diesel engines and auto mufflers as well, if sodium and sulfur are present.

A second type of hot corrosion could also occur at relatively lower temperatures, around 700 to 800°C, when $Na_2SO_4$ reacts with the elements in the base alloy, for example, nickel or cobalt, to dissolve the protective surface oxide barrier. The presence of chromium can be very beneficial in preventing this type of low-temperature hot corrosion by forming an alternative protective oxide, $Cr_2O_3$. Some aircraft turbine blades made of Ni-base superalloy were used to be coated with aluminide in earlier days. However, the aluminide coating has limited resistance to sulfur-enhanced oxidation, that is, hot corrosion. Later, chromium was added to the aluminide coating to improve resistance to hot corrosion and still retain the oxidation resistance. Chromium additions are typically made by diffusing chromium into the part surface prior to applying an aluminum coating.

The sensitivity of hot corrosion to alloying element, operation temperature, and service environment makes the reverse engineering of a part subject to hot corrosion challenging. The compositions of both the coating material and the base alloy need to be carefully identified.

### 4.4.7 Metal Embrittlement

When evaluating material durability and predicting the part life cycle, one of the most challenging tasks is to minimize the unexpected abrupt failure. This type of failure often occurs without noticeable precursors because of the subtle crack initiation process and the rapid crack propagation rate, such as the failures resulting from hydrogen embrittlement or cryogenic embrittlement. When a part is expected to serve under the conditions that have potential to cause embrittlement, the evaluation of these embrittlement effects on part performance is essential in reverse engineering.

The absorption of hydrogen into an alloy lattice can result in brittle failure for some alloys, for example, ferritic and martensitic steels, when they are under stress. The hydrogen lowers the bonding force of the metal lattice at the crack tip and locally embrittles the metal; consequently, the metal fails before yielding occurs. This phenomenon is referred to as hydrogen embrittlement. It often occurs in a humid environment, or an environment with the presence of sulfide, for example, oil well operations, which induces the evolution of hydrogen atoms. In contrast to stress corrosion cracking that usually results from anodic dissolution, hydrogen embrittlement is caused by cathodic polarization that introduces hydrogen atoms, and is reversible when the absorbed hydrogen is released.

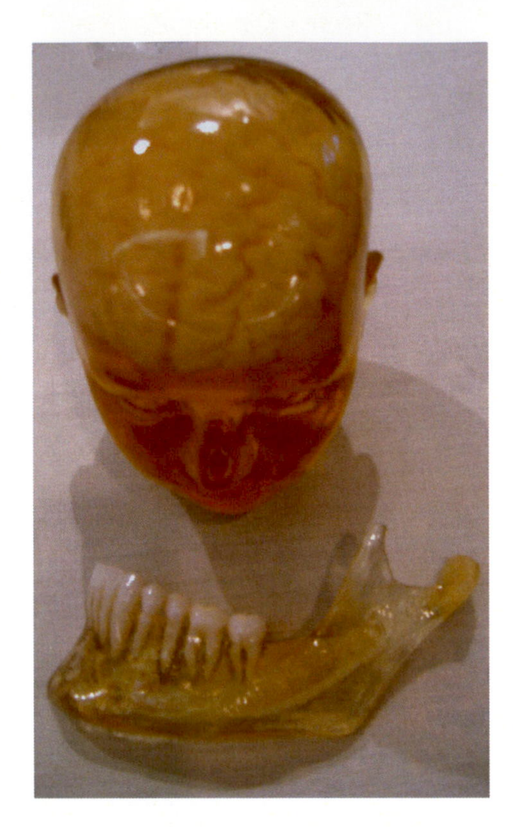

**FIGURE 1.7**
Prototype models in the medical field.

**FIGURE 2.4**
(a) ASTO II in operation.

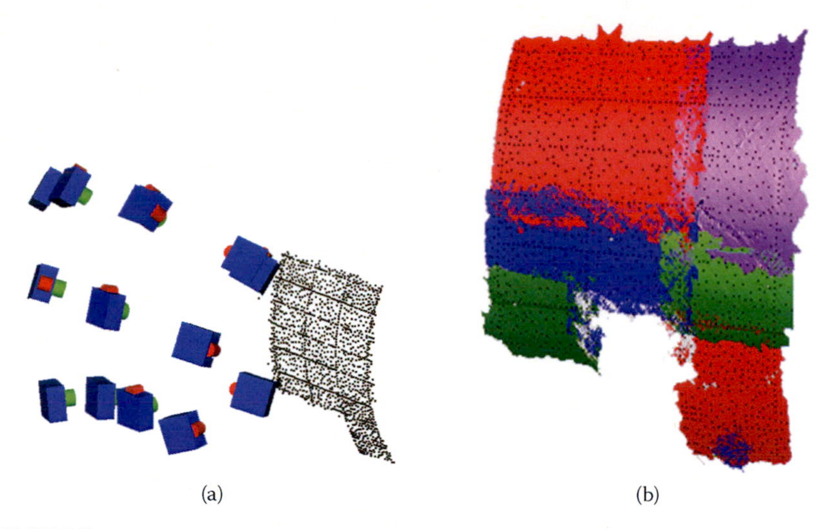


**FIGURE 2.7**
(a) Photogrammetry. (b) Scan compilation. (Both reprinted from 3DScanCo/GKS Global Services. With permission.)

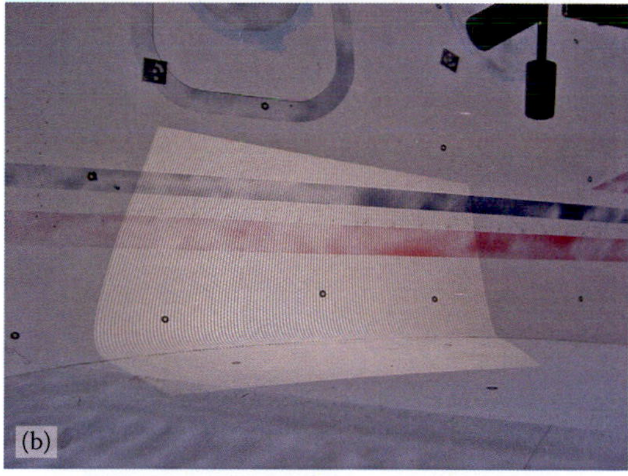


**FIGURE 2.17**
(a) A Falcon-20 aircraft with reference marks under scanning. (b) Fringe patterns. (Both reprinted from Capture 3D. With permission.)

**FIGURE 4.18**
Corrosion due to electrochemical reaction.

**FIGURE 4.19**
General corrosion observed on a mining cart.

**FIGURE 4.20**
Localized corrosion.

**FIGURE 4.21**
A box with distinct macro characteristics of exfoliation corrosion.

**TABLE 5.1**

Periodic Table of Elements

| IA | IIA | IIIB | IVB | VB | VIB | VIIB | VIII | VIII | VIII | IB | IIB | IIIA | IVA | VA | VIA | VIIA | VIIIA |
|---|---|---|---|---|---|---|---|---|---|---|---|---|---|---|---|---|---|
| 1 H | | | | | | | | | | | | | | | | | 2 He |
| 3 Li | 4 Be | | | | | | | | | | | 5 B | 6 C | 7 N | 8 O | 9 F | 10 Ne |
| 11 Na | 12 Mg | | | | | | | | | | | 13 Al | 14 Si | 15 P | 16 S | 17 Cl | 18 Ar |
| 19 K | 20 Ca | 21 Sc | 22 Ti | 23 V | 24 Cr | 25 Mn | 26 Fe | 27 Co | 28 Ni | 29 Cu | 30 Zn | 31 Ga | 32 Ge | 33 As | 34 Se | 35 Br | 36 Kr |
| 37 Rb | 38 Sr | 39 Y | 40 Zr | 41 Nb | 42 Mo | 43 Tc | 44 Ru | 45 Rb | 46 Pd | 47 Ag | 48 Cd | 49 In | 50 Sn | 51 Sb | 52 Te | 53 I | 54 Xe |
| 55 Cs | 56 Ba | 57–71 La–Lu | 72 Hf | 73 Ta | 74 W | 75 Re | 76 Os | 77 Ir | 78 Pt | 79 Au | 80 Hg | 81 Tl | 82 Pb | 83 Bi | 84 Po | 85 At | 86 Rn |
| 87 Fr | 88 Ra | 89–103 Ac–Lr | 104 Rf | 105 Db | 106 Sg | 107 Bb | 108 Hs | 109 Mt | 110 Ds | 111 Rg | | | | | | | |

| | | | | | | | | | | | | | | | |
|---|---|---|---|---|---|---|---|---|---|---|---|---|---|---|---|
| Lanthanides series | 57 La | 58 Ce | 59 Pr | 60 Nd | 61 Pm | 62 Sm | 63 Eu | 64 Gd | 65 Tb | 66 Dy | 67 Ho | 68 Er | 69 Tm | 70 Yb | 71 Lu |
| Actinide series | 89 Ac | 90 Th | 91 Pd | 92 U | 93 Np | 94 Pu | 95 Am | 96 Gm | 97 Bk | 98 Cf | 99 Es | 100 Fm | 101 Md | 102 No | 103 Lr |

| | | | |
|---|---|---|---|
| Non-metals | Transition metals | Rare earth metals | Halogens |
| Alkali metals | Alkali earth metals | Other metals | Inert elements |
| Elements generally detectible by ICP-AES | | | |

**FIGURE 5.28**
Surface hardening and hardness measurement. (Reprinted from Rolinski, E., et al., Heat Treatment Progress, September/October, p. 23, 2006. With permission.)

**FIGURE 5.15**
Aluminum casting macrostructure.

**FIGURE 7.5**
Corrosion observed around a bolt joint.

**FIGURE 7.6**
Uniform corrosion on a steel tube.

**FIGURE 7.9**
A cutaway view of an automobile engine.

**FIGURE 7.10**
A cutaway view of the combustion and turbine sections of a jet engine.

The failures associated with hydrogen embrittlement observed in ferrous materials, particularly high-strength steels, often occur without warning and cracks can propagate rapidly. As a result, they can be catastrophic. The cracks of hydrogen embrittlement are usually intergranular and initiate at the sites with the highest tensile stress. It is highly advisable in reverse engineering to conduct a precautionary analysis on hydrogen embrittlement on a part operating in an environment susceptible to hydrogen embrittlement because of its abruptness and unpredictability.

Liquid metal embrittlement is another catastrophic brittle failure mode and deserves proper consideration in reverse engineering. A normally ductile metal can fail rapidly when it is coated with a thin film of liquid metal such as cadmium. The necessary conditions required for liquid cadmium embrittlement of steel are pure, unalloyed cadmium in contact with steel under tensile stress at temperatures in excess of 320°C. This temperature condition limits the cadmium embrittlement only to the parts exposed to relatively high temperatures, such as the steel aircraft engine compressor disk. The time to failure is generally a function of the temperature and the stress of the exposed part. To prevent cadmium embrittlement, the steel part is first coated with a layer of nickel, followed by a cadmium outer coating. The nickel and cadmium react to form an alloy with higher melting temperature than pure cadmium, thus immobilizing the cadmium and preventing cadmium embrittlement.

The mechanical behaviors of materials at cryogenic temperatures are complex and vary from alloy to alloy. Certain alloys show excellent durability at cryogenic temperatures and are referred to as cryogenic alloys. The yield and tensile strengths of these structural cryogenic alloys will increase as the temperature decreases. For example, plastic deformation on stainless steels such as 301 and 304 at cryogenic temperatures causes partial transformation to martensite, which strengthens these alloys. The effects of low-temperature exposure on ductility and toughness of cryogenic alloys usually depend on alloy composition and structure. Most face-centered cubic (FCC) metals, such as 2024 and 7075 aluminum alloys and IN718 nickel-base superalloy show better tensile and yield strengths and fracture toughness with comparable ductility at cryogenic temperatures; the fatigue crack growth rate is either equal to or lower than the rate at room temperature for IN718. Significant increases in yield and tensile strengths are observed for Ti–6% Al–4% V as the temperature is reduced from room temperature to cryogenic temperatures. However, in contrast to IN718, the fatigue strength of Ti–6% Al–4% V is significantly weaker when the test temperature is reduced from room temperature to cryogenic temperatures. Cryogenic embrittlement is noticeably observed in some metals, for example, carbon steels, at temperatures below –150°C, which space vehicles can be exposed to at high altitude and in outer space. When reverse engineering any parts for a cryogenic service, the effects of temperature and cryogenic embrittlement in particular have to be carefully evaluated.

The environmental effects on mechanical properties once again demonstrate that the interrelationships between various mechanical properties are material specific and rely on many factors, from temperature to humidity. A higher yield strength for one alloy under one environmental condition might imply better fatigue resistance; however, specific supporting data are required to draw any inference to any other alloy in different circumstances. In summary, the demonstration of equivalent material durability and part life limitation in reverse engineering requires part-specific substantiation data.

## References

Colavita, M., and De Paolis, F. 2001. Corrosion management of Italian Air Force fleet. In *Life management techniques for ageing air vehicles conference proceedings*. RTO-MP-079 (II). Neuilly-sur-Seine: NATO Research and Technology Organization.

Dieter, G. E. 1986. *Mechanical metallurgy*. New York: McGraw-Hill.

Kuruvilla, A. K. 1999. *Life prediction and performance assurance of structural materials in corrosive environments. A state of the art report in AMPT-15*. Rome: AMPTIAC.

Kvernes, I. A., and Kofstad, P. 1972. *Met. Trans.* 3:1518.

NTSB. 2005. http://www.ntsb.gov/Pressrel/2005/051222a.htm (accessed December 24, 2009).

Rice, R. C. 1988. *Fatigue design handbook*. Warrendale, PA: SAE.

Sawai, T., Matsuoka, S., Abe, T., et al. 2003. Method of evaluating high fatigue strength material in high tensile strength steel and creation of high fatigue strength material. U.S. Patent 6546808.

Walker, K. 1970. *Effects of environment and complex load history on fatigue life*. ASTM STP 462. West Conshohocken, PA: ASTM International.

# 5

# *Material Identification and Process Verification*

Material identification and process verification are essential to reverse engineering. This chapter will discuss the techniques used to analyze chemical composition, microstructural characteristics, grain morphology, heat treatment, and fabrication processes. The chemical composition of a material determines its inherent properties. The microstructural characteristics are closely related to a material's mechanical properties. Grain morphology reveals the grain size, shape, texture, and their configuration in a material. These material characteristics are often analyzed simultaneously. For example, during an electron probe microanalysis, elemental chemistry is analyzed to identify alloy composition; at the same time, a micrographic image will also be taken to understand the phase transformation that leads to verification of heat treatment and the manufacturing process. The evolution of constituent phases in an alloy is a direct consequence of the prior manufacturing process this alloy has experienced. The identification of these phases by their compositions and quantifying their amounts in an alloy will help engineers verify the manufacturing process used to produce the part.

The end product of material identification and process verification is usually the confirmation of a material specification that is called out by the OEM in its production. Theoretically speaking, all the characteristics listed in a material specification should be tested and verified before it can be called equal to the specification of an OEM design. However, in real-life reverse engineering practice, usually only select characteristics are tested and compared. The characteristics that are tested are determined by their criticalities to the part functionality. The data that are specified in a typical engineering material specification will be reviewed in the next section to establish a foundation and create guidelines for future discussions.

## 5.1 Material Specification

Several institutes of various professions have published material specifications. For instance, the Society of Automotive Engineers International (SAE) publishes Aerospace Material Specifications (AMS) that specify both the

products, such as engineering materials, and the processes whereby the products are fabricated. The AMS are the most frequently cited material specifications in aviation industries. In 1905 the Society of Automobile Engineers was founded, and in 1916 it joined with the American Society of Aeronautic Engineers and the engineers in other closely related professionals to form the Society of Automotive Engineers. The term *automotive* originated from Greek *autos* (self) and Latin *motives* (of motion), and this is a professional society that focuses on modern machinery that steers with its own power. SAE has since played pivotal roles in the advancement of the automobile and aerospace industries (SAE, 2008).

### 5.1.1 Contents of Material Specification

The contents of a material specification depend on the purpose and application of this specification. A typical AMS on a product such as an engineering alloy is identified with a Title section on the first page, followed by eight other sections: Scope, Applicable Documents, Technical Requirements, Quality Assurance Provisions, Preparation for Delivery, Acknowledgment, Rejections, and Notes.

The Title section reveals the revision history of this AMS, the type of alloy, highlights of the material characteristics, nominal composition, and heat treatment condition.

The Scope section covers product form, such as sheet, strip, and plate, and the primary applications of this material, such as "typically for parts requiring strength and oxidation resistance up to 816°C (1,500°F)."

The Applicable Documents section lists all the relevant documents that form part of this specification. Two SAE publications are listed in AMS 5663, which has a composition similar to that of commercial 718 nickel alloy—AMS 2261, "Tolerances, Nickel, Nickel Alloy, and Cobalt Alloy Bars, Rods, and Wire," and AMS 2269, "Chemical Check Analysis Limits, Nickel, Nickel Alloys, and Cobalt Alloys"—along with seven other AMS publications. Also, two ASTM publications, ASTM E8M, "Tension Testing of Metallic Materials (Metric)," and ASTM E10, "Brinell Hardness of Metallic Materials," are listed, along with seven other ASTM publications. Therefore, to claim the conformance of AMS 5663 to the OEM design data, the comparative tensile properties should be evaluated in accordance with ASTM E8M (or its equivalence), which is part of the design document.

The following information is included in many of the Technical Requirements sections: composition, melting practice, condition of utilization, heat treatment, properties, quality, and tolerance. This is one of the core sections of many material specifications. The acceptable chemical composition is usually tabulated with the minimum and maximum elemental contents specified. The acceptable analytical methods are listed in the subsection of the composition. In AMS 5663, the weight percentage of constituent elements is required to be determined by wet chemical methods in accordance

with ASTM E354, or by spectrochemical methods. For lead, bismuth, and selenium, the analytical methods of APR 1313 will be utilized. The acceptance of other analytical methods should be approved by the stakeholders in advance. In other words, the composition determined by energy dispersive X-ray analysis (EDXA) is typically not acceptable in reverse engineering to claim conformance to AMS 5663, unless otherwise agreed upon by all stakeholders. In reference to elemental composition variations, AMS 5663 requires compliance with the applicable requirements of AMS 2269. Strictly speaking, failing to meet any of these requirements can be a justifiable cause for rejection in reverse engineering.

The melting practice directly affects the quality of an alloy. A specific melting practice is required for high-quality material, as is identified in its material specification. A reverse engineered product must demonstrate that it has the same melting practice as the OEM does. To claim conformance to AMS 5663, the following melting practice has to be demonstrated. The alloy should be multiply melted using a consumable electrode practice in the remelting cycle, or it should be induction melted under a vacuum. If consumable electrode remelting is not performed in a vacuum, electrodes produced by vacuum induction melting should be used for remelting. It is worth noting that a double-remelting ingot is not equivalent to a triple-remelting ingot in reverse engineering either.

The mechanical property of a material is a function of its manufacturing process, and therefore the final product form. The available product forms for a material also reflect its formability and machinability, which in turn depend on heat treatment and other prior treatments. It is not uncommon to have several material specifications with the same chemical composition but different product forms and heat treatment conditions, and therefore different material properties. The available product forms or conditions listed in the Condition subsection provide additional information for determining which material specification best fits the OEM part. For example, a material specification that provides sheet and plate product forms is a better fit for a "heat shield" part used in a turbine engine than another material specification that only provides product forms in bar and wire.

One of the most challenging tasks in reverse engineering is to decode an OEM's heat treatment schedule. The precise reverse engineering of a heat treatment process is virtually impossible due to the multiple parameters involved in heat treatment, such as temperature, time, atmosphere, and quench medium. It is further complicated by the fact that often several different heat treatment schedules can produce similar material properties, but none can produce exactly the same properties of the OEM part. Many aging treatment schedules can produce the same hardness number for 2024 aluminum alloy, but with different microstructures and fatigue strengths. Both AMS 5662 and 5663 have the same nominal chemical composition and same melting practice, and provide the same product forms, but they have different heat treatments. In AMS 5663, the precipitation heat treatment is applied

to the alloy after the solution heat treatment; however, in AMS 5662, the alloy will only be subject to solution heat treatment, although it is precipitation hardenable.

Specifically required properties are described in detail in the Properties subsection. These properties usually include various microstructural features, mechanical properties, and resistance to environmental degradation. Two of the most commonly referred to microstructural features are grain size and second phases. The average grain size is usually measured by the intercept method of ASTM E112, which is a linear measurement. In reverse engineering it is advisable to adopt the same method of measurement of grain size for a direct comparison whenever feasible, even if a different method might give a more accurate measurement. The presence of second phases can drastically change the properties. In most nickel alloys, the Laves phase is detrimental. Both AMS 5662 and 5663 require a microstructure free of this phase, and with an acceptable amount of the acicular phase. Unless the "acceptable amount" is otherwise specified, the acceptability of a microstructure can only be determined by a direct comparison between the OEM and the reverse engineered parts. Whenever other microstructral features are specified, such as grain texture or recrystallization, they should be complied with as well.

Hardness provides a first order of approximation of mechanical strength. However, great caution is required to extrapolate mechanical properties directly from hardness. First, hardness is measured using a variety of scales, each representing different material characteristics, and there are no precise conversions among them. Second, the relationships, if any, between hardness and other mechanical properties are usually empirical and lack supporting scientific theories. These relationships are material specific with limited applicability. In reverse engineering, a hardness comparison should always be in the same scale whenever feasible. Conformance to a material specification based on hardness is an estimate at best.

What tests are required and what properties are relevant in reverse engineering? The short answer is that all the properties specified in the material specification are relevant for an accurate conformance. The best reverse engineering practice in material identification is to make a checklist, including all the relevant material characteristics and properties, and compare them item by item. This list is different for each and every material specification. For AMS 5663 it will include hardness, tensile properties at room temperature, tensile properties at 649°C (1,200°F), and stress rupture properties at 649°C (1,200°F). The reported tensile properties should include tensile strength, yield strength, elongation, and reduction, and the tensile test should be conducted on specimens of three orientations: longitudinal, long transverse, and transverse. A word of caution: many material specifications only list the required minimum tensile properties, as shown in both AMS 5662 and AMS 5663. These two specifications require identical tensile properties despite different heat treatments. AMS 5663 requires solution treatment followed by

precipitation hardening, while AMS 5662 only requires solution heat treatment. If a tested tensile strength meets the AMS 5663 minimum requirement, then it does literally satisfy the specification requirement. However, unless a direct comparison with an OEM part, it cannot be decisively concluded that it has a tensile strength equivalent to that of the OEM part. The OEM part might have a tensile strength above the minimum requirement. In reverse engineering, the baseline material properties for comparative analysis are the test results directly measured from the original part, not a material specification.

The tolerance requirements depend on part shape, dimensions, and other factors, such as material flexibility and deformability. In AMS 5662 and AMS 5663, the requirements of tolerances are simply summarized as "all applicable requirements of AMS 2261." However, different requirements might be required in other cases.

The Quality Assurance Provisions section summarizes the responsibility of inspection and classification of tests. Ideally, each heat or lot is tested, and their respective microstructures examined to ensure high quality control. The sampling and testing should also comply with proper procedures, such as AMS 2371: "Quality Assurance Sampling and Testing Corrosion and Heat-Resistant Steels and Alloys Wrought Products and Forging Stock." Any product not confirming to the specification should be rejected, and another alternate material should be considered in reverse engineering.

Specifications on specific subjects are also published. For example, AMS 2242 and AMS 2262 focus on tolerances. They cover established manufacturing tolerances applicable to various product forms made of different alloys. These specifications provide a good reference and guidance on reverse engineering where manufacturing tolerances are often of great concern. Another example, AMS 2248, focuses on chemical analysis limits. It defines limits of variations for determining acceptability of chemical composition of a variety of parts, and provides a valuable reference in alloy composition determination, where the acceptability of variation limits often challenges engineers. Justifications are required to adopt any tolerance or composition if it is out of the scope of the specified ranges.

The best way to confirm a material specification in reverse engineering is by direct comparison of each and every characteristic listed in the specification. However, an alternate method of compliance might be acceptable upon approval or mutual consent.

The material specification goes beyond composition identification and manufacturing process verification. It also extends to packing and identification. AMS 2817 covers procedures that will provide protection for preformed packings of O-rings of elastometric materials from contamination by foreign materials prior to installation, and ensure positive identification. Part packing and identification, though of an administrative nature, also play a crucial role in a reverse engineering project to avoid preinstallation contamination or damage.

### 5.1.2 Alloy Designation Systems

Many alloy designation systems have been developed by various organizations such as SAE International and ASTM International. Different alloy codes and standards are also published in different countries, such as British Standards, German DIN, Swedish Standards, Chinese GB, and Japanese JIS. DIN stands for Deutsches Institut für Normung in German, and German Institute for Standardization in English. It is the German national organization for standardization. The DIN EN number is used for the German edition of European standards. A Swedish Standard is usually designated with a prefix SS. The GB standards are the Chinese national standards issued by the Standardization Administration of China. GB stands for Guobiao, a phonetic transcription of the word *National Standards* in Chinese. JIS stands for Japanese Industrial Standards. It is published by the Japanese Standards Association.

It is of great advantage to have a universally unified alloy code system; that is why the Unified Numbering System (UNS) was proposed. This system consists of a prefix letter and five digits designating a material composition. For example, the prefix *S* is used to designate stainless steels. UNS S31600 is the unified code in the Unified Numbering System for one of the most widely used stainless steels, which is designated as SAE316 by SAE International, 316S31 in British Standards, and SUS 136 in Japanese JIS. However, in the European system, it is designated with a DIN EN number of 1.4401, and given a name of X5CrNiMo17-12-2, while the Swedish Standards system designates it as SS2347. A comprehensive cross-reference system between the UNS and other alloy code systems is yet to be established. From a reverse engineering perspective, the biggest concern is whether two nominally equivalent stainless steels coded in different systems are actually identical. A UNS number alone does not constitute a full material specification because it establishes no requirements for material properties, heat treatment, product form, and quality. Several material specifications are published for stainless steel 316: AMS5524, ASTM A240, and ASTM A666. Great caution needs to be exercised when drawing any inference from cross-references based on different designation systems or codes.

## 5.2 Composition Determination

### 5.2.1 Alloying Elements

Most engineering alloys contain multiple constituent elements to achieve the desired metallurgical and mechanical properties. Superalloys Rene N6 and CSMX-10M are composed of as many as twelve or thirteen microalloying elements to enhance their properties at elevated temperatures (Durand-Charre, 1997). These alloying elements are added for specific purposes and

targeted applications in alloy development. Both Rene N6 and CSMX-10M are advanced single-crystal nickel-base superalloys used for gas turbine components.

Most commercial alloys contain small amounts of various elements. A specific element added to improve alloy properties is called an alloying element. On the other hand, an element that exists in the alloy but is not intended by design is called an impurity. The effects of alloying elements on metallurgical and mechanical properties are complex and material specific. The following types of questions relating to alloy composition are frequently asked in reverse engineering. Can an alloy with less than 0.1% aluminum compared to the OEM alloy composition be used to reproduce the OEM part? Is the 0.05% tungsten detected in the OEM alloy a negligible "trace element" that was accidentally mixed into the alloy, or an alloying element that is purposely added into the OEM alloy? Are two alloys considered equivalent if one contains 0.1% more carbon, while another has a 0.1% higher zirconium content? To answer these questions, an understanding of alloying element effects is required. Although this level of knowledge of materials science is essential in reverse engineering, a detailed discussion of the effects of alloying elements can be overwhelming, as exemplified below.

In the Ni-base superalloys, aluminum and chromium help provide good corrosion and oxidation resistance. The additions of refractory elements niobium, rhenium, molybdenum, or tungsten to these superalloys aim to reduce the coarsening rate of gamma prime ($\gamma'$) precipitate, which is a stable ordered face-centered cubic (FCC) intermetallic precipitate with a composition of $Ni_3(Al, Ti)$. These gamma prime precipitates are coherent with the surrounding gamma ($\gamma$) phase matrix, and very difficult for dislocations to penetrate. They therefore can improve the high-temperature properties of Ni-base superalloys, which is essential when using them in jet engine components such as nozzle guide vanes or turbine blades (Jena and Chaturvedi, 1984). Carbon, boron, and zirconium are added to polycrystalline Ni-base superalloys as grain boundary strengtheners. Boron and zirconium segregate at grain boundaries, and reduce grain boundary energy. As a result, they improve creep strength and ductility by preventing grain decohesion. Carbon and other carbide ($M_{23}C_6$ or MC) formers, like chromium, molybdenum, tungsten, nobelium, tantalum, titanium, and hafnium, also strengthen the grain boundaries because they tend to precipitate there, and hence reduce the tendency for grain boundary sliding. Though an optimal quantity of intermittent carbides along the grain boundaries can impede sliding and enhance mechanical strength, excess carbides will form a continuous chain of carbides and fracture paths along the grain boundary, and therefore weaken the alloy. However, in the single-crystal Ni-base superalloys, such as Rene N6 and CSMX-10M, their effects are less critical because of the elimination of the grain boundary. It is also reported that zirconium or boron does not influence the castability of IN792 during directional solidification when added individually. IN792 is a Ni-base superalloy strengthened

by gamma prime precipitates and used for turbine engine components. However, when both zirconium and boron are present in the alloy, high hot tearing susceptibility was observed, particularly at higher zirconium concentrations (Zhang and Singer, 2004). Hot tearing is an intergranular cracking that occurs along the grain boundaries. It is a casting defect observed in some Ni-base superalloys such as IN792 in investment cast or during directional solidification. Hafnium is usually added to the Ni-base superalloys to avoid the problem of hot tearing. Unfortunately, the addition of hafnium will also induce other effects that are detrimental. First, hafnium is a reactive element that reacts with mold and can form brittle inclusions. Second, hafnium lowers the incipient melting point, and thus the solution treatment temperature, and therefore weakens those Ni-base alloys that obtain their strength by precipitation hardening. Figure 5.1 shows the microstructure of a Ni-base superalloy that has been strengthened by the precipitates on the grain boundary. However, the precipitation of certain phases is also known to be detrimental. Figure 5.2 shows the presence of the deleterious sigma phase (white blocky particles) on the grain boundary that weakens the mechanical properties.

Each defined alloy has a specified alloy composition along with certain unspecified elements accidentally introduced into the alloy during production. The report of these unspecified elements is permitted according to ASTM Standard A751. However, it is neither practical nor necessary to specify limits for every unspecified element that might be present.

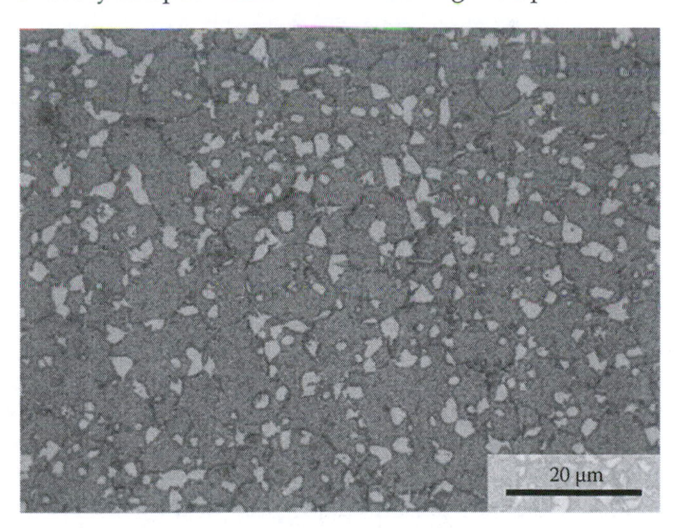


**FIGURE 5.1**
Grain boundary strengthening precipitation in Ni-base superalloys. (Reprinted from Mitchell, R., Department of Materials Science and Metallurgy, Nickel-Base Superalloys Group, University of Cambridge, http://www.msm.cam.ac.uk/UTC/projects, accessed March 2, 2009. With permission.)

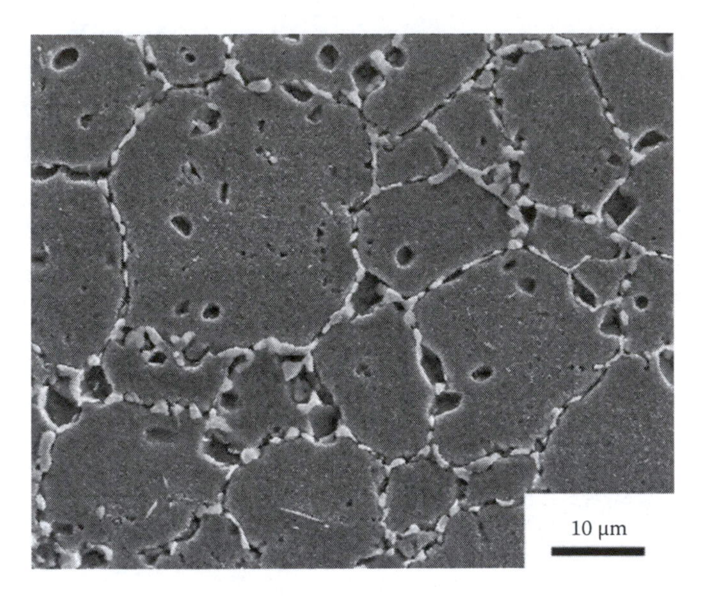


**FIGURE 5.2**
Precipitation of deleterious sigma phase (white blocky particles) on grain boundaries. (Reprinted from Mitchell, R., Nickel-Base Superalloys Group, Department of Materials Science and Metallurgy, University of Cambridge, http://www.msm.cam.ac.uk/UTC/projects, accessed March 2, 2009. With permission.)

The mechanical strength of Ni-base superalloys can also be improved by adding the following alloying elements as solid-solution strengtheners in both gamma and gamma prime phases: cobalt, iron, chromium, niobium, tantalum, molybdenum, tungsten, vanadium, titanium, and aluminum. Their respective quantities are properly controlled to make sure they will not induce undesirable precipitation, particularly to avoid certain brittle phases such as Laves and sigma (Bhadeshia, 2003). Recent superalloys are alloyed with rhenium to increase the strength and elastic modulus of the matrix (Durst and Göken, 2004). Rhenium is a very expensive addition but leads to an improvement in the creep strength and fatigue resistance (Erikson, 1996).

Many alloying elements in Ni-base superalloys are only of small quantities despite their critical contributions to the superalloy's properties and applications. The carbon content is usually from 0.02 to 0.2 wt%, boron from 0.005 to 0.03 wt%, and zirconium from 0.005 to 0.1 wt%. To reproduce these alloys without knowing their original design details, reverse engineering needs to accurately analyze the alloy chemical composition, particularly the quantitative analysis of the critical elements that appear only in trace amounts.

Sometimes the perception of a part can cause a lot of confusion about its true chemical composition. For instance, a U.S. five-cent coin is commonly referred to as a nickel. However, 75% of this coin is copper; only 25% is nickel. The "tin can" that is widely used for food storage is actually made of steel coated with tin. All the alloy identifications have to be based on analytical

data, not by the part popular nickname or perception. The typical methods of identifying a material can be classified into three categories. First, a material can be identified by its physical features, such as refractory index or thermal conductivity. However, this method is usually not used for alloy elemental analysis. Second, a material can be analyzed based on the electric charge-transfer phenomenon using electrochemistry or mass spectrometry. This technique is widely used in analytical chemistry. Lastly, a material can be analyzed using spectroscopy whereby the absorption, emission, or scattering of electromagnetic radiation is analyzed to determine the material chemical composition. The following sections will discuss the alloy chemical composition analysis methods that are widely used in reverse engineering.

### 5.2.2 Mass Spectroscopy

Mass spectrometry analyzes the chemical composition of a sample based on a mass spectrum. A mass spectrum is an intensity vs. mass-to-charge ratio (usually referred to as $m/z$) plot representing the constituent component profile of the sample. The following paragraph briefly explains the process of generating a mass spectrum.

The ions from the individual elements of the sample are extracted into a mass spectrometer and separated on the basis of their mass-to-charge ratio. A detector then receives individual ion signals proportional to their respective concentration to generate a mass spectrum. Inductively coupled plasma–mass spectrometry (ICP-MS) is a type of mass spectrometry where the sample is ionized by the inductively coupled plasma. The plasma used in ICP-MS is made by ionizing argon gas ($Ar \rightarrow Ar^+ + e^-$) with the energy obtained by pulsing an electrical current in wires surrounding the argon gas. The high temperature of the plasma will ionize a portion of the sample atoms to form ions ($M \rightarrow M^+ + e^-$) so that they can be detected by the mass spectrometer. ICP-MS can quantitatively determine chemical concentrations up to parts per tribillion by proper calibration with elemental standards, or through isotope dilution based on an isotopically enriched standard. The ICP-MS can analyze elements with atomic masses ranging from 7 to 250. This range encompasses lithium to uranium. The ICP-MS usually has an analytical resolution from nanograms per liter to 100 mg per liter. Unlike atomic absorption spectroscopy, which can only measure a single element at a time, ICP-MS has the capability to scan for all elements simultaneously. ICP-MS is widely used in the medical and forensic fields, specifically toxicology.

### 5.2.3 Inductively Coupled Plasma–Atomic Emission Spectroscopy

An inductively coupled plasma (ICP) is a very high temperature, up to 8,000K, excitation source that efficiently desolvates, vaporizes, excites, and ionizes atoms. ICP sources are used to excite atoms for atomic emission spectroscopy and to ionize atoms for mass spectrometry. Inductively coupled

plasma–atomic emission spectrometry (ICP-AES), also known as inductively coupled plasma–optical emission spectrometry (ICP-OES), is one of the most popular methods in elemental chemical analysis in reverse engineering. Most elements can be quantitatively measured using ICP-AES up to parts per billion. The exact delectability of an element is instrument specific. The elements that can usually be detected by ICP-AES are enclosed with a heavy border in the periodic table of elements in Table 5.1. Several common elements, such as hydrogen, boron, carbon, nitrogen, and oxygen, cannot be accurately analyzed by ICP-AES. An appropriate interstitial gas analytical technology is required to analyze the gaseous elements such as hydrogen, nitrogen, and oxygen in the alloys. For small quantities of trace elements, such as boron and carbon in Ni-base superalloys, glow discharge mass spectrometry (GDMS) is an acceptable technology for their quantitative measurement.

In ICP-AES analysis, the liquid sample (i.e., solution) is nebulized into an inductively coupled plasma; it has sufficient energy to break chemical bonds, liberate elements, and transform them into a gaseous atomic state for atomic emission spectroscopy. When this happens, a number of the elemental atoms will be excited and emit radiation. The wavelength of this radiation is characteristic of the element that emits it, and the intensity of radiation is proportional to the concentration of that element within the solution. The ICP-AES is used for both qualitative element identification and quantitative chemical composition determination.

Atomic emission spectroscopy (AES) and atomic absorption spectroscopy (AAS) use the emission and absorption of light for elemental composition measurement, respectively. In an AES analysis, all atoms in a sample are excited simultaneously, and can be detected at the same time using a polychromator with multiple detectors. This is the major advantage of AES compared to AAS, which uses a monochromator and therefore only one single element can be analyzed at a time.

### 5.2.4 Electron Specimen Interaction and Emission

The interaction between an electron and a specimen is what makes X-ray analysis and electron microscopy possible, and these two analytical techniques are often used in collaboration with each other in reverse engineering. A brief review of electron specimen interaction and the subsequent emission will be discussed in this section, which will benefit the later discussion on material identification utilizing these techniques. When the energetic electrons in the microscope strike the specimen, a variety of reactions and interactions will occur, as shown in Figure 5.3. The electrons emitted from the top of the specimen are utilized to analyze the bulk samples in scanning electron microscopy (SEM), while those transmitted through the thin or foil specimens are used in transmission electron microscopy (TEM).

When the incident electrons strike a sample, they will scatter primarily in two different modes: elastic or inelastic. In elastic scattering the electrons

**TABLE 5.1** **(See color insert following p. 142.)**

Periodic Table of Elements

| IA | IIA | IIIB | IVB | VB | VIB | VIIB | VIII | VIII | VIII | IB | IIB | IIIA | IVA | VA | VIA | VIIA | VIIIA |
|---|---|---|---|---|---|---|---|---|---|---|---|---|---|---|---|---|---|
| 1 H | | | | | | | | | | | | | | | | | 2 He |
| 3 Li | 4 Be | | | | | | | | | | | 5 B | 6 C | 7 N | 8 O | 9 F | 10 Ne |
| 11 Na | 12 Mg | | | | | | | | | | | 13 Al | 14 Si | 15 P | 16 S | 17 Cl | 18 Ar |
| 19 K | 20 Ca | 21 Sc | 22 Ti | 23 V | 24 Cr | 25 Mn | 26 Fe | 27 Co | 28 Ni | 29 Cu | 30 Zn | 31 Ga | 32 Ge | 33 As | 34 Se | 35 Br | 36 Kr |
| 37 Rb | 38 Sr | 39 Y | 40 Zr | 41 Nb | 42 Mo | 43 Tc | 44 Ru | 45 Rb | 46 Pd | 47 Ag | 48 Cd | 49 In | 50 Sn | 51 Sb | 52 Te | 53 I | 54 Xe |
| 55 Cs | 56 Ba | 57–71 La–Lu | 72 Hf | 73 Ta | 74 W | 75 Re | 76 Os | 77 Ir | 78 Pt | 79 Au | 80 Hg | 81 Tl | 82 Pb | 83 Bi | 84 Po | 85 At | 86 Rn |
| 87 Fr | 88 Ra | 89–103 Ac–Lr | 104 Rf | 105 Db | 106 Sg | 107 Bb | 108 Hs | 109 Mt | 110 Ds | 111 Rg | | | | | | | |

| | | | | | | | | | | | | | | | |
|---|---|---|---|---|---|---|---|---|---|---|---|---|---|---|---|
| Lanthanides series | 57 La | 58 Ce | 59 Pr | 60 Nd | 61 Pm | 62 Sm | 63 Eu | 64 Gd | 65 Tb | 66 Dy | 67 Ho | 68 Er | 69 Tm | 70 Yb | 71 Lu |
| Actinide series | 89 Ac | 90 Th | 91 Pd | 92 U | 93 Np | 94 Pu | 95 Am | 96 Gm | 97 Bk | 98 Cf | 99 Es | 100 Fm | 101 Md | 102 No | 103 Lr |

| | | | |
|---|---|---|---|
| Non-metals | Transition metals | Rare earth metals | Halogens |
| Alkali metals | Alkali earth metals | Other metals | Inert elements |

Elements generally detectible by ICP-AES are enclosed with a heavy border.

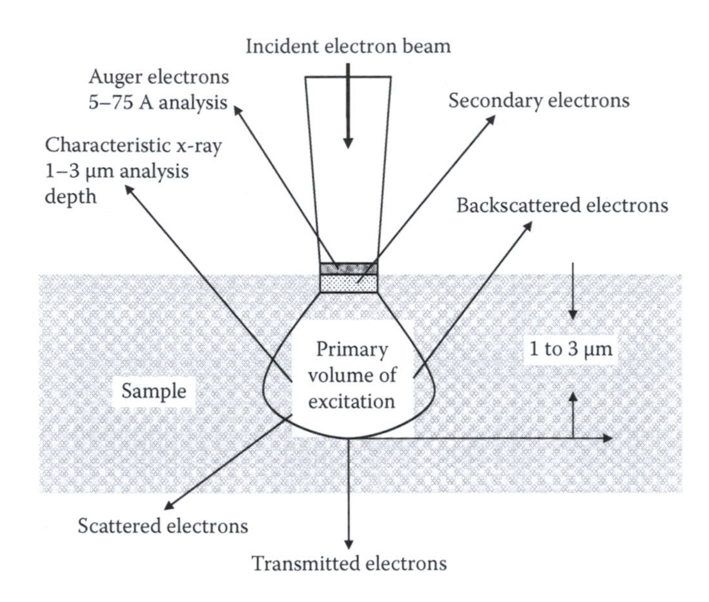


**FIGURE 5.3**
Electron specimen interaction and emission.

scatter away with little (<1 eV) or no change in energy, like the backscattered electrons. Backscattered electrons are scattered backward from the specimen when the incident electron elastically collides with an atom in the specimen. The intensity of backscattered electrons varies directly with the specimen's atomic number. This makes higher atomic number elements brighter than lower atomic number elements in a backscattered electron image, as illustrated in Figure 5.4. During inelastic scattering, energy is transferred to other electrons in the specimen, and the kinetic energy of the incident electron decreases. When the incident electron passes near an atom in the specimen, it will impart some of its kinetic energy to a lower-energy electron. This interaction causes an energy loss and path change of the incident electron and the ionization of the electron in the specimen atom. The ionized electron then leaves the atom with a very small kinetic energy, as little as 5 eV, and is referred to as a secondary electron. In other words, secondary electrons are the electrons emitted from the specimen by inelastic collisions between the specimen and the incident electrons. The electrons scattered inelastically also play an important role in electron energy loss spectroscopy (EELS) for elemental composition and atomic bonding state analysis.

The intensity of secondary electrons is very dependent on topography. Due to their low energy, only the secondary electrons that are very near the surface (<10 nm) can exit the specimen and be detected. A secondary electron detector can be either positively biased or negatively biased. A positively biased secondary electron image shows more topographical features, while a negatively biased secondary electron image has better contrast.

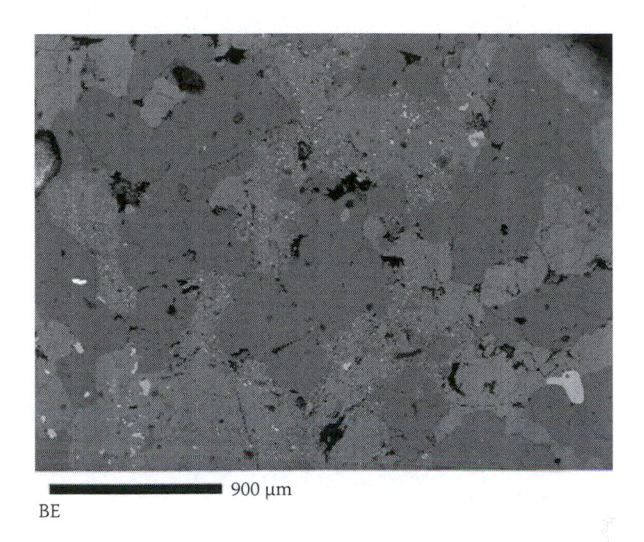


**FIGURE 5.4**
Backscattered electron image.

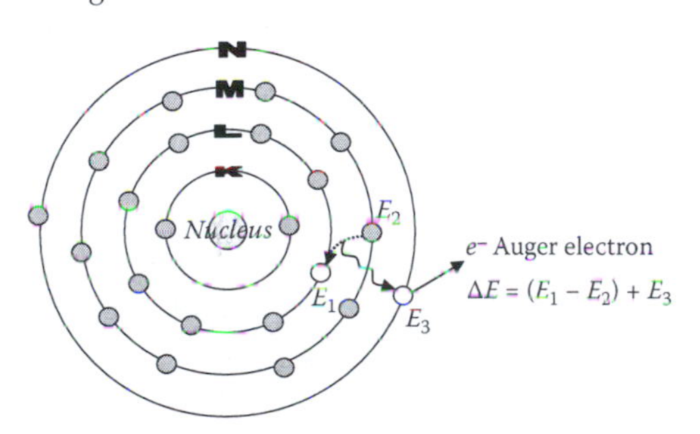


**FIGURE 5.5**
Auger electron emission.

During the secondary electron emission, an electron from the inner shell is emitted from the atom and generates a vacancy in this shell. Another electron in the higher-energy shell from the same atom can fall to this vacancy, resulting in a release of energy, usually in the form of a photon. However, the released energy can also excite another electron in an outer shell and eject it from the atom. This second ejected electron is referred to as an Auger electron. Figure 5.5 schematically illustrates the Auger electron emission process in a titanium atom where $E_1$, $E_2$, and $E_3$ are the electron potential energies in their respective atomic shells, L, M, and N. Conventionally, the potential energy of an electron is set to zero at infinity, and the electrons bound to the atomic orbital have negative potential energy. The binding energy of the ionization energy is referred to as the energy required to free an electron from its atomic shell or orbital, typically reported as positive values. Therefore, the electron potential energy and binding energy are of

the same numerical value but with opposite signs. It shows that when an electron in the M shell with a potential energy of $E_2$ jumps down to the L shell with a potential energy of $E_1$, the released potential energy, $E_2 - E_1$, excites one of the outer electrons in the N shell with a potential energy of $E_3$ to eject it from the atom as an Auger electron. Auger electrons are only emitted from the specimen surface, and have very low kinetic energies, $E = (E_2 - E_1) - E_3 = (E_2 - E_1) + E_3$. Therefore, the resulting Auger electron energy spectra can be used to identify the element and the surface information about the specimen. Today Auger electron spectroscopy is one of the most effective surface analytical techniques for determining the composition of the surface layers of a specimen in reverse engineering.

### 5.2.5 X-Ray Analysis

Another by-product of secondary electron in electron microscopic analysis is X-ray fluorescence. When an electron from the inner shell is emitted from the atom during the secondary electron emission, and another electron from the outer shell falls to its vacancy, the released energy might be emitted as X-ray to balance the total energy of the atom. Figure 5.6a and b schematically illustrates the emission processes of the K-line and L-line X-rays from a titanium atom, respectively, where $E_0$, $E_1$, $E_2$, and $E_3$ are the electron potential energies

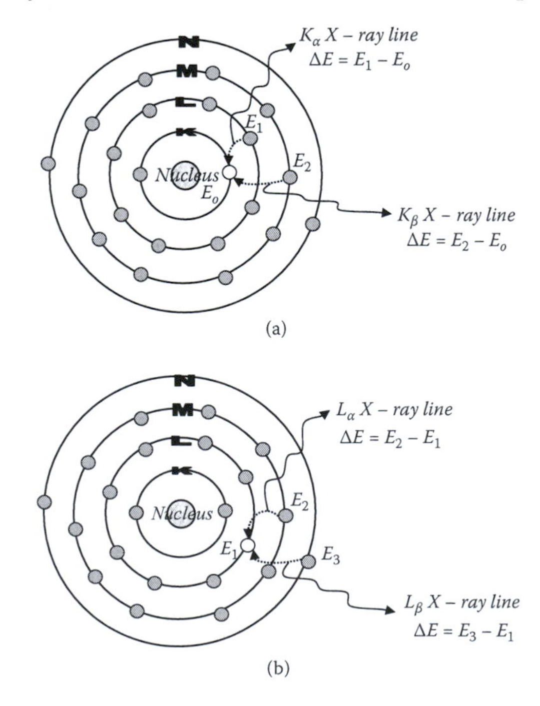


**FIGURE 5.6**
(a) Emission of K-series X-rays. (b) Emission of L-series X-rays.

in shells, K, L, M, and N, respectively. When the vacancy is in the K shell, as shown in Figure 5.6a, the emitted X-rays are referred to as K-line X-rays in the X-ray spectra. The emitted X-ray is further distinguished as $K_{\alpha}$ X-ray if the electron is dropped from the L shell to the K shell, and $K_{\beta}$ X-ray when the electron falls from the M shell to the K shell. The emitted X-rays are referred to as L-line X-rays if the vacancy is located in the L shell, as illustrated in Figure 5.6b. Similarly, the L-series X-rays are further distinguished as $L_{\alpha}$ or $L_{\beta}$ X-rays depending on whether the electron falls from the M or N shell.

The energy dispersive X-ray analysis (EDXA), also referred to as energy dispersive X-ray spectroscopy (EDS), is usually used in conjunction with SEM. An SEM sample has to be conductive. A nonconductive sample will accumulate the incident electrons on its surface and repel other electrons that follow. This effect is referred to as discharging and results in poor imaging. For a nonconductive sample, a thin layer of carbon or gold is applied. A word of caution: Only carbon coating is suitable for EDXA. The gold coating enhances the SEM imaging, but will absorb the incident X-ray and weaken the analytical capability because gold is a heavy metal with an atomic number of 79, while carbon only has an atomic number of 6. EDXA offers a convenient nondestructive method for preliminary chemical composition determination in reverse engineering. The electron beam, typically with an energy of 10 to 20 keV, strikes the sample surface and stimulates X-ray emission. The energy of the X-rays emitted depends on the individual atomic structure of each element, and forms a characteristic X-ray histogram or profile. Figure 5.7a is the EDXA spectrum of a paint chip. It shows multiple characteristic X-ray lines of the same elements, such as Fe, Pb, Ca, and Ti, because X-rays originated from different shells have been emitted. This is a real-life example of paint analysis. The EDAX provides a quick analysis of how many layers of paint there are and the composition of each layer in reverse engineering a piece of artwork. This technology is applied for environment protection as well. If an old house was painted over with a layer of new paint, an EDAX examination over the cross section can verify if the old lead-containing paint was removed. The quantity and energy of the X-rays can be measured by an energy dispersive spectrometer, and the constituent elements can be semiquantitatively measured. A quantitative analysis is achievable but requires delicate calculations and comparative corrections with standards. These standards are the materials containing a known concentration of an analyte. The primary standards are usually extremely pure and stable. They provide a reference to determine unknown concentrations or to calibrate analytical instruments. The National Institute of Standards and Technology provides a wide variety of standard reference materials for validating and calibrating analytical instruments. When the atomic number of an element decreases, this element's detectability gets progressively worse. Any element below sodium (Na) that has an atomic number of 11 in the periodic table of elements cannot be detected by standardless analysis.

Spectral resolution and detection limit are two important parameters in composition analysis. Spectral resolution is the capability of an analytical

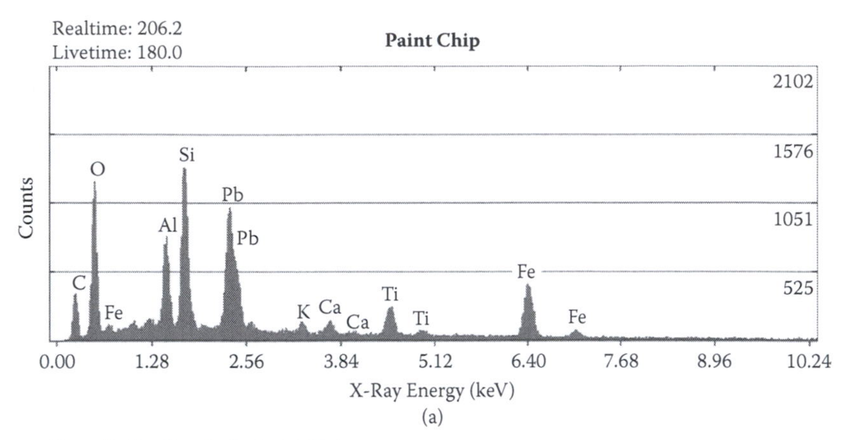


Quantitative Results for Paint Chip
Analysis: Bulk Method: Standardless
Acquired 03-Jun-2009, 20.0 keV@10 eV/channel

| Element | Weight % | Std. Dev. | MDL | Atomatic % | k-Ratio | Intensities | Probability |
|---|---|---|---|---|---|---|---|
| C | 12.64 | 0.92 | 3.72 | 24.99 | 0.0885 | 1534.2 | 0.88 |
| O | 31.53 | 1.29 | 0.37 | 46.82 | 0.1137 | 6518.7 | 0.93 |
| Al | 8.05 | 0.99 | 0.76 | 7.09 | 0.0482 | 5107.8 | 1.00 |
| Si | 12.89 | 0.74 | 0.48 | 10.90 | 0.0853 | 9313.7 | 0.99 |
| K ? | 0.78 | 0.29 | 1.63 | 0.47 | 0.0064 | 524.5 | 0.00 |
| Ca ? | 1.19 | 0.41 | 1.21 | 0.70 | 0.0103 | 802.7 | 0.98 |
| Ti | 3.11 | 0.88 | 0.99 | 1.54 | 0.0264 | 1579.4 | 0.99 |
| Fe | 13.05 | 0.81 | 0.61 | 5.55 | 0.1178 | 4345.9 | 0.98 |
| Pb | 16.76 | 0.92 | 0.65 | 1.92 | 0.1268 | 7659.4 | 0.00 |
| Total | 100.00 | | | | | | |

? These elements are statistically insignificant.

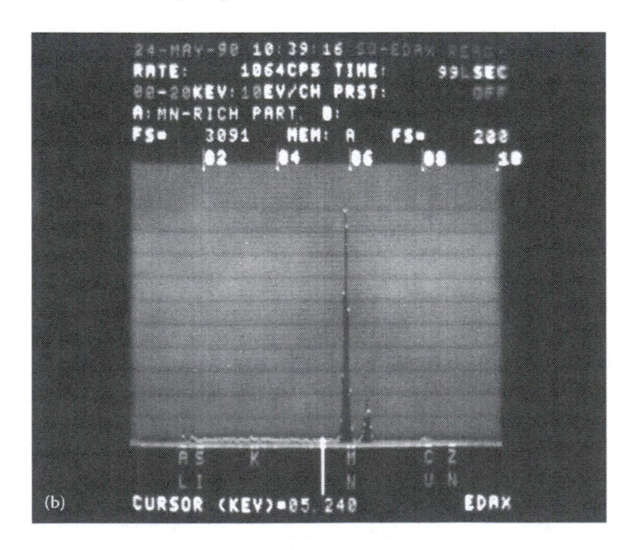


**FIGURE 5.7**
(a) EDXA spectrum of a paint chip. (Reprinted from SEMTech Solutions, Inc. With permission.) (b) EDXA analysis on a Mn-rich particle.

instrument to separate two test data or areas. The detection limit is the smallest quantity an analytical instrument can detect. For EDXA, the electron beam is used and the spectral resolution can be as fine as just 1 micron. For a laser beam that is commonly used in mass spectroscopy, the spectral resolution might be as large as 20 microns. However, mass spectroscopy has a detection limit usually in the range of 1 to 2 ppm, which is much better than the typical range of 50 to 100 ppm for EDXA.

An EDXA detector is used to convert X-ray energy into voltage signals, and separate the characteristic X-rays of different elements. This information is then sent to an analyzer for further analysis and display. Figure 5.7b is an EDXA on a manganese-rich particle in an aluminum alloy. The data are reported as a plot of X-ray intensity or counts on the vertical axis vs. energy, usually in keV, on the horizontal axis. Each peak corresponds to an individual characteristic X-ray from different elements, which reflect their respective identities. The height of the peak, the area under the peak, and the full width at half maximum all have their respective roles in quantitative analysis. The quantity of each element can be semiquantitatively estimated by comparing the relative peak-height ratio or the area under the peak to a standard. The peaks in wavelength dispersive X-ray spectroscopy (WDS) usually are the narrow Lorentzian distributions, and their heights are often measured to reflect the intensity for quantitative analysis. However, the peaks in EDXA are closer to Gaussian distributions, and the areas under the peak are often measured to reflect the intensity for quantitative analysis. The full width at half maximum intensity of the peak accounts for the spectral resolution. EDXAs are subject to some limitations. For example, multiple peaks, such as Mn-$K_a$ and Cr-$K_b$, might closely overlap and make it difficult to resolve them. It is also worth noting that the traditional SiLi detector used for EDXA is often protected by a beryllium (Be) window, and the absorption of the soft X-rays by beryllium could preclude the detection of elements below sodium. The EDXA analytical capability increases in a windowless system. However, it generally cannot detect the presence of elements with an atomic number of less than 5. In other words, EDXA has difficulty detecting hydrogen, helium, lithium, and beryllium.

WDS is another technique utilized for elemental chemical analysis in reverse engineering. EDXA and WDS are usually used in conjunction with SEM, or an electron probe microanalyzer (EPMA). EPMA is a nondestructive elemental analysis technique, similar to SEM but with a more focused analysis area. It works by rastering a micro volume of the sample with an electron beam typical of an energy level of 5 to 30 keV. It then collects the induced X-ray photons emitted by the various elemental species and quantitatively analyzes the spectrum with precise accuracy, up to ppm. In contrast to EDXA, WDS analyzes the electron diffraction patterns based on Bragg's law and has a much finer spectral resolution and better accuracy. WDS also avoids the problems associated with artifacts in EDXA, such as the false peaks and the background noise from the amplifiers. The noise intensity that appears in

most EDXAs partially results from the interaction between the incident electrons and the outermost-shell valence electrons of the sample atoms, which slows down the speed of the incident electrons and releases their kinetic energy to form the background noise. WDS is a high-quality technique commonly used for quantitative spot analysis. EDXA shows a spectrum of elements of a sample simultaneously, as illustrated in Figure 5.7a and b. WDS, however, can only read a single wavelength and analyze one element at a time. The X-ray intensity in any quantitative analysis should be corrected for the matrix effects associated with atomic number (Z), absorption (A), and fluorescence (F), the so-called ZAF factor. In reverse engineering applications, EDXA and WDS are best used as complementary analytical tools. EDXA can be used first to scan the general chemical makeup of an unknown sample, and then WDS is applied to more accurately conduct a quantitative analysis of specific constituent elements of the sample.

## 5.3 Microstructure Analysis

The chemical composition shows what an alloy is made of. Microstructure is an alloy's footprint that traces back its heat treatment history and fabrication flow path, and shows how this alloy is manufactured. The microstructure evolution is a complex process of thermodynamics and kinetics. The laws of thermodynamics determine the existence of a specific phase in an alloy. The Gibbs' phase rule will determine how many phases can coexist during a solidification process when a molten metal is cooling down from the liquid to solid state. The principles of kinetics determine the rate of a kinetic process, and calculates the incubation time of nucleation and the grain growth rate afterwards. The identification of various phases and a quantitative measurement of their respective percentages provide vital information on the manufacturing process and heat treatment that this alloy has experienced.

### 5.3.1 Reverse Engineering Case Study on Ductile Iron

Ductile iron has been widely utilized for various applications in human society for several thousand years since ancient China and other civilizations. It is still used for crankshafts and axle gears in automobiles, and in many other industries, such as railroad and construction today. Its distinct microstructure and versatility makes it an interesting case study in reverse engineering to demonstrate that valuable information can be extracted from alloy microstructure. This case study will also highlight the roles of other subjects discussed in this chapter, such as material specifications and mechanical properties. Figure 5.8 shows the microstructure of a typical ductile iron with nodular graphite surrounded by ferrite in a matrix of pearlite.

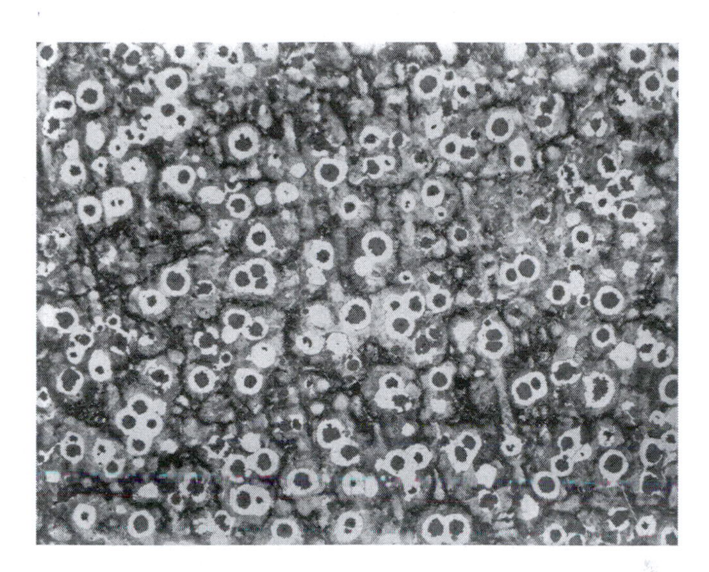

**FIGURE 5.8**
Microstructure of a typical ductile iron with nodular graphite surrounded by ferrite in a matrix of pearlite.

It provides a wealth of information on alloy classification and heat treatment parameters. Based on its distinctive microstructure, if a part is made of ductile iron, its verification can be easily confirmed in reverse engineering. The nominal chemical composition of ductile iron is listed in Table 5.2, though a chemical analysis is required to accurately determine its exact composition. Compared with steel, which contains less carbon and silicon, usually 0 to 2% each, ductile iron has noticeably higher carbon and silicon contents, which along with magnesium will lead to the formation of nodular graphite, as shown in its trademark microstructure. In fact, the weight percent (wt%) of carbon and silicon contents is used as an index for ferrous alloy classification, which is referred to as carbon equivalent (CE), CE = wt% C + 0.33 wt% Si. Seven different types of graphite morphology are defined in ASTM A247 (ASTM, 2006a). The graphite appears in thin flakes like potato chips in gray iron; but in malleable iron, it appears in a massive bulk form like popcorn. Most carbon is consolidated as iron carbide and pearlite in white iron. These microstructure morphologies provide convenient and convincing evidence for alloy identification as ductile iron, gray iron, malleable iron, or white iron, and the affiliated heat treatment schedules that produced them.

More advanced and accurate elemental analyses might be required for the critical part made of ductile iron. Each element has its own specific effect on ductile iron, as detailed below, and its content might need to be determined precisely to meet the specific quality control requirement. Silicon is a heat-resistant element and is usually added for high-temperature strength and to induce graphite and ferrite formation. However, the toughness could deteriorate to an unacceptable level when silicon is higher than 2.6%. Manganese

**TABLE 5.2**

Typical Chemical Composition (Weight Percentage) of Ductile Iron

| Major Elements | | Trace Elements | |
|---|---|---|---|
| Carbon | 3.6–3.75 | Chromium | 0.06 max |
| Silicon | 2.2–2.6 | Cerium | 0.005–0.015 |
| Manganese | 0.35 max | Tin | 0.01 max |
| Phosphorus | 0.02 max | Vanadium | 0.02 max |
| Sulfur | 0.02 max | Titanium | 0.02 max |
| Magnesium | 0.030–0.050 | Lead | 0.004 max |
| Copper | 0.10–0.80 | Aluminum | 0.05 max |
| Nickel | 0.10–2.00 | Other | <0.01 |
| Molybdenum | 0.0–0.20 | | |

*Source:* Data from the Ductile Iron Society.

promotes the formation of pearlite and enhances hardness and strength, but it decreases the annealability and increases segregation. Usually the content of phosphorus is kept as low as possible, since a 0.01% increase in phosphorus could reduce ductility by 1%. The toughness could also deteriorate to an unacceptable level when the content of phosphorus is higher than 0.02%. Sulfur could impede the formation of nodules and "de-ductile iron" the alloy. Magnesium is a deoxidizer: It is added to remove sulfur and produce nodules. However, excessive magnesium could cause graphite to explode and produce shrinkage and spikes in graphite that are sometimes undesirable. Chromium produces carbide. The addition of cerium could tie up tin, vanadium, titanium, lead, and aluminum to isolate these undesirable elements.

The properties of ductile irons of similar compositions are heavily dependent on processing, heat treatment, casting section size, and the subsequent microstructure. The reverse engineering of ductile iron is beyond just identifying its composition. To confirm the grade and properties of the ductile iron used for the OEM part, more comparative analyses are needed. Several standards have been documented for the properties and specifications of ductile irons. ASTM A897 and its metric version, 897M, list the mechanical property requirements of austempered ductile iron in different grades, as summarized in Table 5.3 (ASTM, 2006b). The properties described in metric and U.S. customary units are rounded up for easy comparison. Hardness is only listed in ASTM A897 for reference, and is not mandatory. The critical properties of each grade of ductile irons are dependent on their intended applications. For example, 1600/1300/–grade ductile iron has an ultimate tensile strength of 1,600 MPa (232 ksi) and a yield strength of 1,300 MPa (189 ksi). It does not specify tensile elongation and Charpy impact toughness because it is primarily used for gear and wear resistance applications. Grade 1400/1100/1 sacrifices some wear resistance to improve ductility and toughness. It has an ultimate tensile strength of 1,400 MPa (203 ksi) and a yield strength of 1,100 MPa (160 ksi), and requires a tensile elongation of 1%. It is also used for similar

**TABLE 5.3**

Mechanical Property Requirement of Austempered Ductile Iron

| Grade | Ultimate Tensile Strength, MPa (ksi) | Yield Strength, MPa (ksi) | Elongation (%) | Brinell Hardness Number[a] | Charpy Impact Value, Joule (ft-lb)[b] |
|---|---|---|---|---|---|
| 750/500/11 | 750 (109) | 500 (73) | 11 | 269–321 | 110 (81) |
| 900/650/9 | 900 (131) | 650 (94) | 9 | 269–341 | — |
| 1050/700/7 | 1,050 (152) | 700 (102) | 7 | 302–375 | 80 (59) |
| 1200/850/4 | 1,200 (174) | 850 (123) | 4 | 341–444 | 60 (44) |
| 1400/1100/1 | 1,400 (203) | 1,100 (160) | 1 | 388–477 | 35 (26) |
| 1600/1300/— | 1,600 (232) | 1,300 (189) | — | 402–512 | — |

*Source:* ASTM A897-06 and A897M-06.

[a] Typical range; not a required specification.

[b] Average of the highest three test values of four test samples of unnotched Charpy bars tested at 295 ± 4K (70 ± 7°F).

applications. With the same chemical composition, different properties will result from different heat treatments, such as austempering. Austempering is a heat treatment process that heats ductile iron castings to the austenitizing temperature, usually between 816 and 927°C (1,500 and 1,700°F), depending on the grade, and holds at this temperature long enough to dissolve the carbon in austenite. The part is then quickly quenched to 232 to 399°C (450 to 750°F). The quenching time is usually just a few seconds to ensure a sufficient cooling rate. It is critical to avoid the formation of pearlite around the carbon nodules during quenching, as this would reduce mechanical properties. Afterwards, the part is held at the austempering temperature for isothermal transformation to form a microstructure of acicular ferrite in carbon-enriched austenite. Figure 5.9 schematically illustrates a typical austempering process with unspecified parameters, such as austenitizing temperature, holding time, quenching medium, and cooling rate. To duplicate an equivalent part using reverse engineering, these heat treatment parameters should mirror the parameters in the OEM's process as much possible. The reference documents that verify an OEM's heat treatment parameter, such as the austempering temperature, are not always available in reverse engineering. They sometimes rely on corporate knowledge, information from a professional society such as the Ductile Iron Society, or test results.

Through proper control of the austempering conditions, design engineers can produce a range of properties for austempered ductile iron (Keough, 1998). For instance, for high ductility, good fatigue and impact strengths at the sacrifice of yield strength, a higher austempering temperature is used. A test showing a yield strength of 496 MPa (72 ksi) implies that the austempering temperature is most probably around 399°C (750°F) using the ductile-iron-grade scale as a reference. On the other hand, a lower austempering temperature is used for applications requiring a higher yield strength,

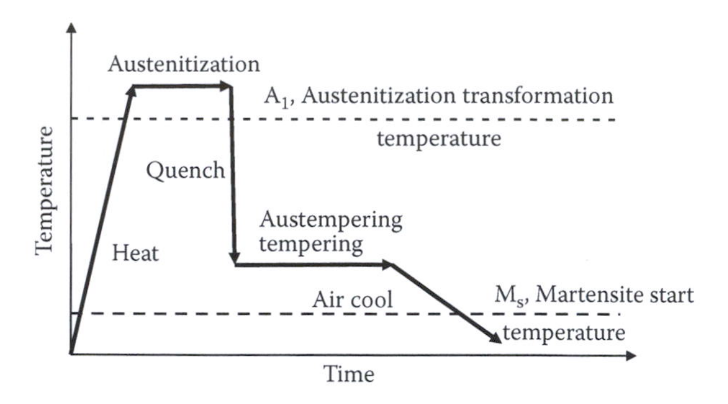


**FIGURE 5.9**
Schematic of a typical austempering process.

hardness, and good wear resistance. If the test shows a yield strength of 1,379 MPA (200 ksi), the austempering temperature is most likely around 260°C (500°F). Based on the mechanical properties of the OEM parts, the engineer can use either the documented reference data or the actual test results from simulated samples to figure out the austempering temperature. The tensile property requirements of ductile iron castings are listed in ASTM A536 (ASTM, 2009).

The property requirements of automotive ductile iron castings are summarized in SAE J434C. In contrast to ASTM A897, the characteristics of ductile iron are primarily specified by hardness and (micro)structure in SAE J434C for automotive industries, and automotive engineers often specify the hardness value in their procurement. This is technically acceptable because both the ultimate tensile strength and yield strength of ductile iron are usually linearly proportional to the Brinell hardness number up to 450. However, the quantitative relationship between hardness and ultimate tensile strength is unique for ductile irons, and not universal. These specifications are the key references used to determine the heat treatment schedules, such as austempering. They also provide hints to the manufacturing process and cast iron grades in reverse engineering. The International Organization for Standardization (ISO) lists the property specifications for spheroidal graphite or nodular graphite cast iron in ISO 1083-2004 (ISO, 2004). Like SAE specifications, ISO standards specify the respective (micro)structure and the hardness value for each grade. Generally, European ductile iron contains relatively less silicon (around 2.1%) than American ductile iron (ranging from 2.2 to 2.6%). European standards are also usually more specific regarding Charpy impact values. These subtle differences make cross-referencing between different material specifications challenging in reverse engineering.

The rest of this section will focus on the three most commonly used analytical techniques for microstructure identification and analysis in reverse engineering: light microscopy, scanning electron microscopy, and transmission electron microscopy.

### 5.3.2 Light Microscopy

To decode a part using reverse engineering often requires engineers to look at the minute details of the part at high magnification. The word *microscope* literally originates from the Greek *mikrós* (small) and *skopeîn* (to look at). Microscope is an instrument for examing objects that are too small to be seen by the naked eye. The first invented, and still the most commonly used today, microscope is the optical microscope. An optical microscope magnifies the image based on the theory of optics. Two terms, *resolution* and *magnification*, are usually used to describe the analytical power of a microscope. Resolution refers to the capability to identify two separate spots located closely together. If two objects are closer than the microscope resolution, then they will blur together in the microscope image. The resolution of a traditional light microscope is restricted by diffraction effect, light wavelength, and the characteristics of the lenses. The best resolution is usually limited to around 0.2 μm. The magnification is the numerical ratio between the image size and the actual object size. When the image is projected to a larger screen, the image will be further magnified, but the resolution remains the same. Generally speaking, a microscope with higher magnification lenses is usually constructed to provide better resolution. By proper design, the resolution of a light microscope will increase from 0.7 to 0.2 microns when the magnification increases from 10× to 100×. However, the field of view, that is, the area that can be examined by the microscope, decreases with increasing magnification. A field of view of 1,000 $\mu m^2$ for a 10× lens will decrease to 100 $\mu m^2$ when the magnification increases to 100×. Modern light microscopes equipped with a laser scanner, digital camera, and imaging software, such as a confocal microscope, can further improve the resolution and provide stereo three-dimensional images.

### 5.3.3 Scanning Electron Microscopy

Scanning electron microscopy is an indispensable analytical instrument in material identification and analysis. Figure 5.10 shows a modified SEM, the JEOL JXA-733 Superprobe. It is equipped with one EDXA in front and five WDS circling around both sides and the back. The EDXA is usually operated at the cryogenic liquid nitrogen temperature, while the WDS operates at room temperature. The electron gun on top of the SEM emits high-energy monochromatic electrons when energized by a high voltage, typically 120 keV for most applications. Electrons are usually generated by thermionic emission from a tungsten filament, or a single-crystal lanthanum hexaboride ($LaB_6$). Alternatively, electrons can also be generated by field emission from a sharp tungsten tip. These electrons are condensed and focused by a series of electromagnetic lenses and apparatuses when they travel down the SEM column, and finally hit the sample. A phosphor screen is used for direct sample

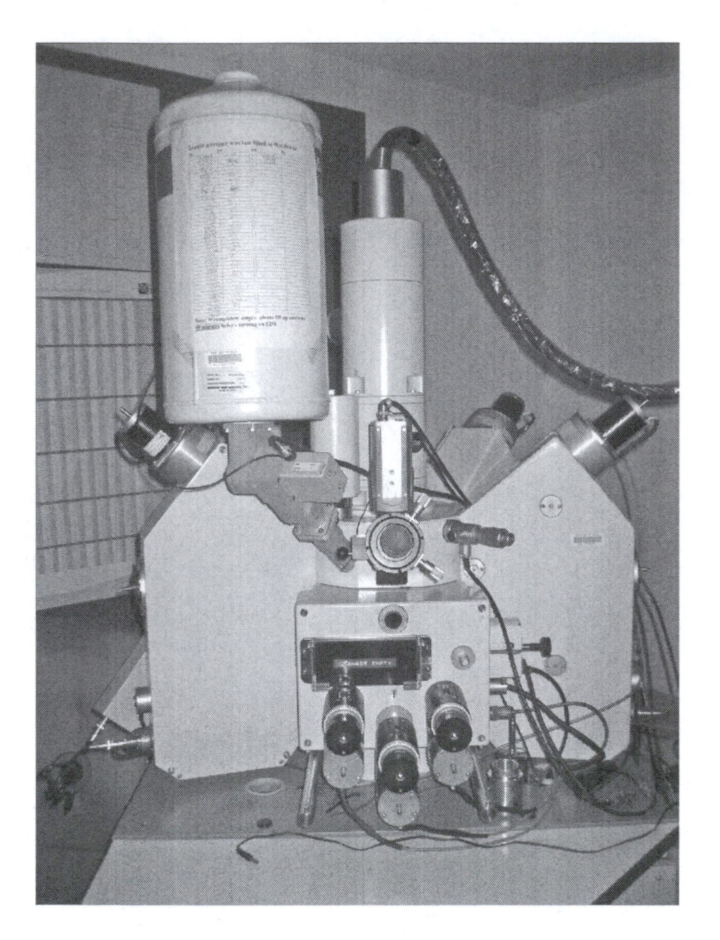

**FIGURE 5.10**
A scanning electron microscope.

examination. The photo or image recording system is either film-based or digitalized with a charge-coupled device (CCD).

In modern materials science and failure analysis, SEM is used for a broad range of applications to analyze solid materials. In reverse engineering an SEM provides the following data of a sample material: topographic features, grain morphology, and alloy composition. Topography illustrates the sample surface features, including the fractography of a fractured surface. Figure 5.11a is SEM fractography of ductile iron failed by tension at room temperature showing nodular graphite spreading out in the cross-sectional matrix. Figure 5.11b shows a transgranular tensile fractography of an aluminum alloy at room temperature. Grain morphology reveals the size, shape, orientation, and texture of grains, in either its original form or a chemically etched form. Though an SEM usually operates in a vacuum on the order of $10^{-5}$ to $10^{-6}$ torr, it is very user friendly, and only minimal sample preparation

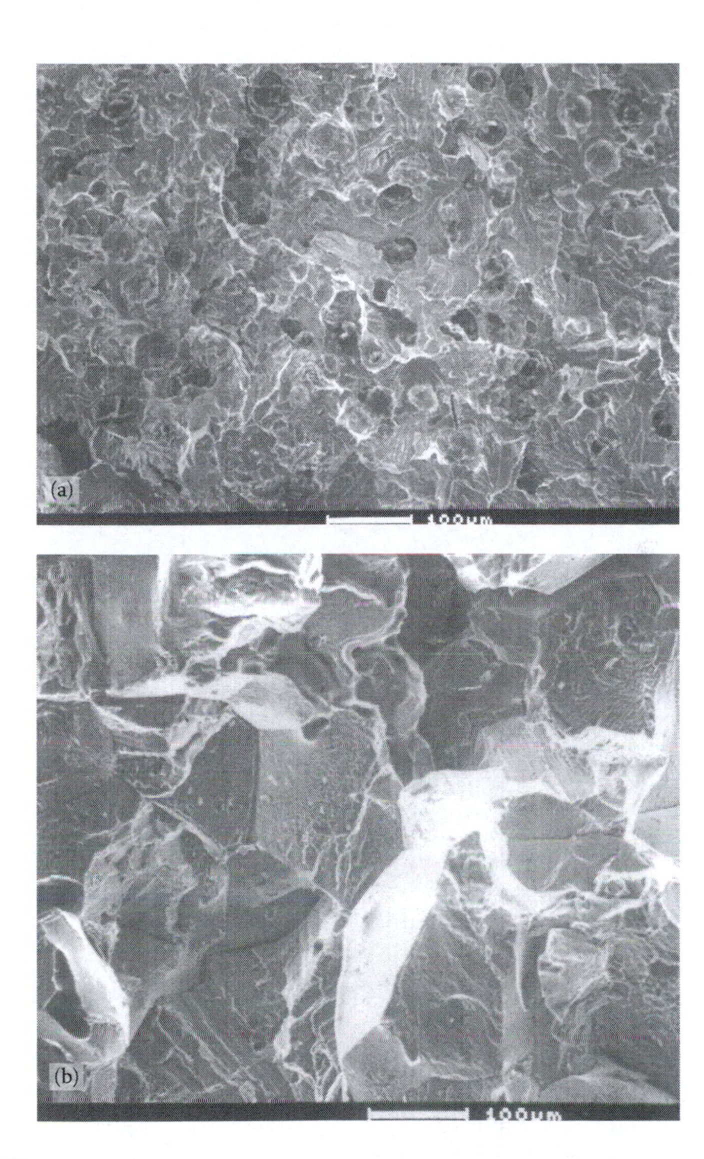

**FIGURE 5.11**
(a) SEM fractography of ductile iron. (b) Intergranular tensile fractography of an aluminum alloy.

is required. The sample size can be as large as 10 cm wide and 40 mm high, which allows direct observation of many real-life samples without cutting them apart. The short wavelength of electrons permits high magnification and resolution. Compared to the typical magnification of approximately 1,000× for conventional light microscopy, SEM can provide magnifications of up to 100,000×, and reach a resolution up to a few nanometers. An SEM also provides great depth of field, allowing complex, three-dimensional objects to remain sharp and in focus. These images, which show detailed features, are

valueless in failure analysis where detailed fractography is often required. The semiquantitative data obtained from an SEM is usually adequate for most applications. It is widely used in pathologic, forensic, metallurgical, and environmental analyses. The three most commonly used SEM imaging techniques—secondary electron imaging, backscattered electron imaging, and X-ray imaging—are discussed below in detail.

The derivative electrons, such as secondary, backscattered, and Auger electrons, and characteristic X-rays emitted from the sample by the electron sample interactions are used for microstructure analysis, elemental composition identification, and phase verification. Secondary electrons are used to image morphology and surface topography of the samples, as shown in Figure 5.11a and b. In contrast to the secondary electron three-dimensional imaging, the backscattered imaging is two-dimensional. Backscattered electron images are used to illustrate composition contrast in multiple phase samples for quick revealing of element and phase distributions. The heavier element or the compound with the higher average atomic number will scatter back more electrons, and appear brighter in the backscattered image, as demonstrated in Figure 5.12a. It is a multicomponent sample. The constituent elements of this sample can be mapped out by X-ray elemental imaging.

The backscattered electron image only shows the contrast difference reflecting the average Z factor that is primarily determined by the atomic number in a phase, as shown in Figure 5.12a. The affiliated X-ray imaging illustrates a composition map showing the spatial distribution of each element of interest in a sample, as shown in Figure 5.12b to e. Figure 5.12b is the calcium X-ray elemental image that illustrates the calcium distribution in the area backscattered electron imaged in Figure 5.12a. Figure 5.12c to e shows the iron, oxygen, and aluminum X-ray images of the same area, respectively. The X-ray composition image can also display integrated element distributions in a textural context by compositional zones, and provides a comprehensive "big picture" of all different elements over the same area. This elemental composition information is crucial in phase identification in reverse engineering. An X-ray composition map can be composed by either EDXA or WDS by progressively rastering the electron beam point by point over an area of interest. It is like scanning the sample and creating a bitmap image pixel by pixel based on chemical elements. Resolution is determined by the rastering electron beam size, scanning speed, and elemental concentration. In many applications, sufficient information can be acquired to map an adequate element image by EDXA. This is typically a faster approach, but sacrifices resolution and detection limits. The elements in low concentration may fail to respond and be missed. An accurate element map usually requires the utilization of WDS in collaboration with an electron microprobe.

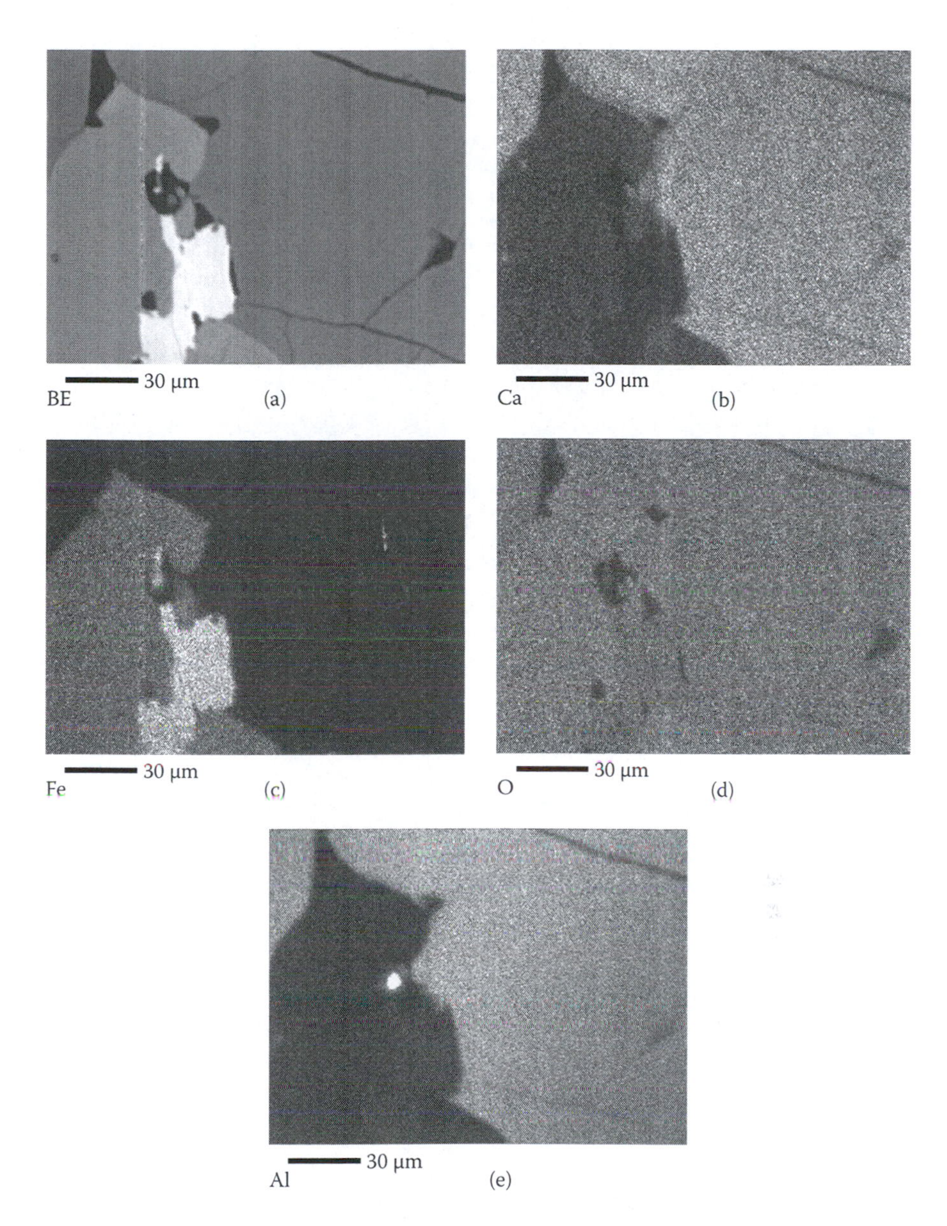


**FIGURE 5.12**
(a) Backscattered SEM image of a multicomponent sample. (b) Calcium X-ray elemental image. (c) Iron X-ray elemental image. (d) Oxygen X-ray elemental image. (e) Aluminum X-ray elemental image.

### 5.3.4 Transmission Electron Microscopy

Since the introduction of transmission electron microscopy in the 1930s, it has become an immensely valuable and versatile technique for the characterization of materials. TEM's high magnification and resolving power allow

for examination of the fine details, down to the atomic level, on the order of a few angstroms ($10^{-10}$ m). TEM provides the following data of particular interest to reverse engineering: grain morphology, crystallographic structure, and chemical composition. A typical TEM is composed of three systems: illumination, image forming, and signal collection. The illumination system includes an electron gun that is the source of electrons, and condenser lenses that focus the electron beam. Both TEM and SEM have their electrons excited from similar electron guns with a high voltage. The higher the voltage, the shorter the electron wavelength, and the better the resolution will be. However, the electron beam will damage the TEM specimen when the kinetic energy becomes too high. The electrons in most TEMs are energized to an energy level around 200 keV. The imaging system consists of an objective lens, intermediate lenses, a project lens, and an aperture for selected area diffraction. The objective lens forms the image, and the intermediate lenses provide the options to get a microstructure image or a diffraction pattern. The signal collection system is equipped with detectors and cameras. To avoid the charging effect that results from the accumulation of electrons on the specimen surface, a nonconductive TEM sample is usually coated with carbon or gold to make it conductive for better imaging.

Compared to SEM, where reflecting electrons are utilized to examine a bulk material surface, TEM does not discern any topographic information because it examines thin films with transmitted electrons. The thin-film sample preparation is more complicated, and the TEM operation requires more training as well. The interpretation of a TEM image requires a good understanding of electron microscopy and the structure of the material.

Different TEM operation modes, such as bright filed, dark field, and diffraction, reveal various data for reverse engineering applications. The bright field is the most common mode. It exhibits the sample using bright-field imaging. Figure 5.13a is a bright-field TEM micrograph that shows $\delta'$ ($Al_3Li$) precipitates in an Al-Mg-Li alloy. The image contrast is formed directly by occlusion and absorption of electrons in the sample. Thicker regions of the sample, or regions with a higher atomic number, will appear dark, while the regions with little electron and sample interaction will appear bright to show a bright field. In the dark-field operation mode, only the scattered electrons are utilized for imaging, as illustrated in Figure 5.13b. It is a dark-field TEM image showing $\delta'$ ($Al_3Li$) precipitates in an Al-Cu-Mg-Li alloy. The areas with little electron–sample interaction reflect no scattered electrons and appear dark. Dark-field imaging is a very useful technique for second phase identification in reverse engineering. The unique analytic capability of TEM allows it to reveal detailed microstructural features. TEM images show the profiles of precipitates, second phases, and dislocation networks, and allow semi-quantitative analyses on these observations. They also show the details of grain morphology, from grain size and grain boundaries to recrystallization. This information is crucial in material characterization, and essential to material identification and process verification.

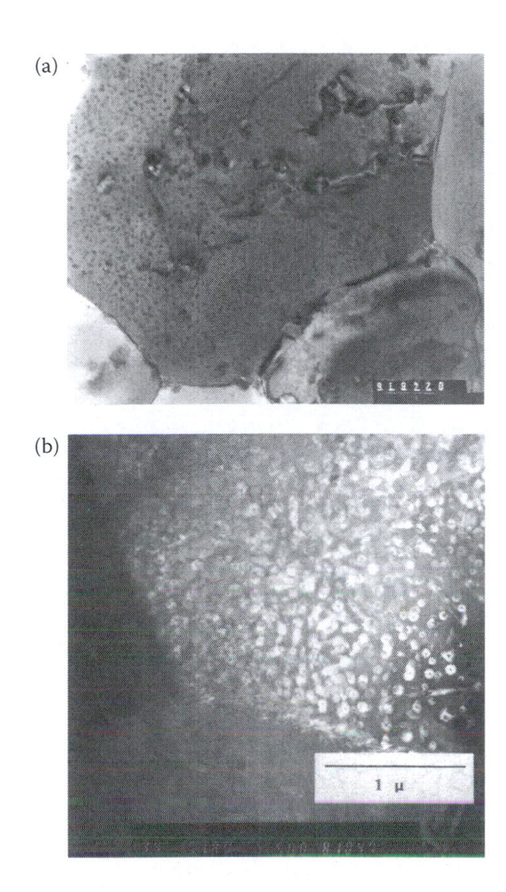

**FIGURE 5.13**
(a) Bright-field TEM image showing $Al_3Li$ precipitates. (b) Dark-field TEM image showing $Al_3Li$ precipitates.

In the imaging mode, the intermediate lens is focused on the initial image formed by the objective lens, and an image is illustrated on the viewing screen. While in the diffraction mode, the intermediate lens is focused on the diffraction pattern formed in the back focal plane of the objective lens, and the viewing screen of the microscope shows a diffraction pattern. In the diffraction mode, the elastically scattered electrons are utilized. The incident electrons might be elastically scattered during their interaction with the specimen atoms. They continuously move forward along a deflected path without losing any energy, and then transmit through the specimen. All the incident electrons are monochromatic with the same energy and wavelength. When a monochromatic electron beam is projected onto a crystalline material, diffraction can occur in accordance with Bragg's law. Bragg's law is mathematically described by Equation 5.1:

$$n\lambda = 2d \sin\theta \tag{5.1}$$

where $n$ is an integral number, $\lambda$ is the electron wavelength, $\theta$ is the incident angle of the electron beam, and $d$ is the spacing between two atomic planes. Diffraction occurs when the distance traveled by the electron beams reflected from successive crystallographic planes differs by a complete number $n$ of wavelengths. This constructive interference through diffraction forms a pattern of diffraction spots. Each spot corresponds to a specific atomic spacing, that is, a crystallographic plane that separates from one another by a distance $d$, which varies according to the crystallographic structure of the material. For any crystal, crystallographic planes exist in a number of different orientations—each with its own specific $d$-spacing. For a single crystal, the diffraction pattern is an array of spots, and each spot corresponds to a specific crystallographic plan, as shown in Figure 5.14a. For a polycrystalline metal, diffraction rings instead of spots are observed. Each diffraction ring corresponds to a specific crystallographic plane. The diffraction ring is a continuity of spots that are the continuous orientation of the same crystallographic plane. Figure 5.14b shows the electron diffraction rings of a polycrystalline

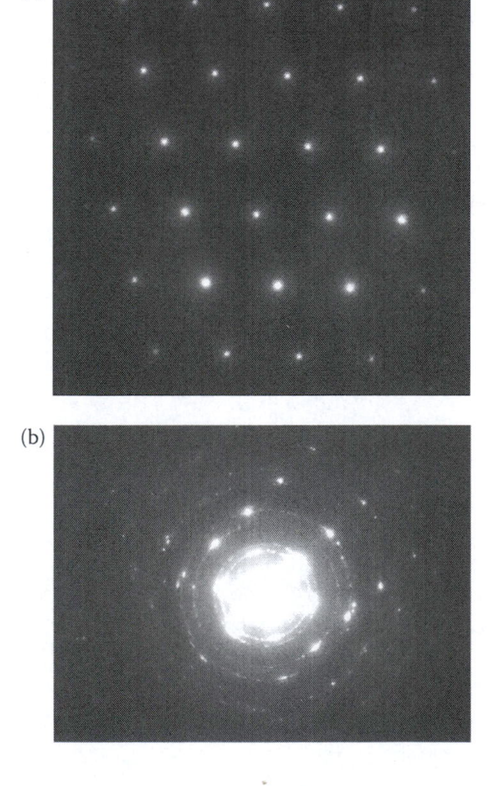


**FIGURE 5.14**
(a) Electron diffraction of a single crystal. (Reprinted from University of Cambridge. With permission.) (b) Electron diffraction rings of a polycrystalline aluminum alloy.

aluminum alloy. The diffraction pattern reveals a wealth of information on the sample material at the atomic scale. This information helps to understand the evolution of phase transformation and the prior thermal history the material has experienced to decode its heat treatment schedule and manufacturing processes in reverse engineering.

An elemental composition analysis is also feasible in an analytical TEM based on the physics of chromatic aberration of electrons when they pass through the thin sample. The interactions between the passing electrons and the constituent elements result in various levels of energy loss. An electron energy loss spectroscopy then forms an image showing a characteristic elemental map of the sample based on the atomic absorption of these interactions.

Table 5.4 summarizes the applications and limitations of the common analytical techniques used in reverse engineering for material identification and process verification.

## 5.4 Manufacturing Process Verification

The design of a machine component starts with material selection, followed by manufacturing process pick, and ends with functional test to ensure its performance. Quality control and cost management are some other factors that also require proper attention. From a reverse engineering perspective, chemical composition analysis is used to determine the material that was selected for the original part. The analytical techniques and material specifications for this process have been discussed in previous sections of this chapter. The material properties and their evaluation are discussed in Chapter 3. The functional test of finished components and their system compatibility will be discussed in Chapter 7. The following sections will focus on manufacturing process verification in reverse engineering. The four primary parameters in manufacturing are temperature, time, force, and atmosphere, such as the solution heat treatment temperature, aging treatment time, force used in forging or press, and atmosphere in which the part is cast or welded. The macrostructural appearance, such as surface finish, and microstructural features, such as grain morphology of the finished parts, usually provide the first evidence of how they are made. Therefore, reverse engineering, a manufacturing process, often starts with macrostructure examination and microstructure analysis to determine the temperature, time, force, and atmosphere that the OEM used to produce the original parts.

The manufacturing process covers all the steps and phases to create a part from raw material to finished product. It starts with melting and solidification, then product forming, machining, and joining, followed by heat treatment

**TABLE 5.4**

Applications and Limitations of Analytical Techniques

| Technique | Application | Limitation |
|---|---|---|
| *Composition Analysis* | | |
| Inductively coupled plasma/optical emission spectroscopy (ICP-OES) | Primarily for the determination of the major component concentrations | Some common elements such as hydrogen, boron, carbon, nitrogen, and oxygen are not detectable |
| Glow discharge mass spectrometry | Trace element analysis | Specialized technology |
| Interstitial gas analysis | Determination of included gases, e.g., hydrogen, nitrogen, and oxygen, in a solid sample up to 1 ppm | Specialized technology |
| DEXA | Quick scanning of sample constituent elements | Qualitative or semiquantative, elements with an atomic number less than 11 (sodium) usually not detectable |
| WDS | Accurate quantitative element analysis | Single element detection |
| *Microstructure Imaging* | | |
| Secondary electron image | Topographical analysis | Lack of composition analysis capability |
| Backscattered electron image | Surface analysis and elemental mapping | Two-dimensional |
| X-ray composition image | Multielement composition mapping showing the spatial distribution of the elements in a sample | Element of low concentration might be not detectable, particularly by EDXA |
| *Microscopy* | | |
| Transmission electron microscopy (TEM) | Grain morphology, crystallographic structure (diffraction), and chemical composition (analytical TEM) | Sample preparation is time-consuming, and data analysis is complicated, while area of analysis is small and might lack representation |
| Scanning electron microscopy (SEM) | Topographical features, fractography, grain morphology, and chemical composition (when equipped with EDXA or WDS) | Surface analysis only |
| Optical microscopy | General applications of microstructural analyses at low magnifications | Magnification is limited to 1,500×, no other analytical capability besides surface examination |

and surface treatment until the final product is complete. To reverse engineer the manufacturing process is a very challenging task. The inherent complexity virtually prohibits a full replication of the original process. Advanced technologies have made the verification process relatively easier today. The end-product performance test also provides reasonable assurance of equivalent functionality between the OEM and reverse engineered parts. The three most utilized manufacturing processes to produce mechanical components are casting, product forming, and machining. Casting is a thermal process that melts and solidifies raw material. Product forming is a deformation process that molds raw material into a part by force. Machining is a shaping process that removes "extra" material to produce the part. To assemble individual parts together, many joint methods are used, such as soldering, blazing, and welding. To improve the strength and performance of the product, many engineering treatments, such as heat treatment and surface treatment, ranging from coating to shot peening, are commonly applied. These processes and treatments and the challenges of replicating them in reverse engineering are discussed below.

### 5.4.1 Casting

The casting process has two primary phases: melting and solidification. This process produces either an ingot with a specified composition for further application, or a semifinished part directly from the molten metal. A high-quality ingot requires multimelting in a controlled environment. A triple vacuum arc remelting (VAR) process is required to produce an aerospace-grade titanium alloy. This process (re)melts and solidifies the ingot in a vacuum or inert gas atmosphere three times to minimize the brittle alpha phase. The size of the ingot could also affect the product properties. Some forged titanium products starting with a larger titanium ingot size have a higher tendency to show deficiency in dwell time fatigue resistance. Different casting processes, such as sand casting, permanent mold casting, and investment deficiency in casting, will produce parts with distinct macro- and microstructures, and therefore different properties due to their different solidification processes. These distinct structural features provide valuable information in reverse engineering to verify the casting method used for the original OEM product. For example, Figure 5.15 shows the macrostructure of the aluminum base for an electric power transmission post. It has a course columnar grain morphology growing inward just beneath the edge and a coarse equiaxed grain morphology in the center. This is a typical direct chill cast structure where three layers of grain configurations are observed: fine equiaxed grains on the chilled surface and coarse equiaxed grains in the center, while columnar grains appear in between these two regions. The macrostructure of the aluminum base, as shown in Figure 5.15, reveals two out of three characteristic structural features of direct chill casting. It is also noted that direct chill casting is one of the most frequently used casting processes

**FIGURE 5.15** **(See color insert following p. 142.)**
Aluminum casting macrostructure.

for wrought aluminum alloys. However, a definite verification of the casting method requires more microstructure analysis.

The effectiveness of microstructure analysis in reverse engineering can be further exemplified by earlier micrographs (Figure 3.1a and b). Even though the figure shows two vastly different microstructures, both samples are aluminum alloys, and each has the same chemical composition. The different microstructures result from different solidification processes. Figure 3.1a shows a typical equiaxed microstructure and provides some evidence that the material was processed by traditional casting. The fine microstructure of Figure 3.1b indicates that this material was solidified at a much higher cooling rate, which can only be achieved by a rapid solidification process. The directionality of the microstructure in Figure 3.1b further implies that it was processed by extrusion or another similar product-forming process. The product-forming process will be discussed in more detail below.

### 5.4.2 Product Forming

Product forming is the shaping of the raw material, such as an ingot or billet, into a product form, such as a turbine disk or axial shaft, by forging, rolling, or extrusion. The deformation of solid raw material during product forming usually results from the force applied to the solid when it is heated up, like a blacksmith working on a horseshoe. The process parameters from the applied force to operating temperature are often traceable because of their "imprinted" effects on the product characteristics and properties. The grain texture reveals the direction of the external force during rolling, extrusion, or forging. The microstructural features, such as subgrain structure

from dynamic recovery, grain shape, and size, all provide a hint in reverse engineering to decode the prior forming process that the original part was subject to. The forming process can also drastically affect the product properties, thus providing more information about the fabrication of the original part using reverse engineering.

### 5.4.3 Machining and Surface Finishing

Machining refers to the manufacturing process that shapes the part geometry by removing material from the workpiece. The traditional machining processes remove materials either by cutting, such as turning, milling, and drilling, or abrading, such as grinding. The advanced machining processes remove materials electrically or chemically instead of mechanically, such as electrical discharge machining (EDM), laser cutting, and chemical etching. In recent years the technologies of laser cutting and EDM have made great strides. They are increasingly becoming the favorite methods for complex, fragile, and tiny parts. In the medical filed, EDM is a particularly preferred method. Many of the cooling holes in jet engine turbine blades are laser drilled.

A cutting process involves a close interaction between the workpiece and the cutting tool. The mechanics of the cutting process are complicated and yet to be fully understood. This process involves mechanical force, material yielding, elastic and plastic deformation, chip formation, and breaking. Choosing a cutting tool and cutting speed is often based on corporate knowledge and experience. Due to the lack of theoretical modeling that can trace the exact cutting process utilized by the OEM, the analytical data based on the finished part become pivotally critical when reverse engineering a cutting process. The postmachining examination is primarily focused on macro appearance and micro characteristics. The key parameters are surface finishing, subsurface microstructure, grain morphology, and surface residual stress. The precise quantitative measurements of these parameters can be very challenging. Making the task even more complex, many of the footprints that are critical to substantiate the operating conditions and cutting parameters used for an OEM part might have been eliminated during the sequence of processes. The microstructure analyses are addressed in other sections of this book. The following will focus on surface finishing, that is, surface roughness and surface residual stress analysis.

A nice surface finishing conjures up the image of a shiny part that brightens the appearance and often the value of a part. During part fabrication, raw materials will be stamped, ground, machined, or heat treated to transform them to the finished parts. These deformation processes shape the material into a part; they also introduce burrs, contamination, scales, and tooling marks on the part surface. The surface finishing can significantly affect part fatigue life; the surfaces of many heavy-duty mechanical springs are electropolished for fatigue life improvement. A good surface finishing also protects the part from corrosion attack by minimizing embedded contamination

on the surface. As a result, surface finishing plays a critical role in machining parts, from the orthopedic industry in the medical field to the jet engine industry in the aviation field. Surface finishing is also essential for the food processing industry to improve products' reaction in making food contact.

In many cases, the undesirable surface imperfections will be either polished up by removing them or covered up by coating. Depending on the applications, both methods are widely used. Though a coating does provide an extra layer of protection, it also causes other concerns, such as delamination. In the medical industry, chipping, peeling or delamination of coatings on any implant, tissue insertion, or surgical tool can pose potentially dangerous conditions for the patient. In the aviation industry, a breaking away of any heat-resistant coating from a jet engine turbine blade can cause severe damage to this blade operated at elevated temperatures.

For decades, chemicals and electricity have been used to improve surface finishing. Each of them will mark its own signature on the part surface that helps to identify the technique used by the OEM later through reverse engineering. For example, electropolishing has been one of the most widely used techniques since the early 1950s. Today, millions of bone screws, plates, and surgical cutting instruments are routinely electropolished as part of the manufacturing process. It is often referred to as reverse plating. In this process, the metal parts are positively charged as anode and immersed into a controlled chemical bath; and when electric current is applied, metal ions are dissolved and therefore removed from the surface. The electroplating uses a similar technique, but the part functions as a cathode, and the process deposits and adds metal ions onto the part instead.

In the United States, the following specifications are commonly used as standards and references in dimensional measurement and surface roughness evaluation:

ASME ANSI B46.1-2002: *Surface Texture, Surface Roughness, Waviness, and Lay*

ASME ANSI Y14.36M-1996: *Surface Texture Symbols*

ANSI Y14.5M: *Dimensioning and Tolerancing*

ASME Y14.38: *Abbreviations and Acronyms for Use on Drawings and Related Documents*

ANSI Y14.3M: *Multi and Sectional View Drawings*

The following international standards are also frequently referenced in surface roughness: International Standards Organization ISO 1302:2002 and Australian Standards AS ISO 1302-2005. These standards measure surface roughness with specifying parameters in various ratings.

Figure 5.16 shows a profilometer for surface roughness measurement. The part surface roughness is commonly measured by one of the three techniques. It can be measured by direct surface contact with a stylus to profile

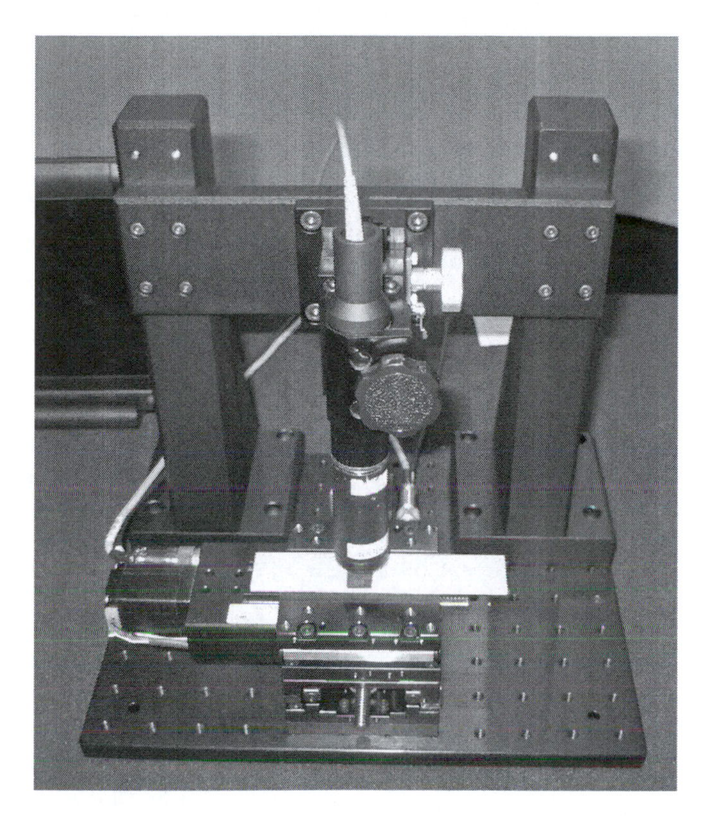

**FIGURE 5.16**
A profilometer for surface roughness measurement.

the surface contours. A confocal microscope can be used to image the surface 3D topography and measure the surface roughness. The surface roughness can also be measured with an interferometer whereby a black-and-white fringe pattern is projected onto the surface and the surface roughness can be measured by the principle of interferometry.

The surface finishing can be quantitatively specified by its roughness and is marked in design drawings as 1/32 or 1/64 in. in many U.S. systems. A simple manual surface roughness tester uses a stylus traveling across the specimen surface. A piezoelectric device installed in this tester will pick up and record the vertical roughness of the surface. The peaks and valleys are then converted to a given roughness scale. The most popular surface roughness scale uses the parameter *Ra* to reflect an average roughness. *Ra* stands for roughness average between the roughness profile and its mean line, as illustrated in Figure 5.17. It is the arithmetic average of the absolute values of the roughness profile heights over a given length in a two-dimensional measurement. In a three-dimensional measurement, the designation *Sa* is used instead of *Ra* to reflect the average roughness profile height over a given surface area. The numerical values of *Ra* and *Sa* can be calculated by

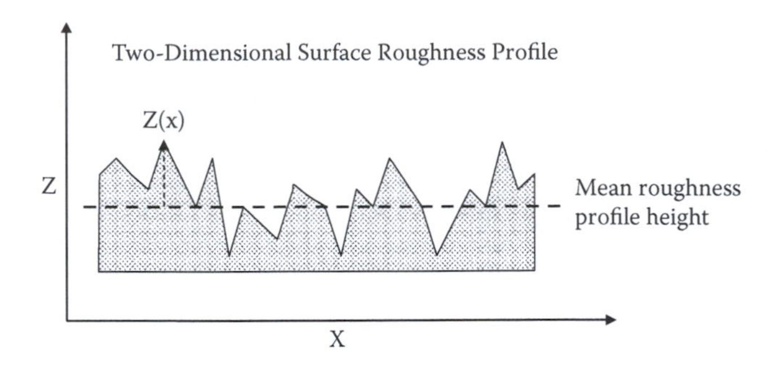


**FIGURE 5.17**
Schematic of two-dimensional roughness average measurement.

Equation 5.2a and b, respectively, where $L$ and $A$ are the length and area of interest, and $Z(x)$ and $Z(x, y)$ are the profile heights at individual surface locations.

$$Ra = \frac{1}{L}\int_0^L |Z(x)|dx \tag{5.2a}$$

$$Sa = \frac{1}{A}\iint_A |Z(x,y)|dxdy \tag{5.2b}$$

Many mechanical components have a specific acceptable surface roughness. For instance, the required surface roughness for automotive bearings usually ranges from an *Ra* value of 0.05 to 0.1 μm (Schmitt Industries, 2009). The surface roughness such as the *Ra* value provides a critical data point in reverse engineering to verify the OEM cutting process. The parts finished by grinding usually have an *Ra* value from 0.05 to 1.6 μm, while a surface finished by horning usually has an *Ra* value from 0.1 to 0.8 μm. Though the *Ra* parameter is easy and efficient in most applications, other surface roughness parameters are also used to meet more specific application requirements.

The specification ANSI/ASME B46.1 also provides guidance for a qualitative roughness comparison using the Microfinish Comparator. The Comparator consists of a series of flat-surface roughness specimens for visual and tactual surface roughness comparisons. It is another tool available in reverse engineering to verify the manufacturing process.

Shot peening is a process whereby a "finished" part is blasted with small metal, glass, or ceramic beads with a controlled mixture of shot, pressure, and exposure time to improve the part's mechanical properties. Shot peening induces a residual compressive stress, which is usually half the ultimate compressive strength or higher in absolute value, at or near the part surface. Metallurgically, a thin layer of part surface is yielded by compressive stress and the metal grains are in a quasi-relaxed state. From a part failure prevention perspective, cracks are generally suppressed in a compressive surface

because fatigue cracks usually initiate at surface imperfections such as burrs and scratches. Shot peening is therefore applied to many mechanical components used in the aviation and automobile industries, such as the turbine blades in jet engines, to improve their fatigue strength and corrosion resistance. The duplication of a shot peening process first requires a measurement of the residual stress on the part surface. SAE Standard J784A, *Residual Stress Measurement by X-Ray Diffraction* (SAE, 1980), provides guidance on a destructive etch/layer method to measure the surface residual stress gradient. X-ray diffraction does not measure stress directly. Instead, it measures the strain imposed on the crystallographic lattice and then calculates the residual stress based on the principles of elasticity. Nondestructive methods have also been introduced by applying radiation beams such as X-rays with various wavelengths on a single surface spot. The penetration depth of the X-ray beam varies as a function of residual stress, and provides a nondestructive measure of the surface residual stress gradient.

### 5.4.4 Joining Process

Mechanical components can be complex in shape and costly to manufacture if they are fabricated together as a single piece. Materials that are weaker and less expensive can be used in sections of the part that carry light loads to reduce cost and weight. It is very common in machine design and manufacturing that several units are first made separately, and then joined together for the finished part to optimize weight, component functionality, and cost-effectiveness by minimizing expensive machining. The three primary permanent joining processes for mechanical components are soldering, brazing, and welding. Reverse engineering these joining processes is particularly important in repairs where broken parts are joined together with replacement materials to restore them back in service. To precisely duplicate these OEM joining processes requires a comprehensive understanding of the interaction between the base material and the filler metal, effects of temperature on alloy phase transformation, and environmental effects on alloy properties.

#### *5.4.4.1 Soldering*

Soldering is the process of making a joint between electrical or mechanical parts with a soft solder, which is an alloy with a low melting point. The two common old-fashioned tin-lead solders, 60% Sn–40% Pb and 50% Sn–50% Pb, melt in a temperature range between 220 and 183°C, which is the eutectic temperature range of tin-lead alloys and their lowest melting range. The typical soldering temperature is usually below 250°C. Environmental and health concerns have recently demanded that lead-free solder be used in many applications. The directive on the restriction of the use of certain hazardous substances in electrical and electronic equipment, commonly

referred to as the Restriction of Hazardous Substances (RoHS) Directive, was adopted in February 2003 by the European Union and took effect on July 1, 2006. It identifies the following six substances as hazardous: lead, hexavalent chromium, cadmium, mercury, polybrominated biphenyls (PBBs), and polybrominated diphenyl ethers (PBDEs). One of the primary environmental concerns that led to the ban of lead-contained soldering was landfilling with lead-contained parts and materials, such as batteries and paints. Because of the relatively insignificant direct environmental effect from using lead in soldering, while a huge economic impact if banned, many exemptions and delays in compliance have been granted for military applications, high-end telecommunication equipment, and medical devices.

Lead is contained in many solder joints on printed circuit boards that are currently used in a lot of machinery. Lead-containing soldering was also used in various electrical connections and other electrical and electronic equipment that might be reverse engineered in the future. It is a challenge to reverse engineer a traditional lead-containing soldering, and replace it with an equivalent lead-free process. The two most popular substitute lead-free soldering alloy series are tin-silver-copper and tin-silver-bismuth-copper alloys. They have higher melting temperatures and will force the soldering temperature to above 250°C. A high soldering temperature can adversely affect the electronic components that are being soldered, particularly if the affiliated parts are made of materials with low melting points, such as plastics. The durability and reliability of some lead-free soldering alloys have raised some concerns. A dendritic microstructure appears on the soldered surface when some lead-free substitute solders, such as Sn-Ag-Cu alloys, are used. This will roughen the surface and make it difficult to discover solder cracks, a common defect in soldering. A long life span of 15 years is usually expected for some commonly used electrical devices, such as the low-voltage circuit breakers used for the protection of low-voltage indoor circuits (Hosogai and Ito, 2005). It has not been proven that a lead-free soldering can provide the longevity demonstrated by traditional lead-containing soldering, primarily due to thermal fatigue. Thermal fatigue is an inherent phenomenon in soldering resulting from thermal stress. It is caused by the different thermal expansion coefficients between the base material and the filler metal under temperature cycling. Many studies have been conducted to investigate the thermal fatigue life of the lead-free solders to test if they have adequate service life margins and acceptable reliability compared to the traditional lead-containing solders. The effectiveness of substituting lead-free soldering for lead-containing soldering by reverse engineering can be more accurately analyzed as more field data are collected in the next few decades.

In reverse engineering, the quality of a soldering technology can be tested to ensure that it is equivalent to or better than the OEM process in accordance with the published guidance. For instance, the effects of soldering on the fracture strength of a component's board-level interconnects can be

evaluated following the industrial specification IPC-9702, *Monotonic Bend Characterization of Broad-Level Interconnects* (IPC, 2004). IPC is an industry association for printed circuit board and electronics manufacturing service companies. In 1999, IPC changed its name from Institute for Interconnecting and Packaging Electronic Circuits to IPC. IPC does not publish material specifications for soldering alloys. The equivalence of a substitute soldering alloy can be evaluated in terms of its characteristics and performance. The following is a sample list of ASTM standards that provide guidelines on the evaluation of critical properties that can be used to verify the substitive soldering alloy's equivalence to the OEM alloy:

ASTM E794, *Standard Test Method for Melting and Crystallization Temperatures by Thermal Analysis*: For liquidus and solidus measurements that are critical temperatures in soldering.

ASTM B193, *Standard Test Method for Resistivity of Electrical Conductor Materials*: For resistivity measurement, one of the most important performance parameters of a soldering alloy.

ASTM E92, *Standard Test Method for Vickers Hardness of Metallic Materials*: For hardness measurement.

ASTM E831, *Standard Test Method for Linear Thermal Expansion of Solid Materials by Thermomechanical Analysis*: For thermal expansion coefficient measurement.

#### 5.4.4.2 Brazing

Brazing is a hard soldering process that joins two parts, often made of dissimilar alloys, with a molten filler metal that flows into the gap between the two parts to be brazed by capillary action. The filler metal used for brazing is commonly referred to as brazing filler metal or brazing alloy. The brazing alloy has a higher melting point than the filler alloy used for soldering. The lead-free solder Sn–3% Ag–0.5% Cu melts around 220°C, while a silver brazing alloy usually has a melting temperature higher than 600°C. Brazing is usually operated at a temperature above the melting temperature of brazing alloy but below the melting temperatures of the joined parts that remain solid during brazing. The heating and cooling processes during brazing alter the composition and microstructure locally, and form a metallurgical bond. However, compared to welding, whereby the two parts to be joined are actually melted and fused together, brazing is processed at a relatively lower temperature, and induces less part distortion. Brazing is particularly suitable for linear joints because the filler alloy naturally flows into the joint area. It is also easier for automation. Brazed joints have great tensile strength, often stronger than those of the original parts. They also

have a very clean, well-finished appearance. This is a fast, economical, and versatile joining method used by many designers and engineers.

The key brazing parameters that require verification in reverse engineering include the types of heat source, operating temperature, heating and cooling rates, operating atmosphere, joint space between parts, type of brazing alloys, part cleaning prior to brazing, and joint cleaning after brazing if flux is used. Many of these operation parameters affect the brazing quality. For example, slow heat cycles generally produce better results than fast heat cycles in brazing. Unfortunately, most of them cannot be directly verified. The following information (Skewes, 2009) helps to explain the operation of brazing, and therefore makes it easier to do an educated analysis on brazing parameters used by the OEM. The heat for brazing is typically provided by a handheld torch, a furnace, or an induction heating system. However, other techniques are also used, such as dip brazing or resistance brazing. The handheld torch is often used in flame brazing for small assemblies and low-volume applications. It is simple, but its quality and repeatability are heavily dependent on the operator's skills. Furnace brazing does not require a skilled operator, and is often used to braze multiple parts at the same time. This method is only practical if the brazing alloy can be pre-positioned. The furnace usually has to remain turned on to eliminate long start-up and cool-down delays, and is not particularly energy efficient. The induction heating system used for brazing provides heat through an induction coil locally to the area being brazed. The joint parts and filler alloy are usually well regulated in position and alignment, and therefore quick and consistent results can be achieved. Dip brazing is used for small wires, sheets, and other components that are small enough to be immersed. The parts are dipped in a molten flux bath, which doubles as the heating agent. Resistance brazing is effective for joining relatively small, highly conductive metal parts. Heat is produced by the resistance of the parts to the electrical current.

What is the exact brazing temperature or the joint space used by the OEM? The brazing temperature is often between the melting points of the part alloys and the brazing alloy, and can be estimated based on the heat source, amount of heat provided, alloys of the joint parts, and brazing alloys. The brazing temperature is usually 30 to 100°C higher than the liquidus of the brazing alloy. The liquidus is the upper melting temperature of an alloy, and can be as high as 750°C for a brazing alloy, depending on the exact composition. Though the joint spacing can be estimated by examining the postbraze joint, the following standard practice also helps in determining its size. In brazing, the brazing alloy is drawn into the joint by capillary action that requires proper spacing between the parts. Usually, the strongest joints are made by allowing just enough space for the filler alloy to flow into the joint area, typically in the range of 0.25 to 1.27 mm (0.001 to 0.005 in.). Wider spacing could result in a weaker joint. It is also important to remember that metals expand and contract at different rates when heated and cooled. When join-

ing dissimilar metals, expansion and contraction rates must be considered and allowed for when the parts are positioned.

The identification of a brazing alloy, particularly its exact chemical composition, is another challenging task in reverse engineering. The melting and solidification of the alloy might have altered the original composition due to evaporation, contamination, and metallurgical transformation. Silver, copper, and aluminum alloys are the most commonly used filler metals for brazing. Silver-base brazing alloys are frequently chosen because of their relatively low melting points. Copper-base brazing alloys have higher melting points but are generally more economical. Depending on the application, the brazing alloy may be in the form of a stick, paste, or preform. A preformed brazing alloy can be positioned around the joint before the heat cycle begins, and is normally the best choice when even distribution and repeatability are paramount considerations. The melting brazing alloy will tend to flow toward areas of higher temperature, so the heat is usually applied to the opposite side of where it is positioned. The heat then helps draw the molten metal down into the joint area. It is virtually impossible to figure out exactly what form of brazing alloy was employed after the brazing. This is one of the examples in reverse engineering where equivalent functionality and performance of the finished part take precedence over the formality of the manufacturing process. Similar arguments can also be made about prior brazing preparation, from clamping fixtures to cleaning procedures. The joint area has to be clean. The brazing alloy will not flow properly if grease, dirt, or rust blocks its path. As critical as they might be, the cleaning procedures used by the OEM are very difficult to retrace after brazing because most of the physical tracks have been erased during brazing.

Some residual evidence might help engineers to figure out the brazing environment, the flux used, and postbrazing cleaning. Nonetheless, this task is as challenging as verifying other brazing parameters. When brazing is done in the open air, the joints are normally precoated with flux, a chemical compound that protects the part surfaces. Advanced analytical techniques help engineers to decode the flux used by the EOM and the operating atmosphere under which the brazing is conducted. The existence of surface oxidation is evidence of an open-air brazing operation. The part brazed in open air will show a heavily oxidized surface, while the one brazed under a controlled environment will show a much cleaner surface with little oxidization. A flux coating helps prevent oxidation when the metal heats up; it protects the filler alloy and improves its flow. As heat is applied to the joint, the flux will dissolve and absorb the oxides that form. A variety of fluxes are available for use at different temperatures, with different metals, and for a variety of environmental conditions. Flux residues are chemically corrosive and may weaken the joint if they are not completely removed. The joint usually will immediately quench in hot water as soon as the brazing alloy is solidified. To remove residual oxidation, the parts can be dipped in hot sulfuric or hydrochloric acid. However, care should be taken

to avoid etching the joint with too strong an acid solution. Parts brazed in a protective atmosphere require no cleaning. From the reverse engineering perspective, this information provides another valuable data point. Brazing can also be conducted in a protective atmosphere such as argon, nitrogen, or in a vacuum. In a vacuum system, parts are heated in a fully enclosed, stainless steel chamber, which can be pumped down to $10^{-6}$ torr, provides tight process control, and produces the cleanest parts, free of any oxidation or scaling. However, vacuum brazing requires an alloy free of volatile elements, such as cadmium.

The applications of the parts also provide some relevant data in reverse engineering. To braze aerospace components, medical devices, and other instruments that require the highest part quality, a high-vacuum environment is usually preferred. In summary, the chemical reactions during the brazing process and some untraceable evidence make the task of reverse engineering brazing very challenging. Engineers should educate themselves on these subjects as much as possible, and make their best estimate based on analytical data, customary practice, and corporate knowledge.

#### *5.4.4.3 Welding*

There are a number of welding technologies that are used in various industries. Traditionally they are named based on either their respective heat source or operating environment. Electron beam welding uses an electron beam to heat and weld two parts, while a friction welding process utilizes the heat generated by friction to join the parts together. Tungsten inert gas (TIG) welding is an electric arc welding process that operates under the shield of inert gas, such as argon, hydrogen, or helium. It is also referred to as gas tungsten arc welding (GTAW). The sources of heat for welding may also be extracted from a gas flame, laser, or plasma, and the protective shielding can be vapor or even slag.

Resistance spot welding is one of the oldest electric welding processes, and is still widely used today. The weld is made by combining heat and pressure. The resistance heat generated by the parts themselves melts the parts and joins them together. The electric power and current, the press pressure, and time all play a critical role in this process, and are the parameters that need to be verified when reverse engineering resistance spot welding. Figure 5.18 shows the macrograph of a resistance spot weld of two plates of 1.5 mm thickness. They are Al-Cu-Mg alloys, solution heat treated, cold worked, and naturally aged. The weldment reveals dendritic grains. The continuous horizontal line on the far left is the original separation line between the plates. The two short lines on the right (inside the weld) are solidification cracks. The micrograph also shows that the weld nugget formed in between the surfaces of two parts instead of on just one side from one workpiece, as usually occurs in gas tungsten arc spot welding. This characteristic is the signature of resistance spot weld. Specific welding processes have distinct features that provide invaluable information

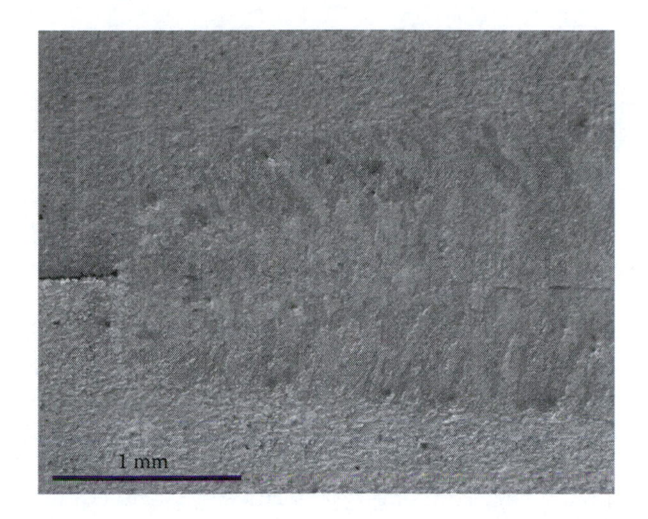


**FIGURE 5.18**
Macrograph of resistance spot weld. (Reprinted from Adamowski, J., http://www.doitpoms.ac.uk/miclib/systems.php?id=2&page=3, accessed February 22, 2009. With permission.)

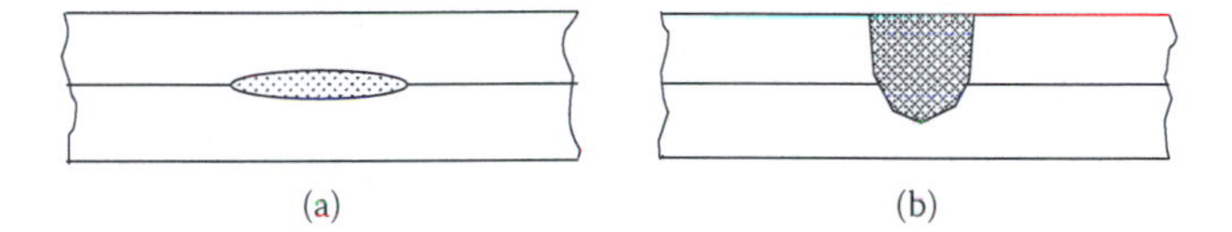


**FIGURE 5.19**
Schematic weld nugget configurations of (a) resistance spot weld and (b) tungsten arc spot weld.

in reverse engineering for welding process verification. Figure 5.19 shows the schematics illustrating the different weld nugget configurations between a resistance spot weld and a gas tungsten arc spot weld.

Table 5.5 highlights the respective characteristics of soldering, brazing, and welding. Compared to soldering and brazing, which are nonfusion processes, the welding process is a fusion process that operates at a much higher temperature. A filler alloy is usually used, and occasionally pressure is also applied in conjunction with heat in welding. In electric arc welding, the arc can produce a temperature of about 3,600°C at the tip. It forms a metallurgical bond between the workpieces and is likely to introduce residual stress and generate a heat-affected hazard zone and distortion. The high temperature also promotes chemical reactions between the base metal and oxygen and nitrogen in the air to form oxides and nitrides that could have adverse effects on the mechanical properties of the part. Therefore, some kind of arc shielding is often used in many arc welding processes. The shield of gas, vapor, or slag provides a protective covering for the arc and the molten pool to prevent molten metal from making contact with air. The details of

**TABLE 5.5**

Comparison among Soldering, Brazing, and Welding

| Parameter | Soldering | Brazing | Welding |
|---|---|---|---|
| Melting temp. of filler, °C (°F) | <450 (840) | >450 (840) | >450 (840) |
| Base metal | Does not melt | Does not melt | Usually melts |
| Flux | Required | Optional | Optional |
| Joint | Mechanical adhesive plus metallurgical bond | Metallurgical bond | Metallurgical bond |
| Distortion and residual stress | Atypical | Atypical | Likely |

electrode type, filler alloy, power supply, electric circuitry, arc temperature, and arc shielding are usually specified in the OEM welding procedures. In an electric arc welding, the electrode is consumable, and the filler alloy is melted and reacts with the base metals to form new metallurgical phases. The arc welding circuitry and operation conditions could be OEM proprietary information. In other words, a large portion of the data is either lost during the process or is protected by law. The postprocess verification of these parameters based on the finished product is very challenging. Reverse engineering any of these joint processes requires specialized expertise and experience. An expert welder might be able to determine the inert gas that is used in the OEM's TIG process based on his or her expert knowledge. In a TIG welding, helium is generally used to increase heat input, and therefore increase welding speed or weld penetration. The use of hydrogen will result in cleaner-looking welds and also increase heat input. However, hydrogen may promote porosity or hydrogen cracking.

### 5.4.5 Heat Treatment

The engineers have to overcome the following two challenges to successfully duplicate the exact heat treatment using reverse engineering. First, different heat treatments can produce almost identical material characteristics. Second, many of the detailed parameters in heat treatment cannot be verified based on a post–heat treatment analysis. The most practical method for reverse engineering is to adopt "the end justifies the means" approach. Since the end expectations differ from part to part, the best-fit analytical technology also varies. This once again highlights the part-specific nature of reverse engineering.

Various heat treatments are applied to mechanical parts for property improvement. A solution treatment followed by an aging treatment is usually aimed to improve part mechanical strength, such as ultimate tensile and yield strengths. The annealing treatment will improve ductility and machinability, and enhance structural stability. More complex heat treatments are

applied to improve fatigue or creep resistance. Fracture toughness is a function of crack length, load condition, and material strength, and can also be enhanced by proper heat treatment. Stress relief is a function-specific heat treatment commonly used to reduce residual stresses induced during fabrication. Normalization is another heat treatment commonly applied to ferrous alloy to refine the grain size and improve the uniformity of microstructure and properties after hot working such as forging or rolling. To achieve these objectives, the normalization treatment will be conducted at a predetermined temperature for a specific time duration, and subsequently cooled in still air at room temperature. Many heat treatment schedules have several heat-up and cool-down cycles to obtain the most optimal microstructure and property. To further complicate the task, the same mechanical properties can be obtained with different combinations of microstructure morphology in terms of grain size and amount of precipitation. To substantiate the equivalency of one alloy with two different microstructures is very complicated in reverse engineering, if ever possible. Sufficient data or adequate tests are required to demonstrate that improvement of one mechanical property, such as fatigue strength, will not adversely affect another property, such as creep resistance, if different microstructures are observed.

Despite the uncertainties, microstructure analysis remains the most convenient and convincing technique available to reverse engineer an OEM's heat treatment. Many engineering materials, such as steel, aluminum, and titanium alloys, will develop a series of distinct microstructure morphology as a result of heat treatment. The following case study uses a titanium alloy as an example because the microstructure evolution of titanium alloys is very sensitive to heat treatment, and they are widely used in many industries. Titanium alloys have been used for simple hammer heads, expensive race car engine valves, delicate artificial knees, and complicated jet engine components. One modern Boeing 777 aircraft uses approximately 58,500 kg of titanium alloys, accounting for about 9% of the aircraft weight. About 15% of all the materials used for a Boeing 787 aircraft are titanium alloys. Many reverse engineering projects involve parts made of titanium alloys. Figure 5.20 shows five sets of microstructure of the same titanium alloy. The microstructure as received is shown in the center, surrounded by four other different microstructures resulting from various heat treatment schedules, as schematically illustrated in Figure 5.21. Figure 5.20a shows primary $\alpha_2$ in $\beta$ matrix with acicular $\alpha_2$ precipitates. Figure 5.20b shows scattered $\alpha_2$ in $\alpha_2/\beta$ matrix, while Figure 5.20c shows equiaxed $\alpha_2$ in $\alpha_2/\beta$ matrix. Figure 5.20d and e shows more microstructure evolutions resulting from heat treatment. Figure 5.20d has a fine $\alpha_2/\beta$ matrix with coarse $\alpha_2$ at the prior $\beta$ grain boundaries. Figure 5.20e is a microstructure of fine Widmanstatten morphology with clearly defined $\beta$ grain boundaries. In reverse engineering, the observed microstructures, as shown in Figure 5.20, and the knowledge of phase transformation of titanium alloy help engineers select the "right" heat treatment schedule from Figure 5.21a to d. At approximately 885°C, the hexagonal closed-packed $\alpha$ phase transforms to

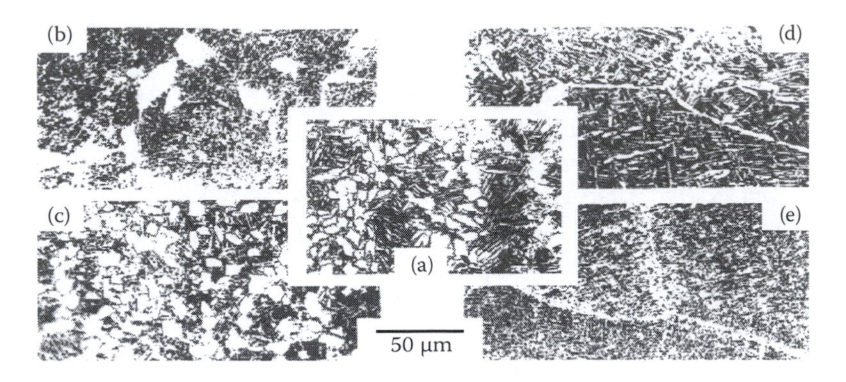


**FIGURE 5.20**
Microstructure evolution of a titanium alloy: (a) primary $\alpha_2$ in $\beta$ matrix with acicular $\alpha_2$ precipitates, (b) scattered $\alpha_2$ in $\alpha_2/\beta$ matrix, (c) equiaxed $\alpha_2$ in $\alpha_2/\beta$ matrix, (d) fine $\alpha_2/\beta$ matrix with coarse $\alpha_2$ at the prior $\beta$ grain boundaries, and (e) fine Widmanstatten morphology with clearly defined $\beta$ grain boundaries.

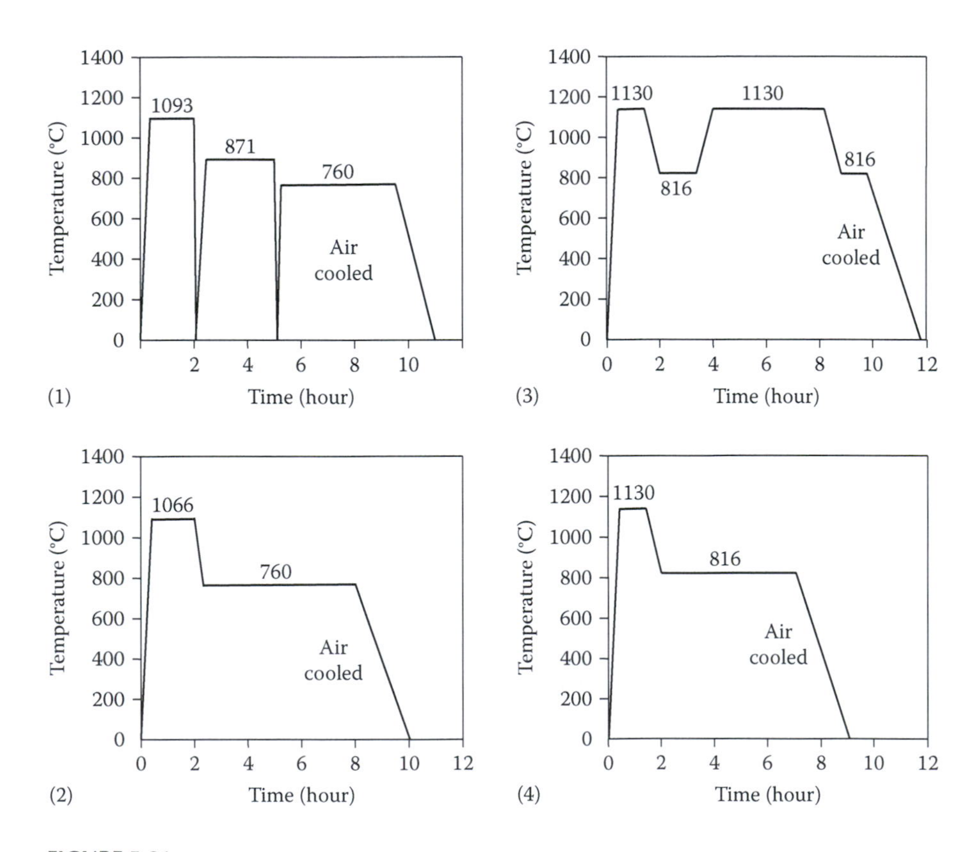


**FIGURE 5.21**
Four different heat treatment schedules (1), (2), (3), and (4) corresponding to the microstructures (b), (c), (d), (e), respectively in Figure 5.20.

the body-centered cubic β phase in pure titanium. The β transformation temperatures (also known as β transus temperatures) of titanium alloys might be higher or lower, depending on their respective alloying elements. Figure 5.21 shows that when the titanium alloy is heat treated above and below the β transformation temperature, various β configurations will develop. The $\alpha_2$ phase is an ordered intermetallic compound in titanium alloys.

However, in most cases the design data, such as Figures 5.20 and 5.21, are not readily available for comparison. It might be required to conduct simulated heat treatments on test coupons and build up the database for the necessary comparative analysis in a reverse engineering project.

It is essential to have a sufficient number of samples to show the full characteristics of the part in reverse engineering heat treatment. For example, directionality might be a feature of a normalized low-carbon steel, but this feature will not be revealed if only one micrograph is taken in the transverse section. Figure 5.22a and b shows the microstructures of a low-carbon alloy with a nominal composition of Fe–0.1% C–1% Mn–0.5% S. The alloy was first hot worked, and then normalized at 1,000°C. Figure 5.22a and b

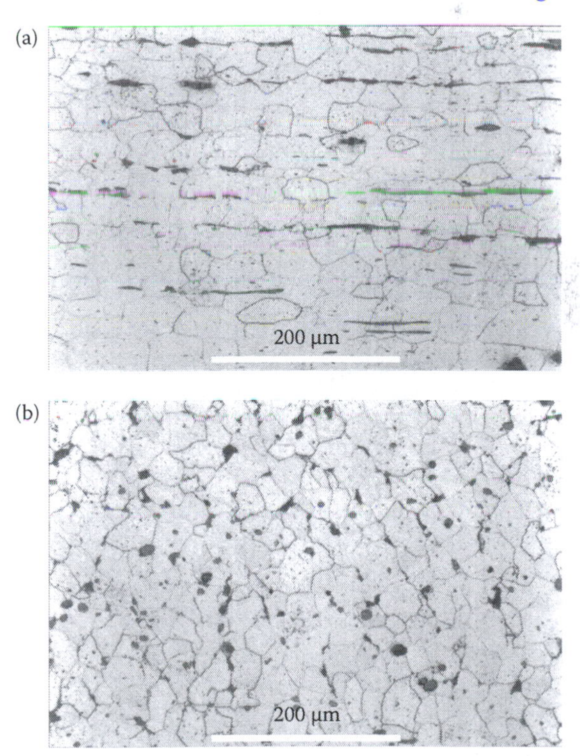


**FIGURE 5.22**
Micrographs of a low-carbon steel: (a) longitudinal view with directionality revealed as stringers of pearlite, and (b) transverse view showing no directionality. (Reprinted from Cochrane, R. F., DoITPoMS Micrograph Library, University of Cambridge, 2002. With permission.)

shows a microstructure consisting mostly of ferrite that has been normalized with scattered recrystallization. However, the pearlite portion of the alloy is unaffected by normalization and, consequently, retains its directionality resulting from prior hot work. This directionality evidenced as stringers of pearlite can be clearly observed in Figure 5.22a, which was taken along the direction of hot work deformation. However, the directionality is completely shielded in Figure 5.22b, which was taken perpendicular to the direction of deformation. In this case, a single micrograph was not able to provide all the information we need in reverse engineering.

The amount of hot or cold work prior to, during, or after heat treatment can significantly affect material microstructure and properties. Therefore, the best practice in reverse engineering to determine the amount of hot or cold work is to trace back the microstructure evolution and the variation of mechanical properties as demonstrated by the following example. Figure 5.23a to d is the microstructure of a 718Plus alloy with the following nominal composition: Ni–17.9% Cr–9.3% Fe–9.0% Co–5.5% Nb–2.7% Mo–1.5% Al–1.0% W–0.7%

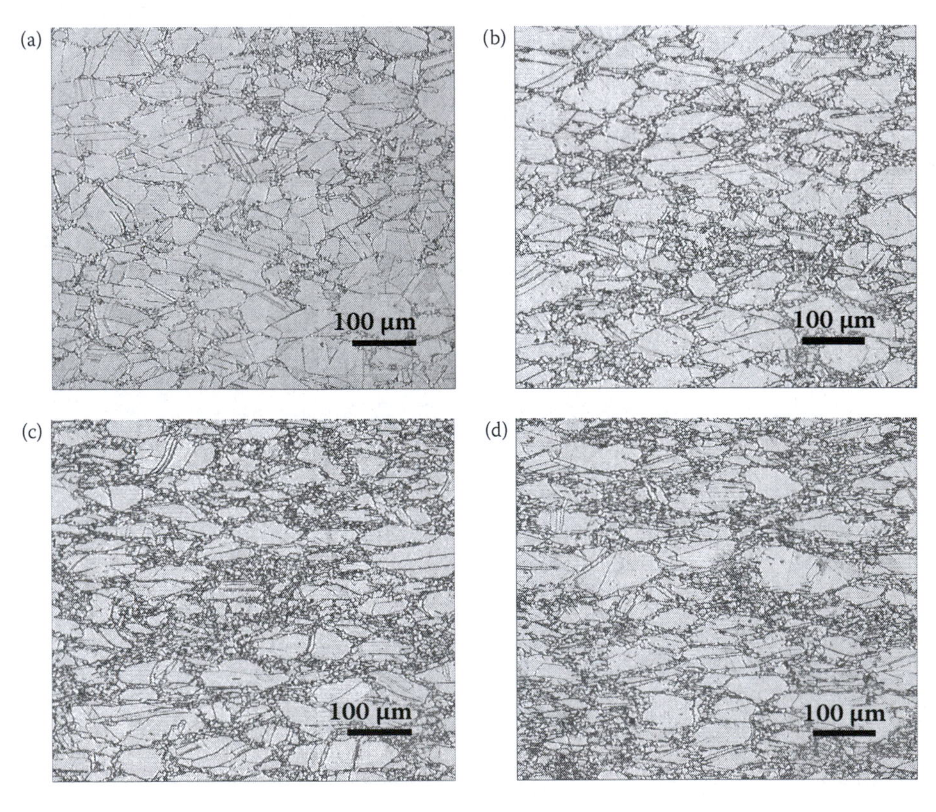


**FIGURE 5.23**
Microstructures of a 718Plus alloy (a) as hot rolled, (b) 16% cold work, (c) 25% cold work, and (d) 35% cold work. (Reprinted from Bond, B. J. & Kennedy, R. L., Superalloys 718, 625, 706 and derivatives, 2005. With permission.)

Ti–0.02% C, but each with different amounts of hot or cold work (Bond and Kennedy, 2005). The material was taken from a production heat produced by the vacuum induction melting (VIM)–vacuum arc remelting (VAR) process. Vacuum induction melting is a process to melt metal in vacuum using electromagnetic induction. This process induces electrical eddy currents in the metal to melt and refine it. Vacuum arc remelting is a secondary melting process to produce highly purified ingots used in the biomedical, aviation, and other fields where closely controlled speciality alloys are used. The combined VIM-VAR process is used to ensure high-quality metals and alloys. To verify that the VIM-VAR process is used for the OEM part, the engineers have to base their judgment on corporate knowledge and industry experience, and the distinct signature of the VIM-VAR process of producing high-quality ingot with precision chemical composition. In this example, the initial ingot of 432 mm in diameter was first pressed to a billet of 203 mm in diameter, then a section was further pressed to 102 mm square and hot rolled to 28 mm round bars. Figure 5.23a shows the microstructure of this alloy as hot rolled. The hot rolled bars were annealed at 982°C, and cold drawn to three levels of cumulative reductions in area, 16, 25, and 35%, followed by direct aging treatment. Direct aging is an aging heat treatment that immediately follows hot working without prior solution heat treatment. Figure 5.23b to d shows the microstructure of the same alloy with 16, 25, and 35% cold work, respectively. In reverse engineering, the difference in microstructure provides an index of the amount of hot or cold work the material has experienced. This index can collaborate with other indices, such as the variations of ultimate tensile and yield strengths as a function of cold work, as plotted in Figure 5.24a. It shows that both ultimate tensile strength (UTS) and yield strength (YS) linearly increase with the

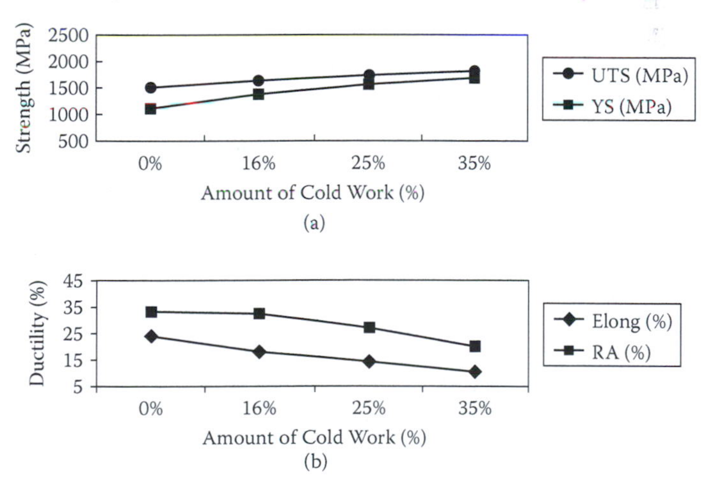


**FIGURE 5.24**
(a) Tensile strengths of a cold-worked 718Plus alloy. (b) Tensile ductility of a cold-worked 718Plus alloy.

amount of cold work. The variations of tensile ductility of these three alloys as a function of cold work are plotted in Figure 5.24b. It shows both tensile elongation and reduction of area decrease with the amount of cold work in an approximately linear relationship as well. Though not all alloys show similar linear relationships between UTS, YS, elongation, and reduction of area with the amount of cold work, the above exemplified approach provides a method in reverse engineering to decode the amount of hot or cold work a material has experienced before. From time to time this baseline information of microstructure and property variations might have to be independently established for comparison in a reverse engineering project.

Hardness measurement is another widely used technology to reverse engineer heat treatment. The relationship between hardness and heat treatment is best exemplified by the fact that the term *hardenability* has been used as a parameter to quantitatively measure the easiness of transformation of steel from austenite to martensite by heat treatment. Austenite and martensite are two metallurgical phases of steel resulting from different heat treatments, each with distinctively different hardness. Hardness is also widely used to monitor the heat treatment of precipitation-hardenable alloys such as aluminum-copper alloys and nickel-base superalloys. Precipitation-hardening heat treatment is one of the most frequently used strengthening mechanisms for engineering alloys. Figure 5.25 shows three simulated aging curves of an aluminum-copper alloy, at 160°C, 180°C, and room temperature, respectively. They present hardness profiles over time as a result of aging treatment. The alloy will reach the peak hardness earlier when aged at higher temperature, while at a lower peak hardness value. The alloy can also be naturally aged, and slowly increase its hardness at room temperature. The solubility of copper in aluminum decreases rapidly from about 5.5% at 548°C to almost zero when the temperature is cooled down to room temperature. If the aluminum-copper is first heated to a temperature around 560°C and held there for approximately 1 hour, most of the copper up to approximately 5.5% will be dissolved into the aluminum matrix to form a homogeneous solid solution. This process is referred to as solution treatment. It is the first part

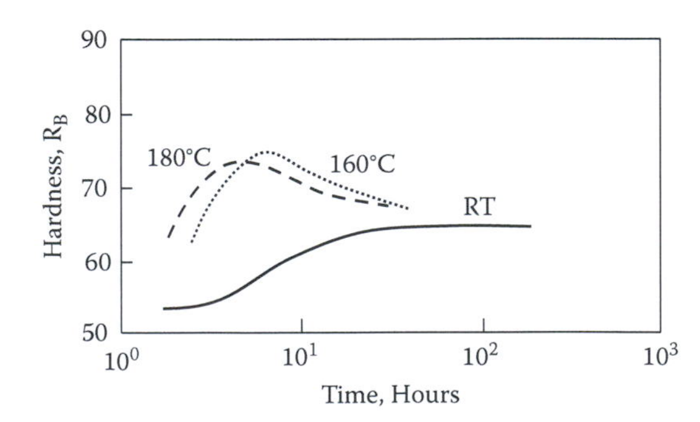


**FIGURE 5.25**
Simulated aging curves of an aluminum-copper alloy.

of the precipitation-hardening heat treatment. Afterwards the alloy will be quenched in water to cool it to room temperature. The rapid cooling prevents copper atoms from being rejected out of the aluminum matrix, and forming a thermodynamically unstable supersaturated aluminum-copper solid solution. The second part of precipitation-hardening heat treatment is aging treatment. It is a process that warms up the supersaturated aluminum-copper solid solution to a temperature usually between 150 and 190°C and holds it there for isothermal transformation. How much and in what form or shape the copper atoms will precipitate out depends on the complex kinetic process. However, the measurement of hardness during the aging treatment provides a simple barometer index that helps engineers study the aging effects. It also provides a useful tool in reverse engineering to track back how the OEM's part was aged. The data in Figure 5.25 show that various combinations of aging time and aging temperature can produce the same hardness. To verify exactly which set of time and temperature was employed, further analysis on the part's microstructure is needed to examine the precipitate morphology.

Hardness is a cost-effective index that reflects a first-order approximation of the prior heat treatment. This method is particularly relevant because of the long list of engineering alloys that are precipitation hardenable, such as 2024 and 7075 aluminum alloys, 17-4 PH stainless steel, maraging steel, Inconel 718, Rene 41, and Waspaloy. In many reverse engineering projects, engineers are working on the parts made of these alloys. However, for a precise reverse engineering of heat treatment, more analyses and substantiation data are required.

### 5.4.6 Specification and Guidance for Heat Treatment

The heat treatment process is a complex and involved process, and is alloy specific. There are many steps and stages in heat treatment, including solution treatment, annealing, stabilization, precipitation, normalization, and stress relief. Fabricators and vendors often establish their own proprietary heat treatment processes and specifications. Many heat treatment options can produce similar material characteristics of the OEM counterpart. However, few can guarantee identical properties. For example, several heat treatment schedules will produce the same hardness number for a steel, but with different fatigue strengths.

SAE AMS 2759, *Heat Treatment of Steel Parts, General Requirements* (SAE, 2008), provides the general requirements for the heat treatment of steel parts. In conjunction with these general requirements in AMS 2759, the following specifications provide further details for specific heat treatment procedures:

AMS 2759/3 establishes the requirements for heat treatment of precipitation-hardening corrosion-resistant and maraging steel parts.

AMS 2759/4 establishes the requirements for annealing, stress reliving, and stabilizing heat treatment of austentic corrosion-resistant steel parts.

AMS 2759/5 provides guidance for martensitic corrosion-resistant steel parts.

Similarly, the technical requirements, quality assurance, applicable reference documents, and other notes related to the surface heat treatment procedure are also detailed in their respective material specifications. For example, AMS 2759/6 covers the surface heat treatment gas nitriding, while AMS 2759/8 covers ion nitriding. Reverse engineering of ion nitriding should refer to AMS 2759/8 for guidance. Surface enhancement treatment, however, is beyond just surface heat treatment, and this subject will be discussed in more detail in the next section.

AMS 2774 should be used when reverse engineering the heat treatment of a part made of wrought nickel or cobalt alloy. A wealth of information and references are included in AMS 2774. Other professional organizations, such as ASTM, also publish various standard specifications related to the heat treatment of nickel and cobalt alloys, for example, ASTM B637, *Precipitation-Hardening Nickel Alloy Bars, Forgings, and Forging Stock for High-Temperature Service* (ASTM, 2006c). The engineers can often take references from the requirements of these specifications, such as ASTM B637, to determine what analyses and tests need to be conducted and what material characteristics are required to be checked before they can claim ASTM B637 is the specification called out in the OEM's original design. ASTM B637 requires that a "chemical analysis shall be performed on the alloy and shall conform to the chemical composition requirement in carbon, manganese, silicon, phosphorus, sulfur, chromium, cobalt, molybdenum, columbium, tantalum, titanium, aluminum, zirconium, boron, iron, copper, and nickel." Therefore, the percentage of all these above-mentioned alloying elements should be quantitatively determined whenever applicable in reverse engineering a part made of nickel or cobalt alloy. ASTM B637 recommends specific procedures for annealing treatment, solution treatment, stabilizing treatment, and precipitation-hardening treatment. It also specifies a material's hardness, tensile strength, yield strength, elongation, reduction in area, and stress rupture properties. To confirm that the OEM parts were heat treated in accordance with the ASTM B637 specifications; first, hardness testing, tension testing, and stress rupture testing have to be performed on the material. Then the tested results should comply with the required hardness, tensile strength, yield strength, elongation, and reduction in area specified in ASTM B637.

The test matrix in a reverse engineering project should be carefully established to provide necessary and sufficient data of material properties to confirm the equivalency of the adopted material and the OEM material specified by the reference material specifications. It is worth noting that there are more than twenty material specifications published for Ti-64 alloy by various

**TABLE 5.6**

Specifications for Ti-64 Alloy

| AMS Specification | 4934 | 4935 | 4936 |
|---|---|---|---|
| Composition | Ti-6Al-4V | Ti-6Al-4V | Ti-6Al-4V |
| Product form | Extrusions and flash welded rings | Extrusions and flash welded rings | Extrusions and flash welded rings |
| Process | Solution heat treated and aged | Annealed, beta processed | Beta processed |

institutions. They all have the same or very close nominal chemical compositions. As illustrated in Table 5.6, three AMS specifications are published for Ti-64 alloy with the same nominal chemical composition and the same product form. They are only distinguished by different heat treatment processes. There are also a large number of material specifications for Inconel 718, 7075 aluminum alloy, and other engineering alloys. The verification of alloy composition alone usually does not provide sufficient and necessary information to identify a specific material specification used by an OEM in their design. Product form, mechanical properties, and microstructure are often required to pin down the "right" material specifications in reverse engineering. All these material characteristics are closely related to heat treatment.

### 5.4.7 Surface Treatment

A proper surface treatment is essential to part performance. However, the technical details of many surface treatment processes are proprietary. Furthermore, the complexity and versatility of these processes make reverse engineering of surface treatment very challenging. The three most commonly used surface treatment techniques are surface heat treatment, protective coating, and shot peening.

#### *5.4.7.1 Surface Heat Treatment*

Surface heat treatment, also known as surface hardening or case hardening, is primarily designed to increase the hardness and wear resistance of a part surface while maintaining the ductility and toughness of the part interior, such as gears, bearings, and stamping dies. Various gears are used in many automotive applications. Figure 5.26 shows the gear mechanism in a Toyota manual transaxle. Figure 5.27a shows the gear applications in a BMW differential, and Figure 5.27b a planetary gear mechanism. Carbonization and nitridization are among the most frequently used methods for surface heat treatment. Traditionally, the carbon or nitrogen is diffused into the part surface through either a gas atmosphere, liquid solution, or solid package at an elevated temperature. Modern technology has advanced beyond the typical three matter states (gas, liquid, and solid) and makes it possible for nitrogen

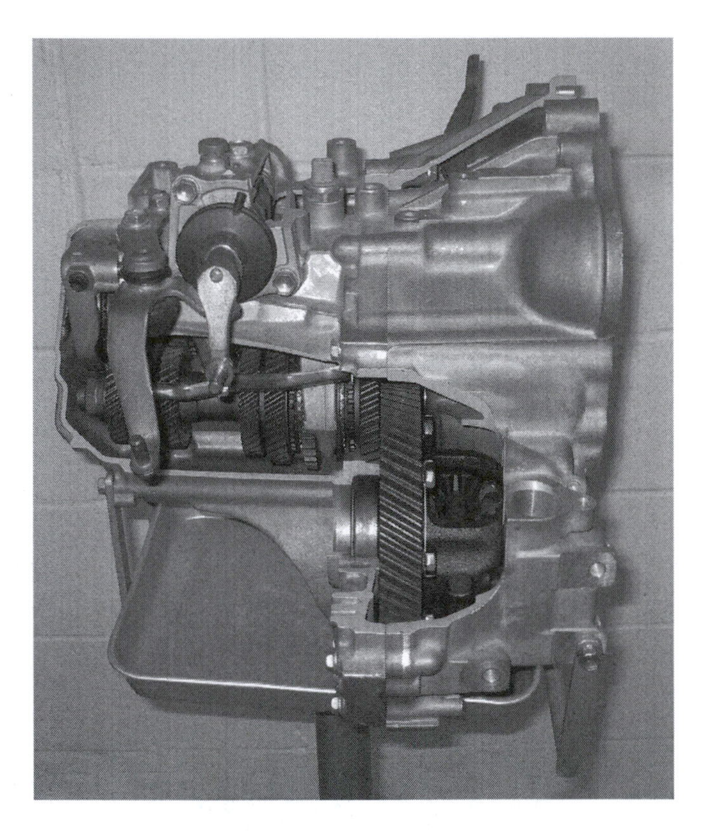

**FIGURE 5.26**
Gear mechanism in a Toyota manual transaxle.

to diffuse into the part surface in a fourth state, "plasma," such as by plasma (ion) nirtriding. In the plasma state, matter exists in its excited form, also known as the ionized form, wherein the outermost electrons of the element are knocked off. Thus, a plasma is usually an ionized gas consisting of neutral and charged particles. The high carbon or nitrogen content at the surface will form carbides or nitrides to increase the surface hardness.

The parameters of the carbonization or nitridization process, such as temperature, methods of infusion, and depth of surface hardening, can be best revealed by examining the microstructure as shown in Figure 5.28 (Rolinski et al., 2006). It is a micrograph showing the surface microstructure of an automotive stamping die made of tool steel and plasma nitridized at 510°C (950°F). The outer surface in the light color is a layer of nitride compound that provides a significantly higher hardness, as reflected in the hardness profile included in Figure 5.28. The Vickers hardness HV0.1, tested with a mechanical load of 0.1 kg, decreases in the darker diffusion zone immediately underneath the surface where a network of nitrocarbides can be observed. The hardness levels off at approximately 0.254 mm (0.01 in.) deep into the surface.

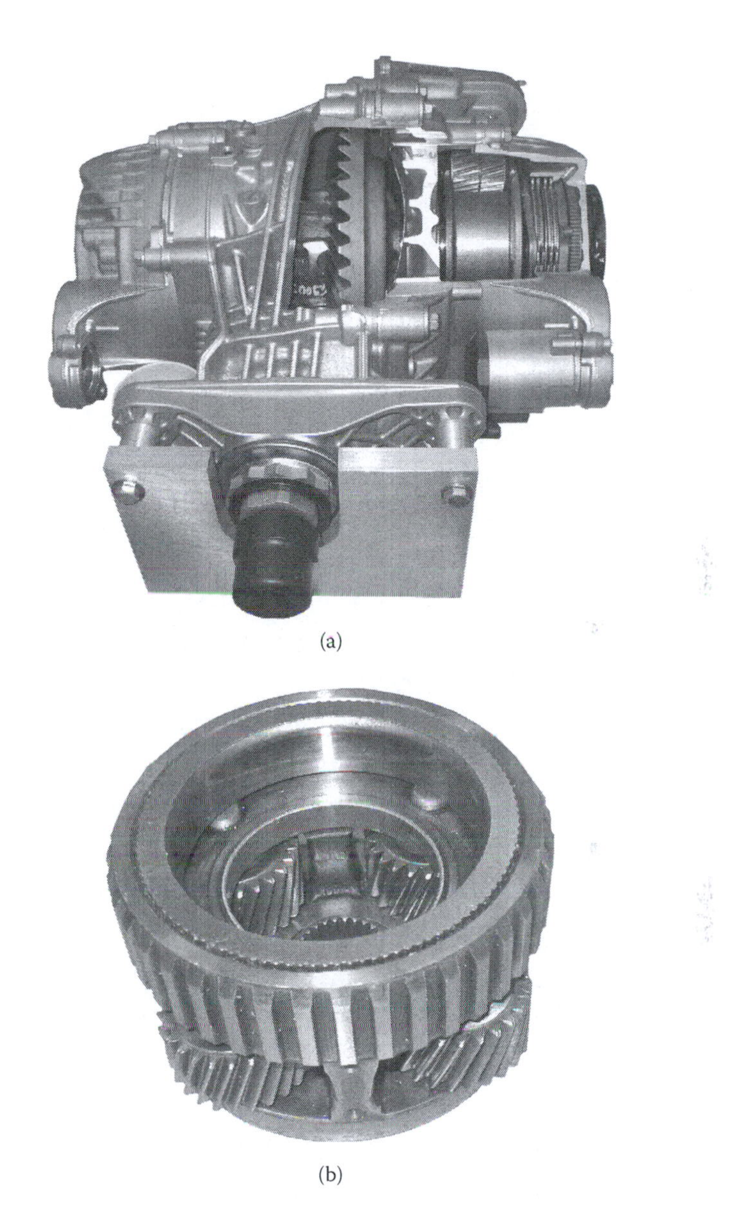

(a)

(b)

**FIGURE 5.27**
(a) Gear application in a BMW differential. (b) Planetary gear in a BMW car.

### *5.4.7.2 Coating*

Generally speaking, surface treatment by coating can be classified into five categories: electroplating, deposition, thermal spraying, anodizing, and galvanizing (also known as hot dipping).

Electroplating is usually further specified by the plating alloy, such as nickel plating and silver plating. Nickel plating is applied to metal parts to

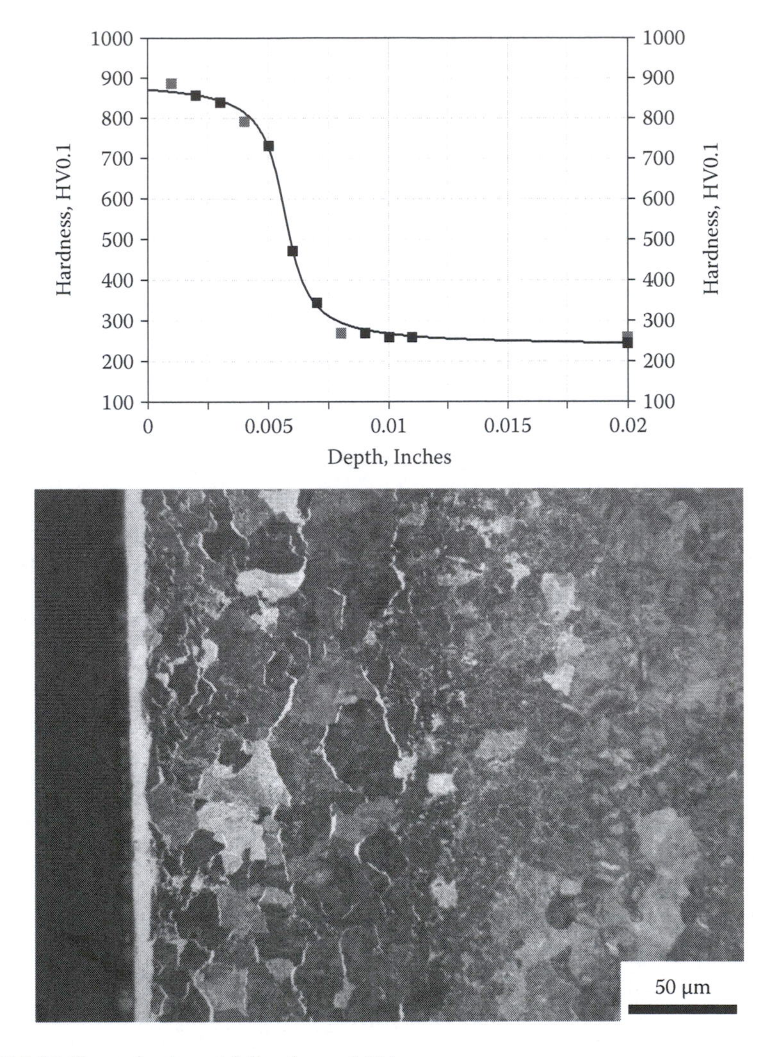


**FIGURE 5.28** (See color insert following p. 142.)
Surface hardening and hardness measurement. (Reprinted from Rolinski, E. et al., *Heat Treating Progress*, September/October, 6:19–23, 2006. With permission.)

provide corrosion and oxidation resistance. It is a process that often needs to be decoded in many reverse engineering projects. For a bearing surface, silver plating is usually applied to prevent galling or seizing. AMS 2410 covers the engineering requirements for silver plating, and helps reverse engineer bearings. However, the silver plating specifications for fasteners made of low-alloy steel, corrosion- and heat-resistant steel, and nickel alloys for high-temperature (up to 1,400°F) applications are specified in AMS 2411. Many fixtures, tools, and hardware are chrome or titanium plated.

The deposition processes also have three primary categories: physical vapor deposition (PVD), chemical vapor deposition, and ion implantation.

These depositions are often applied to the delicate instruments, such as medical and electronic devices. AMS 2403 and 2424 cover the materials, methods, applications, and processes for nickel electrodeposition and the properties thereof this deposition. They provide a wealth of information helping to decode this process in reverse engineering.

Many mechanical parts are coated by thermal spraying. Depending on their respective operating parameters, such as deposition rate, particle speed, and heat source, there are four major methods to applying thermal spraying: high-velocity oxyfuel (HVOF) spraying, plasma spraying, arc spraying, and flame spraying. Table 5.7 summarizes the primary characteristics of these thermal spraying processes and coating characteristics, such as adhesive strength, oxide content, porosity, and deposit thickness. The high particle speed of HVOF provides noticeably higher adhesive strength than other processes. This is because the adhesive strength of most thermally sprayed coatings results directly from mechanical adhesion in lieu of metallurgical bonding. The relatively low porosity observed in HVOF coating is also a direct consequence of the high particle speed of this process. These coating characteristics provide identifiable distinctions of HVOF that are critical for reverse engineering. Other thermal spraying processes, such as plasma spraying, are also applied to various mechanical parts to provide protection from wear, heat, corrosion, and abrasion. AMS 2437 covers the basic engineering requirements of plasma spray coating and provides a good reference to decode an OEM's coating process and to justify an alternative if necessary.

Black oxide coating is a process used typically to increase the antichafing and antifriction properties of carbon and low-alloy steel parts, particularly on sliding or bearing surfaces. The specifications of black oxide coating process are covered in AMS 2485.

For aluminum alloys, anodic treatment is widely applied to protect their surfaces by increasing their corrosion resistance. The anodic treatment also often produces a decorative colored surface as well. The engineering requirements of anodic treatment for aluminum alloys are covered in AMS 2470, 2471, and 2472 for chrome acid process, sulfuric acid process/undyed coating, and surface acid process/dyed coating, respectively. The engineering

**TABLE 5.7**

Comparison of Thermal Spraying Processes and Coating Characteristics

| | Particle Speed (m/s) | Adhesive Strength (MPa) | Oxide Content (%) | Porosity (%) | Deposition Rate (kg/h) | Typical Deposit Thickness (mm) |
|---|---|---|---|---|---|---|
| HVOF | 600–1,000 | >70 | 1–2 | 1–2 | 1–5 | 0.2–2 |
| Plasma | 200–300 | 20–70 | 1–3 | 5–10 | 1–5 | 0.2–2 |
| Arc | 100 | 10–30 | 10–20 | 5–10 | 6–60 | 0.2–10 |
| Flame | 40 | <8 | 10–15 | 10–15 | 1–10 | 0.2–10 |

requirements for producing a chemical film coating to increase the corrosion resistance of aluminum alloy parts are specified in AMS 2473. A word of caution: Adverse effects resulting from surface treatments are also observed. Many parts made of aluminum alloys are anodized to enhance their surface conditions. However, this surface treatment may not be appropriate for aluminum-copper alloys. Application of the anodizing treatment to aluminum-copper alloys might lead to deterioration in fatigue resistance because the copper-rich regions can develop pits during the early stages of the anodizing treatment. These pits will then initiate fatigue cracks. This information does not automatically exclude any parts made of aluminum-copper alloys from anodic treatment. However, understanding this effect certainly helps engineers to reverse engineer these parts. In reverse engineering, one easy and quick way to check whether an aluminum surface is anodized is just polishing off the surface layer and checking its electric conductivity. The aluminum alloy itself is electrically conductive; however, its anodized surface is nonconductive.

The coating material needs to be analyzed by proper surface analytical techniques such as Auger microscopy, or low-energy electron spectroscopy (LEES). The precise measurement of coating thickness is one of the few inherent impossibilities in reverse engineering. Most surface coating processes do not produce a uniform thickness over the part surface. Many surfaces also have multiple coating layers with rugged interfaces between them. The chemical reaction between the adjacent layers and the diffusion of atoms between them can mix the compositions of individual layers and make the identification of the original chemical compositions virtually impossible. As a result, reverse engineering surface coating is usually based on the best available data, and a judgment call that heavily relies on engineering expertise, knowledge, and training.

### *5.4.7.3 Shot Peening*

It is a common practice to induce residual compressive stress on a part surface by shot peening to increase fatigue strength and resistance to stress corrosion cracking. Aircraft landing gear failures primarily result from fatigue and stress corrosion cracking, so shot peening is widely used to improve the performance of landing gear. Accurate measurements of surface residual stress and its gradient profile underneath the surface are essential for reverse engineering shot peening. These measurements can be performed either destructively by etching and removing the surface layer by layer, or nondestructively by monitoring the penetration and reflection of X-rays. To minimize the introduction of new residual stresses to the surface layer, the removal of material is usually performed by electropolishing in small increments as thin as 0.5 μm (0.2 mil).

Today's technology is not able to decode all the detailed shot peening parameters and its dynamic effects, but it does provide abundant

information to help reverse engineer an OEM's shot peening process. The method of X-ray diffraction combined with microstructure analysis along with advanced analytical software can obtain reasonably accurate surface and near-surface residual stress measurements and their gradients. X-ray diffraction is based on the principle of Bragg's law. Figure 5.29 shows the detailed configuration of an X-ray diffractometer. The sample is held in the center goniometer, the X-ray is emitted from a source on the left, and the diffracted beam is collected on the right side. Figure 5.30 is a schematic of X-ray diffraction showing incident X-ray, diffraction cone, and a detector film with X-ray diffraction peaks. The center hole on the film is where the incident X-ray passes through, and it is also the datum for measuring the X-ray diffraction peak positions. The diffraction cone angle, θ, and the X-ray diffraction peaks are a function of material lattice spacing, *d*. This lattice spacing

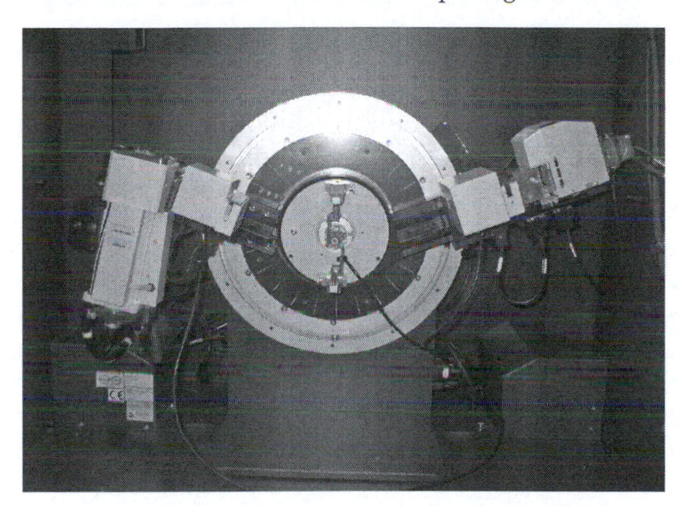

**FIGURE 5.29**
Setting of an X-ray diffractometer.

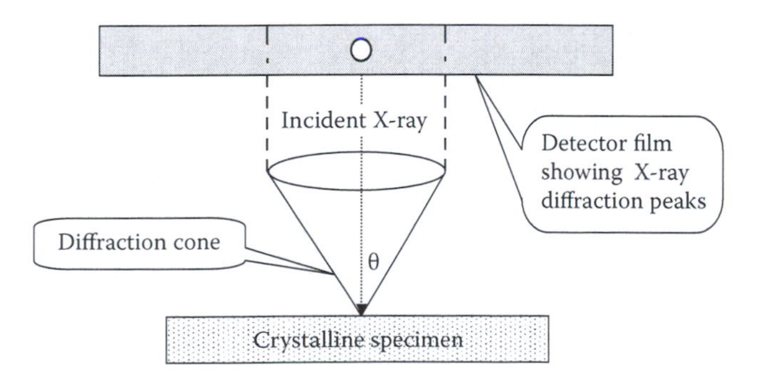


**FIGURE 5.30**
Schematic of X-ray diffraction.

changes when it is subject to stress due to strain deformation. As a result, the amount of residual stress can be measured by X-ray diffraction. Obviously, the X-ray diffraction technique only applies to crystalline specimens because only they have defined crystalline lattice spacing. The test specimen should be bare on the surface, and any galvanizing or paint has to be removed to get a good result. An accurate reading of X-ray diffraction also requires a competent technical analyst. Several standards and guidances on residual stress measurement are published, notably SAE J874a, *Residual Stress Measurement by X-Ray Diffraction* (SAE, 1980); and ASTM E915, *Standard Test Method for Verifying the Alignment of X-Ray Diffraction Instrumentation for Residual Stress Measurement*. ASM Handbook, Volume 11, also provides some general guidance on X-ray diffraction theory and residual stress measurement. Reference information on automatic shot peening is covered in SAE AMS 2430 and 2431. They discuss the engineering requirements and the media for shot peening by impingement of metallic shots, glass beads, or ceramic particles.

## References

ASME ANSI. 2002. *Surface texture, surface roughness, waviness, and lay*. ASME ANSI B46.1-2002. New York: ASME ANSI.

ASTM. 2006a. *Standard test method for evaluating the microstructure of graphite in iron castings*. ASTM A247-06. West Conshohocken, PA: ASTM International.

ASTM. 2006b. *Standard specification for austempered ductile iron castings*. ASTM A897/A897M-06. West Conshohocken, PA: ASTM International.

ASTM. 2006c. *Standard specification for precipitation-hardening nickel alloy bars, forgings, and forging stock for high-temperature service*. ASTM B637-06e1. West Conshohocken, PA: ASTM International

ASTM. 2009. *Standard specification for ductile iron castings*. ASTM A536-84. West Conshohocken, PA: ASTM International.

Bhadeshia, H. K. D. H. 2003. *Nickel-based superalloys*. University of Cambridge. http://www.msm.cam.ac.uk/phase-trans/2003/Superalloys/superalloys.html (accessed December 16, 2008).

Bond, B. J., and Kennedy, R. L. 2005. Evaluation of Allvac® 718Plus™ alloy in the cold worked and heat treated condition. In *Superalloys 718, 625, 706 and derivatives 2005*, ed. E. A. Loria. 203–212, Warrendale, PA: TMS.

Durand-Charre, M. 1997. *The microstructure of superalloys*. Amsterdam: Gordon and Breach Science.

Durst, K., and Göken, M. 2004. Micromechanical characterisation of the influence of rhenium on the mechanical properties in nickel-base superalloys. *Mater. Sci. Eng*. A387–89:312–16.

Erikson, G. L. 1996. *Superalloys*, ed. R. D. Kissinger, et al., 35. Warrendale, PA: TMS-AIME.

ISO. 1976. *Spheroidal graphite cast irons*. ISO 1083:2004. Geneva: ISO.

Jena, J. K., and Chaturvedi, M. C. 1984. The role of alloying elements in the design of nickel-base superalloys. *J. Mater. Sci.* 19:3121–39.

Keough, J. R. 1998. Ductile iron data for design engineers. Section IV. Austempered ductile iron. http://www.ductile.org/didata/section4/4intro.htm#Austempering (accessed January 2, 2009).

Hosogai, S., and Ito, H. 2005. *Responding to the RoHS directive and technology for low voltage circuit breakers*. Technical Report. Mitsubishi Electric ADVANCE. http://global.mitsubishielectric.com/pdf/advance/vol110/110_TR10.pdf (accessed January 2, 2009).

IPC. 2004. *Monotonic bend characterization of broad-level interconnects*. IPC-9702. Northbrook, IL: IPC.

Rolinski, E., Sharp, G., and Konieczny, A. 2006. Plasma nitriding automotive stamping dies. *Heat Treating Progr.* September/October, 6:19–23.

SAE. 1980. *Residual stress measurement by X-ray diffraction*, ed. M. E. Hilley, 62–65. SAE J784A. Warrendale, PA: SAE.

SAE AMS. 2008. *Heat treatment of steel parts, general requirements*. SAE AMS 2759. Warrendale, PA: SAE International.

SAE International. 2008. An abridged history of SAE. http://www.sae.org/about/general/history (accessed December 3, 2008).

Schmitt Industries, Inc. 2009. http://www.schmitt-ind.com/pdf/Roughness.pdf (accessed June 22, 2009).

Skewes, S. 2009. The brazing guide, induction atmospheres. http://www.inductionatmospheres.com/brazing.html?gclid=CL7oo6X68pcCFQEoGgodBjXbDg (accessed January 3, 2009).

Zhang, J., and Singer, R. F. 2004. Effect of Zr and B on castability of Ni-base superalloy IN792. *Metal. Mater. Trans.* 35A:1337.

# 6

# *Data Process and Analysis*

Data process and analysis in reverse engineering are composed of systematic assessment and quantitative evaluation. These are two independent yet interrelated processes. The assessment process identifies and collects relevant data from the earlier work through data acquisition. Data acquisition is a critical but tedious exercise in reverse engineering. In practice, engineers often collect as much data as they can for later analyses. The subsequent evaluation process collates, interprets, and analyzes the data obtained through assessment processes to draw statistical inferences and ensure quality performance of the new part produced by reverse engineering. All the raw data should be attained by creditable methods based on scientific and engineering principles whenever feasible. The reliance on anecdotal data by indirect estimation might lead to further uncertainty in later analyses.

Despite the advancement in statistics and other interpretive methods to present data in various formats and analyze the trends, one of the challenges still confronting many engineers in reverse engineering today is to correlate all the data from multiple sources into a logical conclusion. In reverse engineering we might need to determine the heat treatment schedule from hardness measurements and tensile properties, to decide the fatigue strengths from a set of test results, and to calculate grain size from the measurements based on grain morphology. No matter what techniques are used for data acquisition, the data have to be accurate and verified. Any inference thereby drawn from the data should be able to show a logical collaboration among the collected data. It is particularly important to collate the characteristic signatures in drawing any inference from the data. The surface texture of a component is a signature that provides crucial clues for the machining tools used in manufacturing. Hardness is a signature widely used in reverse engineering to determine the heat treatment of a part.

How many samples or measurements are required to ensure statistical accuracy is one of the most commonly asked questions in reverse engineering. This chapter will discuss this question by introducing the fundamental principles of statistics and their applications in data process and analysis. The reliability theory is closely related to statistics but was independently developed. This chapter will also discuss the applications of reliability theory, which is critical to reverse engineering data process and analysis in many cases.

## 6.1 Statistical Analysis

Reverse engineering is a data-driven, fact-based, high-tech endeavor. All the conclusions are based on measured data and tested results. It is imperative that the data be accurate and properly interpreted. Data acquisition and filing are usually the first steps in a reverse engineering project. These steps mechanically collect and file raw data without any elaboration on the data. Data interpretation and the subsequent conclusion or inferences drawn from the raw data integrate the data with various statistical theories. Assumptions are always introduced in any theoretical analysis, and inevitably concerns of accuracy and applicability of these assumptions can be raised in some cases. The suitability of any statistical analysis and the applicability of a statistical conclusion in reverse engineering are the most essential criteria that need to be clarified beforehand.

In reverse engineering it is very important to properly interpret the statistical data with some basic understanding of statistics. Statistics is a branch of applied mathematics concerning the collection, organization, and interpretation of numerical data. It uses probability theory to estimate population parameters such as dimensions and tensile strength. It helps engineers make correct decisions or conclusions based on limited data. The statistical mean of length in a part dimensional measurement along with other statistical parameters, such as the upper limit, lower limit, and standard deviation, provide valuable information in part geometric shape determination. Statistics, however, is a numerical data analysis with little scientific explanation. An in-depth engineering analysis can only be obtained by integrating statistical data with scientific principles. The average tensile strength obtained through statistics provides a numerical data point reflecting the material strength, but it will not explain what manufacturing process or heat treatment was used to achieve this tensile strength.

Theoretically, any measurement is subject to statistical randomness, and the measurements of the elemental composition of an alloy also show a statistical distribution. The focus then shifts to what is the engineering significance of this statistical randomness. The goal of a composition analysis for alloy identification in reverse engineering is the call-out or establishment of a material specification. The acceptable composition defined in virtually all the material specifications is listed as a range instead of a singular percentage. In other words, despite the potential significant effects of alloy composition on the alloy's characteristics, all engineering alloys have a range of acceptable compositions. A composition range, instead of a singular value, is defined in the material specification due to inherent alloy manufacturing variations and inaccuracy of analytical techniques. Will two or three tests significantly improve the statistical confidence compared to just one single test in alloy identification? From the perspective of alloy identification, the engineering significance of composition variation based on statistical data

obtained from two or three test results can be beneficial but might be marginal. The number of samples required in reverse engineering to identify the material used for the original part is often a judgment call based on part criticality and an engineer's experience and expertise. Nonetheless, whenever feasible, the best practice is to test a large number of samples to establish an alloy composition range based on actually measured data. This practice is recommended for a critical part and precision reverse engineering.

Descriptive statistics and inferential statistics are the two most used statistical analyses. Descriptive statistics summarizes or describes a collection of data, such as the grade point average (GPA) of a student or the batting average of a baseball player. It simply states a statistical fact based on data, while inferential statistics draws inferences about the general population from a selected sample. It generalizes statistical sample data with specified certainty and requires random sampling. Since statistics is primarily a numerical data analysis, caution is usually required when drawing a statistical conclusion. For example, in a class of twenty students, if one student wins a lottery of $10 million and the remaining nineteen students win nothing, statistically speaking, the average winning prize for each student in this class is $500,000. It is also worth noting that statistics tells the statistical chance of an event to happen, but does not specify exactly when it will happen or to whom.

The most important statistical subjects relevant to reverse engineering are statistical average and statistical reliability. Most statistical averages of material properties such as tensile strength or hardness can be calculated based on their respective normal distributions. However, the Weibull analysis is the most suitable statistical theory for reliability analyses such as fatigue lifing calculation and part life prediction. This chapter will introduce the basic concepts of statistics based on normal distribution, such as probability, confidence level, and interval. It will also discuss the Weibull analysis and reliability prediction.

### 6.1.1 Statistical Distribution

A statistical analysis starts with data distribution. The statistical distributions for data of interest are usually quantitatively described by statistical functions. The statistical functions are mathematical equations expressed in terms of statistical parameters.

Discrete distribution and continuous distribution are the two most common statistical distributions. The binomial distribution is a typical example of discrete distribution. The normal (also known as Gaussian) distribution and the Weibull distribution are two examples of continuous distributions. If a coin is tossed forty times and the number of heads-up occurrences is recorded, and this exercise is repeated seventy-five times, then the distribution of the heads-up occurrence frequency observed is binominal. The total number of events in this exercise is seventy-five. The highest possible frequency is forty, and the lowest one is zero. The frequencies of numbers

of occurrence ranging from ten to thirty are tabulated in Table 6.1 based on actual experimental data. Also included in Table 6.1 are the cumulative frequencies at and beyond a specific frequency. A total of twelve times was recorded where the coin landed heads up twenty-one times in any one event; therefore, the corresponding frequency is twelve. However, a total of forty-six times was counted for the coins to land heads up from zero up to twenty-one times. Figure 6.1 shows a plot of the distribution profile, where the $x$-axis is the number of occurrences of a specific frequency of the heads-up landing marked in the $y$-axis.

A binominal distribution describes the behavior of a variable (e.g., the frequency of heads-up occurrences) if the following conditions apply: (1) the population size is fixed, (2) each record is independent, (3) each observation represents one of two outcomes, and (4) the probability of success is the same for each observation. In the above-mentioned coin-tossing example, all four conditions are met. The population size is seventy-five because there are a total of seventy-five records, one from each event when the coin is tossed forty times. Obviously, the records of each event will be independent of the others. The coin can land on either its head or tail each time. Each observation

**TABLE 6.1**

Observation of Coin Tossing

| Number of Occurrences | Frequency | Cumulative Frequency |
|---|---|---|
| 10 | 0 | 0 |
| 11 | 0 | 0 |
| 12 | 0 | 0 |
| 13 | 2 | 2 |
| 14 | 3 | 5 |
| 15 | 2 | 7 |
| 16 | 5 | 12 |
| 17 | 6 | 18 |
| 18 | 3 | 21 |
| 19 | 7 | 28 |
| 20 | 6 | 34 |
| 21 | 12 | 46 |
| 22 | 8 | 54 |
| 23 | 4 | 58 |
| 24 | 4 | 62 |
| 25 | 3 | 65 |
| 26 | 1 | 66 |
| 27 | 3 | 69 |
| 28 | 0 | 69 |
| 29 | 0 | 69 |
| 30 | 0 | 69 |

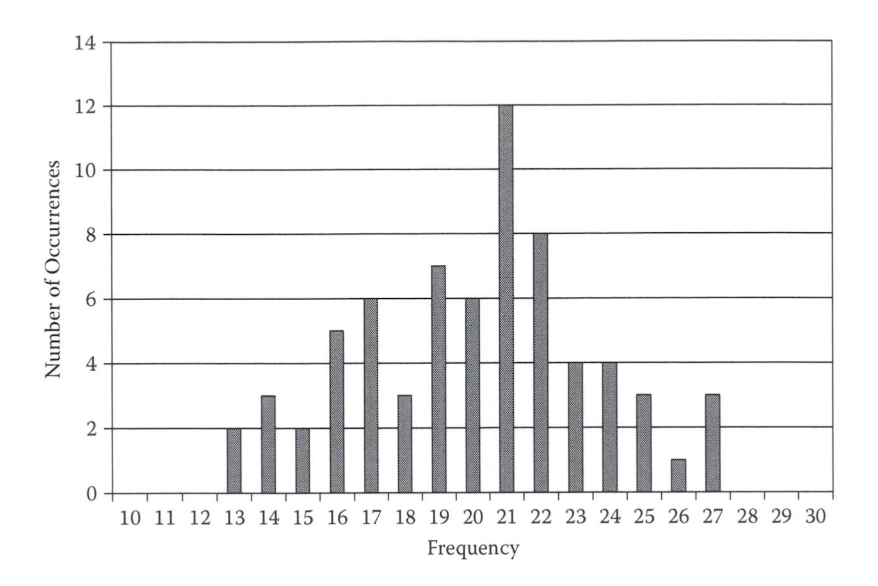


**FIGURE 6.1**
Discrete distribution of coin tossing.

has to be one of two outcomes, either heads up or tails up. If we define the heads-up landing as a success, the probability of success is always the same, 50/50 chance for each tossing.

A variety of data are approximated well by the normal distribution. Mathematically a normal distribution curve can be expressed as Equation 6.1, where $x$ is a variable and $P(x)$ is the quantitative frequency or probability of this variable. A typical standard distribution can be fully defined by two statistical parameters: the mean, $\mu$, and the standard deviation, $\sigma$. The maximum value is at the mean, and the minimum value is at plus and minus infinities. When $\mu = 0$ and $\sigma = 1$, the distribution is referred to as a standard normal distribution.

$$P(x) = \frac{e^{-(x-\mu)^2/(2\sigma^2)}}{\sigma\sqrt{2\pi}} \tag{6.1}$$

When a typical normal distribution is plotted with the horizontal axis as the variable, $x$, and the vertical axis as the frequency or probability of this variable, $P(x)$, it shows a bell-shaped curve that is symmetric about the mean, $\mu$. Schematics of normal distribution curves are shown in Figure 6.2. The width of the curve depends on the standard deviation. The shape of the curve flattens out with increasing standard deviation, implying a broad distribution of data.

When the population size is thirty or larger, the discrete binomial distribution diagram, shown in Figure 6.1, approaches the profile of a continuous

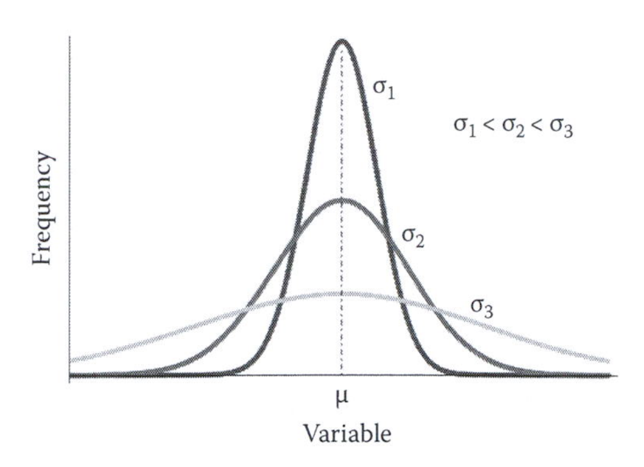


**FIGURE 6.2**
Schematics of normal distribution curves.

normal distribution curve. As a result, a discrete binomial distribution can be approximately represented by a continuous normal distribution in statistical analysis.

### 6.1.2 Statistical Parameter and Function

Statistical functions are mathematically described in terms of statistical parameters. The following statistical parameters are the most commonly used parameters in statistical analysis: population, sample, variate, variance, standard deviation, mean, median, and skewness.

A population in statistics consists of an entire set of objects, observations, or scores that have something in common. For example, a population might be defined as the entire stock of 3 mm diameter bolts made of 316 stainless steel. A sample is a subset of a population. Since it is usually impractical to test every member of an entire population, sampling from a population is typically the best approach available. Inferential statistics draw conclusions about a population from sampling. A variate is a generic random variable. It can be the length of a geometric measurement such as the length of the bolt, or a quantitative property value such as the hardness number of 316 stainless steel. In statistics, a capital $X$ is used to represent a variate in its general form, while a small $x$ is designated for the specific numerical value of the variate. The variance, $\sigma^2$, is a statistical parameter that shows how much the measurements vary from one to another. It is a measurement of data distribution spread. The mean, $\mu$, of a set of measurements is defined as the sum of the total measurements divided by the total number, $N$, of measurements as defined by Equation 6.2, where $x_i$ is the individual value and $\bar{x}$ is the arithmetic mean. The variance can be mathematically calculated by Equation 6.3. For example, in a population that is composed of seven length measurements showing respective lengths of 10, 9, 9, 10, 11, 10, and 11 cm, the

mean and the variance will be 10 and 0.571, as calculated by Equations 6.2. and 6.3, respectively:

$$\mu = \bar{x} = \frac{1}{N}\sum_{i=1}^{N} x_i \tag{6.2}$$

$$\sigma^2 = \frac{\sum_{i=1}^{N}(x_i - \mu)^2}{N} \tag{6.3}$$

The difference between an individual measurement, $x_i$, and the mean, $\mu$, can be either positive or negative. If the values $(x_i - \mu)$ for all $i$ measurements are just simply summed up, two values might cancel out each other when they are of the same absolute value but opposite signs, and the average deviation from the mean would be underestimated. To avoid this error, the difference is squared, $(x_i - \mu)^2$. The variance is the arithmetic average of these squared differences, as defined by Equation 6.3. The standard deviation, $\sigma$, is mathematically defined as the square root of variance, as expressed by Equation 6.4. The standard deviation is a statistical parameter that characterizes data dispersion. It is a measure of data scattering if the measurements are accurate, for instance, the scattering of test scores in a class of fifty students. It is a gauge of error in the measurements in case they are not always accurate, such as the variation of the length measurements of a bolt. If the value of $\sigma$ is small, this means the measurements all have similar values. A larger $\sigma$ reflects a larger variation among measurements, as illustrated in Figure 6.2.

$$\sigma = \sqrt{\frac{1}{N}\sum_{i=1}^{N}(x_i - \mu)^2} \tag{6.4}$$

The median is a statistical parameter representing the middle of a distribution: half the variate values are above the median and half are below the median. The median is less sensitive to extreme variate values than the mean, and therefore a better measure than the mean for highly skewed distributions. The median score of a test is usually more representative than the mean score of the class of how well the average students did on the test. The set of numbers 1, 2, 3, 7, 8, 9, and 12 have a mean of 6 and a median of 7. The mode is the statistical parameter that represents the most frequently occurring variate value in a distribution and is used as a measure of central tendency.

Several different terms are used for the same statistical functions by different statisticians and in different books. The cumulative distribution function, $D(x)$ or $F(x)$, is also referred to as the distribution function, the cumulative

density function, $CDF(x)$, or the probability distribution function; in other words, $D(x) = F(x) = CDF(x)$. While this book recognizes and supports the prerogative of statisticians to use and adopt the terminology of their choice, however, it is more convenient for the readers to have a consistent terminology in this book. For the purpose of consistency, the term *cumulative distribution function*, $D(x)$, will be used throughout this book.

Probability is expressed by a number between 0 and 1 that represents the chance that an event will occur. A probability of 1 means the event will definitely occur. A probability of 0 means the event will never occur. The probability or chance of occurrence is also expressed as a percentage between 0 and 100%. The probability function, $P(x)$, is also referred to as the probability density function, $PDF(x)$, or the cumulative probability function; therefore, $P(x) = PDF(x)$. The term *probability function* will be used throughout this book. The curve for a probability function $P(x)$ of a normal distribution with a mean of $\mu$ and a standard deviation of $\sigma$ can be mathematically described by Equation 6.1, as discussed before. The curve for a cumulative distribution function literally reflects the cumulative effect. The cumulative distribution function, $D(x)$, of a normal distribution is defined by Equation 6.5. It calculates the cumulative probability that a variate assumes a value in the range from 0 to $x$. Figure 6.3 is a plot of the cumulative distribution function curve from the data in Table 6.1.

$$D(x) = \frac{1}{\sqrt{2\pi}} \int_0^x e^{-z^2/2}\, dz \tag{6.5}$$

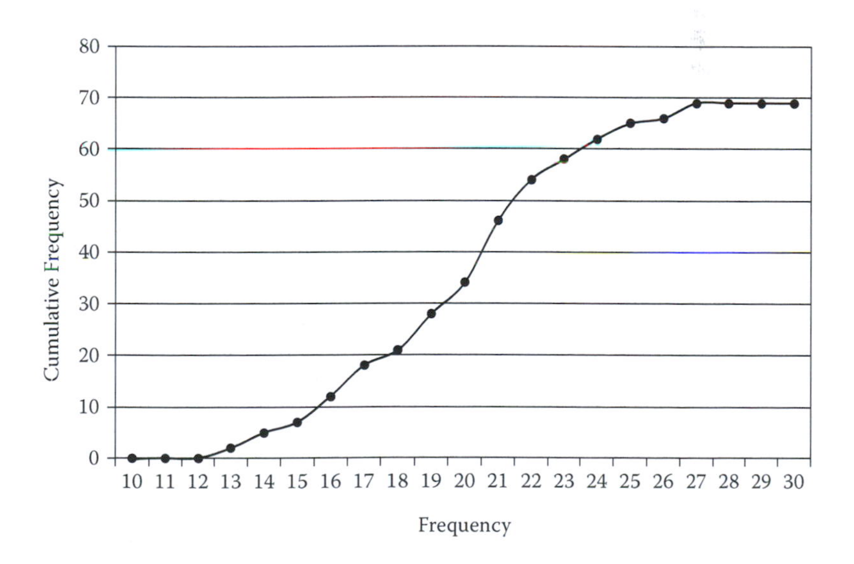


**FIGURE 6.3**
Schematic of a cumulative distribution function.

## 6.2 Data Analysis

### 6.2.1 Statistical Confidence Level and Interval

Engineers are frequently struggling to determine the required statistical sample size that can provide sufficient and enough data in a reverse engineering practice. Currently there are no governmental regulations or industrial guidance that mandates a specific minimum sample size for reverse engineering. Part of the reason is that statistical accuracy is often expressed in terms of confidence level and confidence interval; they in turn depend not only on sample size, but also on population size.

In statistics, the confidence interval is an interval with an associated confidence level generated from repeated sampling from the same population. The confidence interval depends on the number of observations or measurements, and the spread of data. For example, a poll predicts with 95% confidence that a candidate will have a 75 to 80% chance of winning an election. The confidence level is 95% and the confidence interval is between 75 and 80% for this poll. In reverse engineering, an engineer might report his measurement with 99% confidence that there is a 95 to 98% chance the length of a bolt is 6.3 cm. For a specific confidence level and associated confidence interval, the larger the population size is, the smaller the required sample percentage needs to be. Figure 6.4 is a statistical confidence chart of an election survey for a college class president. The *x*-axis is the population size, i.e., the class size of total student number, and the *y*-axis is the survey sample size of the class measured as a percentage. Certainly the survey is 100% accurate and the margin of error is 0% if the survey is conducted with every individual in the entire class (i.e., the entire population), no matter the class size. However, if not every individual is surveyed, the numbers of students that need to be surveyed will be different, depending on the class size, to achieve the same level of certainty. In a class size of 100, to get a 95% certainty with ±5% margin of error, that is, a confidence level of 95% with a

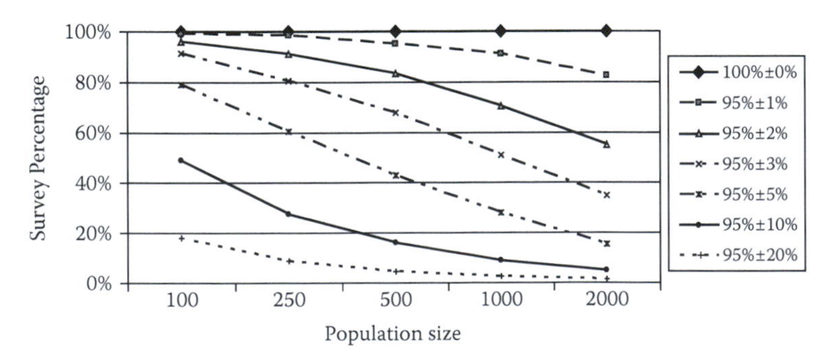


**FIGURE 6.4**
Statistical confidence chart.

confidence interval between 100 and 90%, about 80% of the class needs to be surveyed. The required percentage decreases to approximately 60% when the class size increases to 250 students for the same confidence level and confidence interval, and only less than 20% of the class is required when the class size increases to 2,000. Figure 6.4 also shows that smaller margins of error (narrower confidence intervals) require larger sample sizes. With the same population size of 500, to get the same confidence level of 95% when the margin of error decreases from 5% to 2%, the required sample size will increase from about 43% to 84% of the entire population.

For a normal distribution, the probability that a measurement falls in the interval of $\pm n$ standard deviation, $\pm n\sigma$, of the mean, $\mu$, can be calculated by Equation 6.6. This equation quantitatively calculates the confidence level (the probability) of finding the measurement within a defined confidence interval, from $\mu - n\sigma$ to $\mu + n\sigma$.

$$P(\mu - n\sigma < x < \mu + n\sigma) = \frac{2}{\sqrt{\pi}} \int_0^{n/\sqrt{2}} e^{-u^2} du = erf\left(\frac{n}{\sqrt{2}}\right) \tag{6.6}$$

where $erf(n/\sqrt{2})$ is the error function. The values of confidence levels corresponding to specific confidence intervals of a normal distribution are tabulated in Table 6.2. They are also plotted in Figure 6.5. For normal distribution data such as hardness numbers, statistically only an approximately 68.27% confidence level holds for a narrow confidence interval of $\pm 1$ standard deviation from the mean. The confidence level expands to approximately 95.45% when the confidence interval expands to $\pm 2$ standard deviations. The confidence level will further improve to 99.73% for a confidence interval of $\pm 3$ standard deviations, and so forth.

Conversely, for a given confidence level, the confidence interval can be calculated by Equation 6.7:

$$n = \sqrt{2} erf^{-1}(P) \tag{6.7}$$

**TABLE 6.2**

Confidence Level Corresponding to Specific Confidence Interval of a Normal Distribution

| Confidence Interval | Confidence Level $P(\mu - n\sigma < x < \mu + n\sigma)$ |
|---|---|
| $\pm\sigma$ | 68.27% |
| $\pm 2\sigma$ | 95.45% |
| $\pm 3\sigma$ | 99.73% |
| $\pm 4\sigma$ | 99.994% |
| $\pm 5\sigma$ | 99.999% |

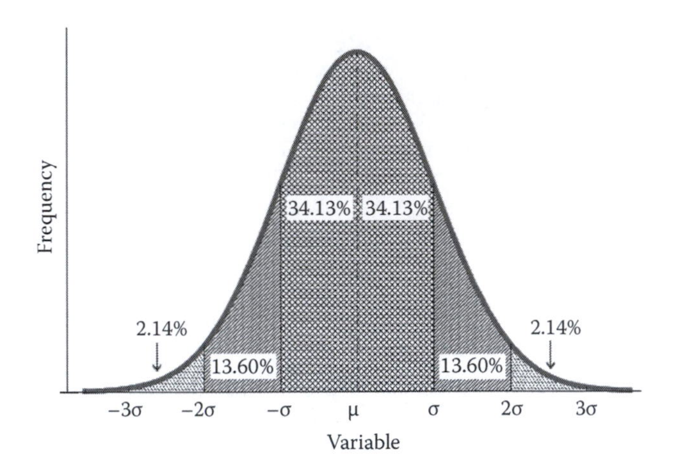


**FIGURE 6.5**
Confidence level and interval of a normal distribution.

where $erf^{-1}(P)$ is the inverse error function, and their values are tabulated in Table 6.3. In a reverse engineering measurement with normal distribution, if an 80% confidence level is required, then the associated confidence interval between ±1.28155σ is a given that has to be accepted. When a 90% confidence level is required, the confidence interval expands to ±1.64485σ. In statistical data, the confidence level and confidence interval are directly affiliated with each other and usually reported side by side as a pair.

Theoretically, any measurement will produce a set of statistical data and should be so reported if feasible. Let us consider a case study whereby an OEM shaft will be duplicated by reverse engineering. The length of the shaft is measured, and the report shows that the measured data have a nominal distribution with a mean of 250 mm, and a standard deviation of 0.5 mm. The following exemplifies a typical statement reporting this measurement in statistical terms based on the calculations detailed in Equation 6.8a and b. "With a confidence level of 95%, the mean shaft length lies within the confidence intervals between 250.98 and 249.02 mm."

$$250.98 \approx 250.97998 = 250 + 1.95996\sigma = 250 + 1.95996 \times 0.5 \tag{6.8a}$$

$$249.02 \approx 249.02002 = 250 - 1.95996\sigma = 250 - 1.95996 \times 0.5 \tag{6.8b}$$

The statistics of test results and measured data might be critical to part reliability analysis in machine design and reverse engineering. In some cases the method of probabilistic fracture mechanics along with the statistical data is preferred over deterministic fracture mechanics in failure prevention analysis. However, a simple average value, that is, 250 mm for the shaft length in the above example, is usually sufficient in most calculations

**TABLE 6.3**

Confidence Interval for Specified Confidence Level of a Normal Distribution

| Confidence Level $P(\mu - n\sigma < x < \mu + n\sigma)$ | Confidence Interval |
|---|---|
| 80% | 1.28155σ |
| 90% | 1.64485σ |
| 95% | 1.95996σ |
| 99% | 2.57583σ |

whenever the deterministic method applies, where a singular value instead of a static distribution is used. The deterministic methodology is usually adopted for its simplicity and effectiveness in reverse engineering.

### 6.2.2 Sampling

Data collection is essential in reverse engineering. However, if the data are not gathered systematically and efficiently, this process can be frustrating, time-consuming, and costly. Sampling is a streamlined method for data collection. It is a common statistics technique used to draw inference for a large population size, such as, to obtain the tensile strength of 2024 aluminum alloy that has been produced by many vendors for many years. The population size is too large to conduct an evaluation of the entire population. Therefore, sampling is the best way to determine its tensile strength statistically.

The sampling distribution is a statistical distribution of the probability function related to a reference statistical parameter. In the above example, a sampling distribution of tensile strength is the statistical distribution of the probability of having 2024 aluminum alloy at a specified tensile strength. The sampling distribution is the first step leading to the determination of the tensile strength of 2024 aluminum alloys.

Sampling in reverse engineering practice is disparaged from time to time because it is often time-consuming, and if it is not done effectively and correctly, the results can be confusing. Judging the quality of a sampling process requires a good understanding of the principles of statistical sampling. It is not always true that more samples are better, because the focus is on quality, not quantity, of the data. Engineers, scientists, and all other stakeholders, such as governmental regulators, have yet to agree on an acceptable method to determine the appropriate sample size in reverse engineering. Nonetheless, understanding and using good sampling techniques can streamline the process, and greatly reduce the amount of data that need to be collected.

In accordance with established statistical theories, if the original population distribution $D(x)$ is a normal distribution, the sampling distribution

of the sample mean, $\bar{x}$, will be a normal distribution as well. The sampling distribution of the sample mean is the distribution of the means of the samples. Statistically, the sampling distribution mean is equal to the population mean, $\mu_{\bar{x}} = \mu$, where $\mu_{\bar{x}}$ and $\mu$ are the sampling distribution mean and the original population distribution mean, respectively. The population mean is also often referred to as the expected value. The standard deviation of a sampling distribution can be calculated by dividing the population standard deviation by the square root of the sample size, $n$, as mathematically expressed by Equation 6.9:

$$\sigma_{\bar{x}} = \frac{\sigma}{\sqrt{n}} \tag{6.9}$$

The equivalence relationship between the sample mean and population mean, and the mathematical relationship between the standard deviations of the sampling and the population distributions are valid regardless of the sample size for normal distributions.

According to the central limit theory, for sufficiently large sample sizes, usually greater than thirty, the sampling distribution of sample means, $\bar{x}$, will be approximately normal even for nonnormal distributions. The central limit theory further states that the sampling distribution mean still approximately equals the population mean, that is, $\mu_{\bar{x}} = \mu$; when the sample size is sufficiently large, even the original population distribution is not normal. The mathematic relationship expressed by Equation 6.9 is also approximately correct for the standard deviations between the sampling and the population distributions when the sample size is sufficiently large. In the following example, an engineer is tasked to determine the length of an OEM shaft in a reverse engineering project. The OEM has manufactured and distributed 5,000 pieces of this shaft. The population size of this shaft is therefore 5,000. It has been documented that the shaft length has a normal distribution. In this example, the shaft length, $X$, is the random variable, that is, the variate in statistical terms. Since the population distribution is normal, the population mean will be equal to the sample mean. The engineer can randomly measure thirty sets of samples with each set having twenty-five pieces of shaft. The sample size of each set is twenty-five, and the measured data will show a normal distribution. The sample mean for each set can be determined; each set has different values, such as 250.5 mm, 249.7 mm, etc. These sample means of each set can be used to plot the sampling distribution. If the mean of the plotted sampling distribution is 250 mm, that should also be the population mean of all 5,000 shafts. If the standard deviation for the sampling mean is 0.5 mm, then the standard deviation of population distribution can be calculated using Equation 6.9a.

The result is 0.1 mm. In summary, the statistical average length of the shaft is 250 mm with a standard deviation of 0.1 mm.

$$\sigma_{\bar{x}} = \frac{\sigma}{\sqrt{n}} = \frac{0.5}{\sqrt{25}} = 0.1 \tag{6.9a}$$

### 6.2.3 Statistical Bias

In reverse engineering, it is mandatory to only utilize a testing machine or measuring device that has been regularly calibrated according to standards and regulations. The American National Standards Institute and other institutions have published a series of standards for periodic calibrations of a variety of instruments. It is also often required that the personnel involved with the test be properly trained and certified at the respective level of the work he or she is performing. All of these requirements are aimed at avoiding potential erroneous data.

One of the potential sources of errors is attributed to statistical bias that primarily results from the following three biased effects: biased sample, biased estimator, and biased measurement instrument. In a statistical analysis, when most samples are chosen from one specific section of the population, the sample population will have the potential to be biased. For instance, if all the 2024 aluminum alloy samples were provided by one vendor in a hardness analysis, the sample would be biased. Furthermore, the source of samples requires carefully scrutiny to avoid hidden biased effects. For instance, the 2024 aluminum alloy samples might be provided by three different suppliers; however, all three suppliers might obtain their materials from the same foundry.

Biased data from measurements might result from uncalibrated instruments. For example, a biased Rockwell hardness tester with a misalignment might constantly produce lower hardness values. It is usually required to check the tester with a standard test block to ensure that the hardness test machine is functioning properly before collecting any test data. Figure 6.6 shows the ASTM hardness test standards for Rockwell hardness testers manufactured by Instron Corporation.

A systematically biased measurement can also result from the mentoring system. The measured average speed on a highway will be lower than the speed monitored otherwise if the data are collected using a police patrol car tailing other automobiles on the highway. The collected data are systematically biased in this case. Figure 6.7 is a schematic of the effects of bias and imprecision of data distribution. Only unbiased and precise data should be used in any reverse engineering analysis.

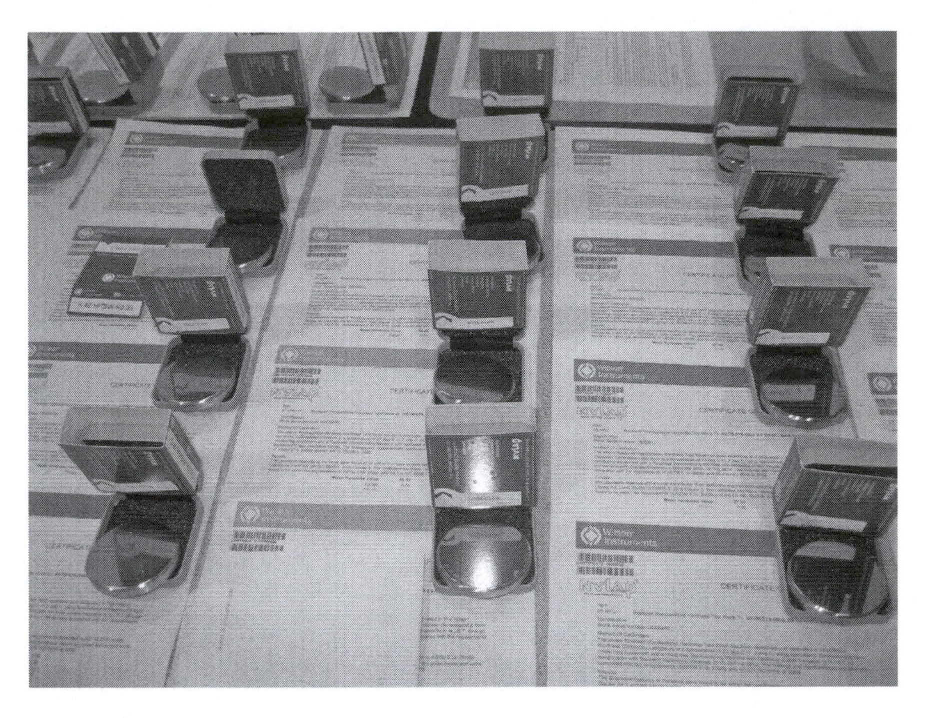

**FIGURE 6.6**
Rockwell hardness test standards.

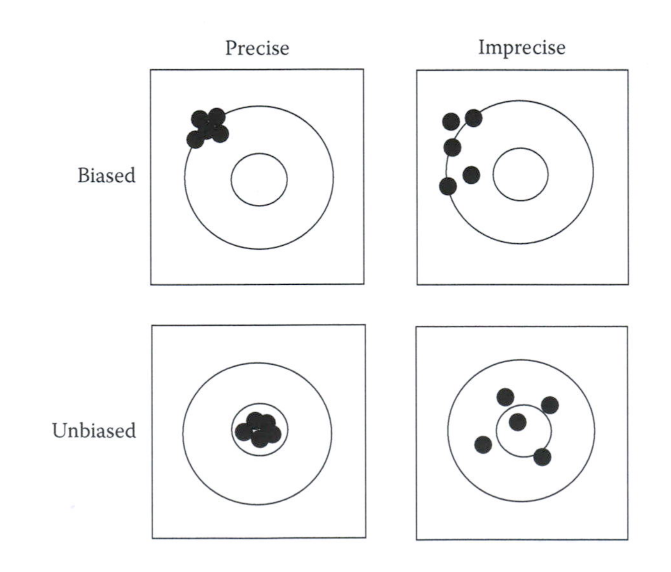


**FIGURE 6.7**
Schematic of statistical bias.

## 6.3 Reliability and the Theory of Interference

There are two aspects of reliability in reverse engineering. Reliability in statistics refers to the consistency or repeatability of a measurement. It relies on the repeatability of the same measurement instrument, and the comparability of similar measurement instruments. It also depends on the repeatability of the operator of the test machine or the device. It does not imply the validity of the data that reflect whether the test result or measurement is what was intended to be obtained. A misaligned Rockwell hardness tester will repeatedly produce consistently biased hardness data. Statistically, the measurement by this tester is of high reliability despite it is not producing valid engineering data.

The second aspect of reliability is referred to as part functionality. The reliability of a part reflects the probability that this part will perform a required function without failure under stated conditions for a stated period of time. Reliability of a part analyzes time to event. Statistically, failure of a part is deemed as a sample event. The objective is to predict the rate of events for a given population (often referred to as failure rate or hazard rate) or the probability of an event for an individual part. Part reliability plays a crucial role in machine design, and in reverse engineering that reinvents the same part. It is an important continued operational safety issue, and also has a significant financial impact because it helps to predict the probability of failure for a part over a period of time.

The reliability theory is heavily dependent on statistics and probability, but was developed apart from mainstream statistics and probability to help insurance companies in the nineteenth century. It also heavily relies on the theory of interference. Part functional reliability will be discussed below in terms of safety margins and Weibull analysis.

### 6.3.1 Prediction of Reliability Based on Statistical Interference

In conventional deterministic mechanics, both part strength and applied stress have singular values. When part strength is larger than applied stress, the part is safe; otherwise, it will fail. However, when part strength and applied stress are presented as statistical distributions, the theory of statistical interference will be used to analyze part reliability. Figure 6.8 exemplifies the statistical mechanics, where both strength and stress are presented as statistical distributions. The mean strength is higher than the mean of the applied stress. The part is safe and needs no further analysis in terms of conventional deterministic mechanics. However, it is statistically possible that the part can become unsafe in the overlapped area where the part is overloaded. This circumstance arises when an abnormally weak part is subjected to an unusually heavy load.

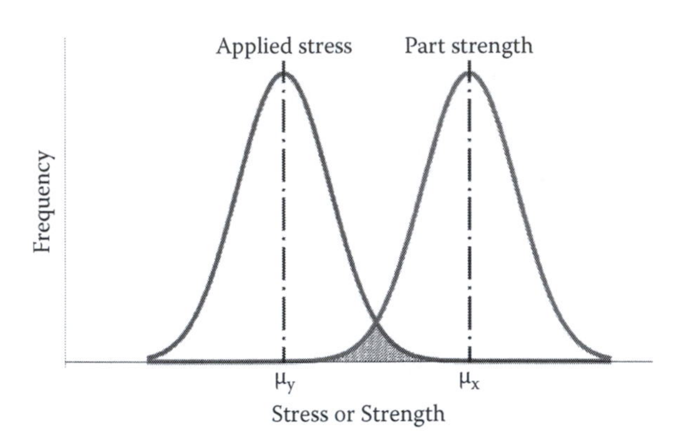


**FIGURE 6.8**
Strength and stress distributions and interference.

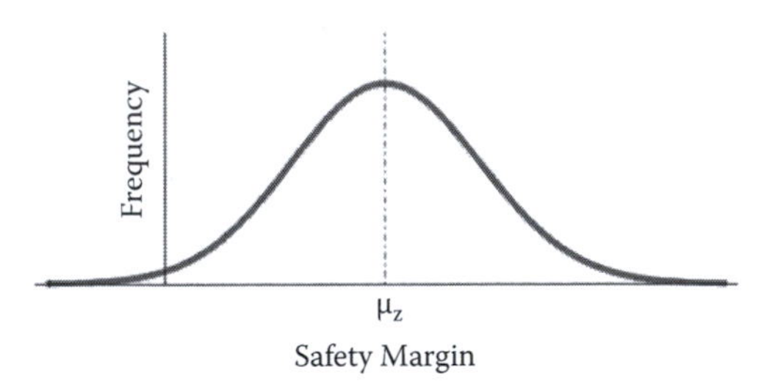


**FIGURE 6.9**
Normal distribution of safety margin.

In Figure 6.8, both part strength and applied stress are normal distributions, the mean of the part strength is designated as $\mu_x$, while the mean of the applied stress is designated as $\mu_y$. By definition, the safety margin, $\mu_{safety}$, will be $\mu_{safety} = \mu_{strength} - \mu_{stress}$. The profile of safety margin is a derived distribution of the two normal distributions of part strength and applied stress. As a result, it is also a normal distribution. Figure 6.9 is a schematic of the normal distribution of safety margin. The $x$-axis is the safety margin, $\mu_{safety}$, with the mean as $\mu_z = \mu_x - \mu_y$, and the $y$-axis is the frequency of occurrence corresponding to a specific safety margin value. The part will fail when $\mu_{safety} = \mu_{stress} - \mu_{strength} < 0$. A vertical line on the far left end delineates this safe-failure boundary in Figure 6.9. This specific boundary safety margin value can be set at zero that will certainly lead to a failure when the safety margin is below this boundary, or a small positive value to build in some conservatism in failure prevention. According to the theories of statistics, the variance of the derived distribution is the sum of the variances of the two base distributions. The standard deviation of the safety margin can therefore be calculated by Equation 6.10, where $\sigma_x$, $\sigma_y$, and $\sigma_z$ are the standard deviations of part strength, applied stress, and safety margin, respectively.

$$\sigma_z = \sqrt{\sigma_x^2 + \sigma_y^2} \tag{6.10}$$

The derived normal distribution of safety margin as plotted in Figure 6.9, and the affiliated derivative parameters such as the mean of safety margin, $\mu_z$, and standard deviation, $\sigma_z$, build the foundation for further analysis of statistical reliability of the part reinvented by reverse engineering. The part is reliable when $\mu_{safety} > 0$, but unsafe when $\mu_{safety} < 0$. Probabilistic mechanics conducts further statistical analysis to calculate the probability for these conditions to occur.

In the following example, a shaft made of steel is reverse engineered. A total of thirty parts are analyzed, and the test data show that their tensile strength has a normal distribution with a mean of 650 MPa and a standard deviation of 10 MPa. This part will be subjected to a tensile stress in service that fluctuates in a normal distribution pattern, with a mean of 550 MPa and a standard deviation of 20 MPa. Therefore,

$$\mu_x = 650 \text{ MPa}$$

$$\sigma_x = 10 \text{ MPa}$$

$$\mu_y = 550 \text{ MPa}$$

$$\sigma_y = 20 \text{ MPa}$$

and

$$\mu_z = \mu_x - \mu_y = 650 - 550 = 100 \text{ MPa}$$

and

$$\sigma_z = \sqrt{\sigma_x^2 + \sigma_y^2} = \sqrt{10^2 + 20^2} = 22.36 \text{ MPa}$$

Based on the above calculations, the safe-failure boundary will lie around $\mu_{safety} = -4.473\sigma_z$, where $\mu_{safety} = \mu_{strength} - \mu_{stress} = 100 - 4.473 \times 22.36 \approx 0$, in the safety margin distribution curve, similar to Figure 6.9. Since this is a normal distribution, there is only a 0.0004% chance that the part will fall into the area left to the delineated boundary. In other words, the reliability of this part is 1 – 0.0004% = 99.9996%. This part is therefore very reliable and safe. The same conclusion can be easily reached using conventional deterministic mechanics by simply comparing the mean of part strength, 650 MPa, to the mean of applied stress, 550 MPa. If the mean of the applied stress increases to 600 MPa, provided all other parameters remain the same, deterministic mechanics will reach the same conclusion: the part is safe because $\mu_{strength} > \mu_{stress}$. In this case, statistical mechanics will predict a lower reliability of only approximately 98.73%, compared to 99.9996%, when the applied stress is 550 MPa. The reliability parameters relevant to the detailed calculations are summarized in Table 6.4.

**TABLE 6.4**

Reliability Analysis

| Parameter | Case 1 | Case 2 |
|---|---|---|
| Parameter/applied stress | $\mu_y = 550$ MPa | $\mu_y = 600$ MPa |
| Mean of part strength | $\mu_x = 650$ MPa | $\mu_x = 650$ MPa |
| Mean of applied stress | $\mu_y = 550$ MPa | $\mu_y = 600$ MPa |
| Mean of safety margin | $\mu_z = \mu_x - \mu_y = 100$ MPa | $\mu_z = \mu_x - \mu_y = 50$ MPa |
| Standard deviation of part strength | $\sigma_x = 10$ MPa | $\sigma_x = 10$ MPa |
| Standard deviation of applied stress | $\sigma_y = 20$ MPa | $\sigma_y = 20$ MPa |
| Standard deviation of safety margin | $\sigma_z = \sqrt{\sigma_x^2 + \sigma_y^2} = 22.36\text{MPa}$ | $\sigma_z = \sqrt{\sigma_x^2 + \sigma_y^2} = 22.36\text{MPa}$ |
| Safety margin boundary, $-n\sigma_z$ | $100 - 4.473 \times 22.36 \leq 0$<br>$n = 4.473$ | $50 - 2.236 \times 22.36 \leq 0$<br>$n = 2.236$ |
| Error function | $erf\left(4.473/\sqrt{2}\right) = erf(3.1629)$<br>$= 0.999992$ | $erf\left(2.236/\sqrt{2}\right) = erf(1.5810)$<br>$= 0.9746$ |
| Reliability | $1 - 0.999992 = 0.000008$<br>$0.000008/2 = 0.000004$<br>$1 - 0.000004 = 0.999996 = 99.9996\%$ | $(1 - 0.9746) = 0.0254$<br>$0.0254/2 = 0.0127$<br>$1 - 0.0127 = 0.9873 = 98.73\%$ |

All the material properties, such as tensile strength and fatigue endurance, can be presented as statistical data. All the loads externally applied to the part can also be represented as statistical data. The above example indicates that a simple conclusion on safety can easily be reached using the conventional deterministic mechanics that only applies a singular value in its calculation. However, when reliability in terms of a quantitative percentage is required, engineers usually use statistical mechanics, where the relevant parameters are expressed in statistical terms.

## 6.4 Weibull Analysis

The Weibull distribution was first formulated in detail by Walloddi Weibull in 1951, and thus it bears his name. It more accurately describes the distribution of life data, such as fatigue endurance, compared to other statistical distributions, such as the normal distribution which fits better for hardness and tensile strength. Weibull analysis is particularly effective in life prediction. It can provide reasonably accurate failure analyses and failure predictions with few data points, and therefore facilitates cost-effective and efficient component testing. Weibull analysis is widely used in many machine design

and reverse engineering practices, where fatigue life is a determining factor in part life prediction.

A statistical distribution is characterized by the location, scale, and shape parameters. A normal distribution is defined only by the location (i.e., mean) and scale (standard deviation) parameters. The standard normal distribution has a location parameter equal to 0 and a scale parameter equal to 1. In addition to the location and scale parameters, the Weibull distribution is further defined by a third parameter, a shape parameter. A normal distribution has a symmetric shape; that is, the mean, median, and mode are identical, whereas a typical Weibull distribution is asymmetric and skewed, with a long tail either on the right or left side. The mean, median, and mode are not coincident at the same location. To assess the suitability of a distribution, either normal or Weibull, for a set of data, a graphical technique is usually utilized. This technique generates a series of probability plots for competing distributions to see which fits best. The probability plot can also estimate the location and scale parameters of the chosen distribution. Engineers are urged to check the suitability of a statistical distribution before applying it to any analysis.

The three-parameter Weibull cumulative distribution function, $F(t)$, that predicts the cumulative probability of failure up to a specific time, $t$, is mathematically expressed by Equation 6.11. The probability density function, $f(t)$, which is a derivative of the cumulative distribution function, is expressed by Equation 6.12:

$$F(t) = 1 - e^{-(t-t_o/\eta)^{\beta}} \tag{6.11}$$

$$f(t) = \frac{\beta}{\eta}\left(\frac{t-t_o}{\eta}\right)^{\beta-1} e^{-\left(\frac{t-t_o}{\eta}\right)^{\beta}} \tag{6.12}$$

where $\eta$ = scale factor or characteristic life, $\beta$ = shape parameter or slope of Weibull plot, and $t_o$ = location parameter or guaranteed life ($t_o = 0$ in a two-parameter Weibull).

The two-parameter Weibull cumulative distribution as described by Equation 6.13a is often converted to a linear equation like Equation 6.13e by simple mathematical manipulations detailed by Equation 6.13b to d. This linear equation allows the use of a simple graphic solution to plot a straight line, with its slope showing the numerical value of $\beta$ in a Weibull probability paper. The Weibull probability paper was widely used to check the applicability of the Weibull distribution model, and also to obtain the values of the parameters $\eta$ and $\beta$ by a graphic method referred to as Weibull plotting, before computing software programs became readily available. Figure 6.10 shows a sample Weibull plotting with the shape parameter, $\beta$, estimated to be 1.4, as illustrated on the upper left corner in the plot. The six data points show a clear linear incremental trend of the unreliability as the time increases, and therefore it verifies that the data follow Weibull statistics.

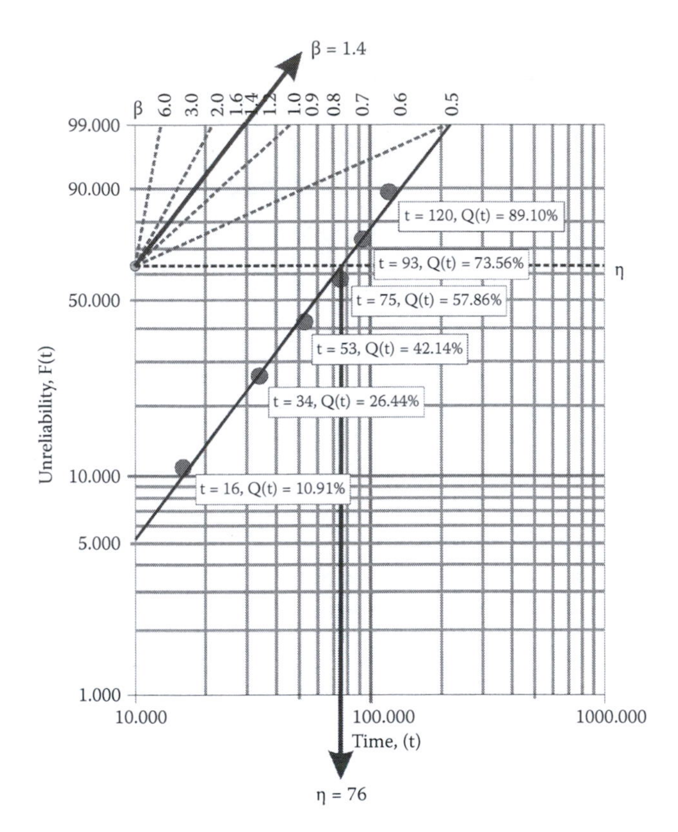


**FIGURE 6.10**
A sample Weibull plotting. (Reprinted from ReliaSoft. With permission.)

Both the abscissa and ordinate of the Weibull probability paper are in logrithmic scales based on Equation 6.13e. Depending on the orders of magnitudes on the abscissa, these papers are referred to as one-, two-, or three-cycle papers. The one-cycle paper has a scale from 1 to 10, the two-cycle paper has a scale from 1 to 100, and so forth. The paper in Figure 6.10 is a three-cycle paper. The time, $t$, or cycles to failure are plotted as the abscissa. The unit for the x-axis can be minutes, hours, or days for time; or any other life dimensions of a part. For example, it can be cycles for fatigue life, meters for the distance traveled, or numbers of print, etc. Any dimension that measures a part life can be used as a unit for the x-axis. This also applies to the following Figures 6.11, 6.12, 6.13a, c, and d. The Weibull cumulative distribution function, $F(t)$, is plotted as the ordinate; it represents the cumulative percentage failed at the time, $t$, or the unreliability, as shown in Figure 6.10. The determination of the $F(t)$ position in the plot often requires further statistical calculation. For example, the value of median rank as the unreliability percentage is often used for the Weibull plotting for censored (i.e., incomplete) data. The scale parameter $\eta$ is referred to as characteristic life because $F(t) = 1 - e^{-1} \approx 0.632$ when $t = \eta$, regardless of

the value of the shape parameter, β. Therefore, the corresponding $t$ value at $F(t) = 0.632$ gives the reading of η. The value of scale parameter η is determined to be 76 in Figure 6.10.

$$F(t) = 1 - e^{-(t/\eta)^{\beta}} \tag{6.13a}$$

$$1/[1 - F(t)] = e^{(t/\eta)^{\beta}} \tag{6.13b}$$

$$\ln[1/[1 - F(t)] = (t/\eta)^{\beta} \tag{6.13c}$$

$$\ln\ln\{1/[1 - F(t)]\} = \beta\ln(t) - \beta\ln(\eta) \tag{6.13d}$$

$$y = \beta x + \alpha \tag{6.13e}$$

where

$y = \ln\ln\{1/[1 - F(t)]\}$
$x = \ln(t)$
$\alpha = \beta\ln(\eta)$

In a reliability analysis the most relevant parameters are reflected by the reliability and hazard functions. The reliability function can be mathematically expressed by Equation 6.14:

$$R(t) = 1 - F(t) \tag{6.14}$$

The hazard function reflects the instantaneous failure rate at a specific time $t$. For a two-parameter Weibull, it is mathematically expressed as Equation 6.15:

$$h(t) = (\beta/\eta)(t/\eta)^{\beta-1} \tag{6.15}$$

In many applications, the Weibull analysis is applied to predict the part reliability or unreliability based on limited data with the help of modern computer technology. The limited data will inevitably introduce some statistical uncertainty to the results. Figure 6.11 shows the unreliability as a function of time in a Weibull plot. The six data points reasonably fit on the straight line and verify that the data are a Weibull distribution. The hourglass curves plotted on each side of the Weibull line represent the bounds of 90% confidence intervals for this analysis. The width of the intervals depends on the sample size; it narrows when more samples are analyzed.

The effects of scale parameter, η, on a Weibull probability density function curve are illustrated in Figure 6.12. A change in the scale parameter, η, has the same effect on the Weibull probability density function profile as changing the abscissa scale. The scale parameter also bears the same unit as the abscissa scale, in terms of either time or cycle. With a constant β value, when η increases, the profile will flatten out, and the mean will move to the right

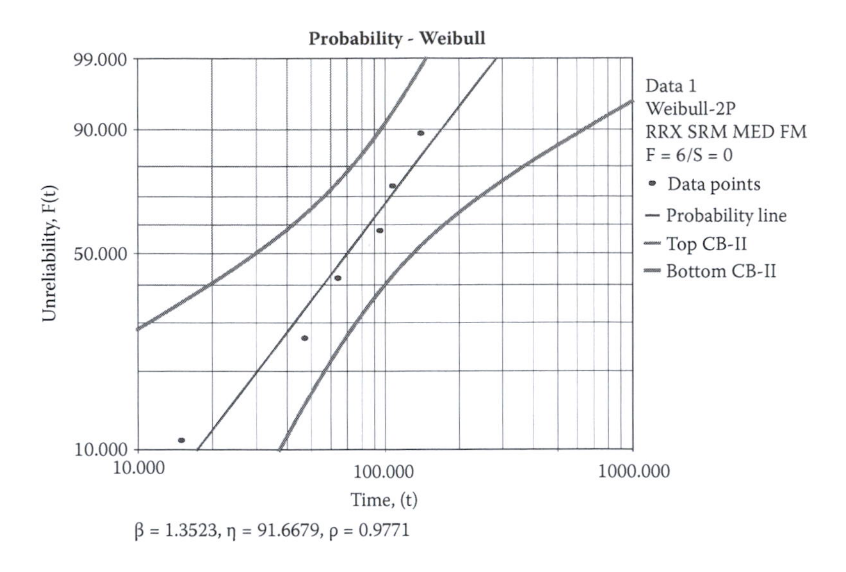


**FIGURE 6.11**
A Weibull plot with the boundaries of confidence level. (Reprinted from ReliaSoft. With permission.)

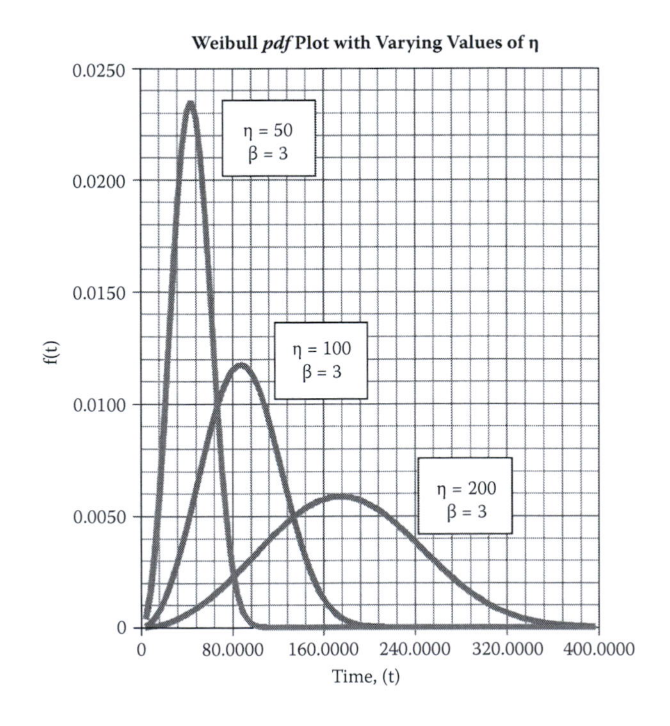


**FIGURE 6.12**
Effect of scale parameter on Weibull probability density function. (Reprinted from ReliaSoft. With permission.)

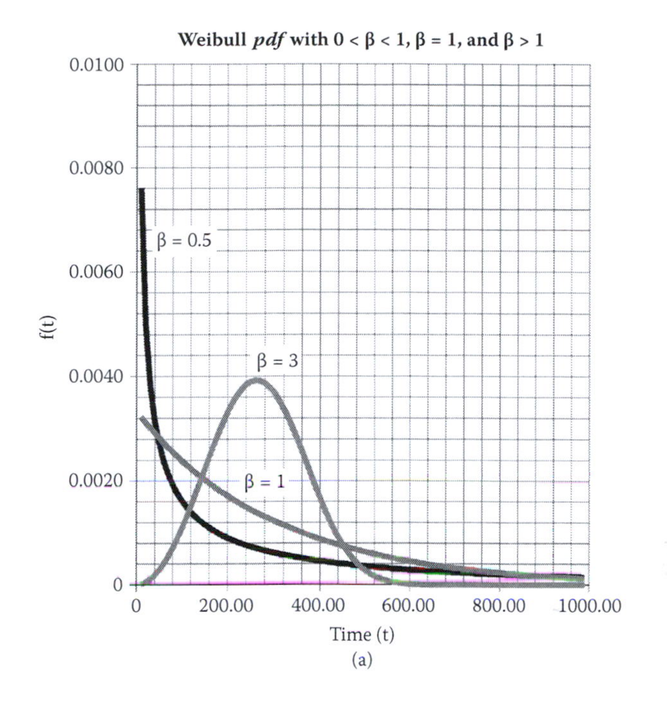

(a)

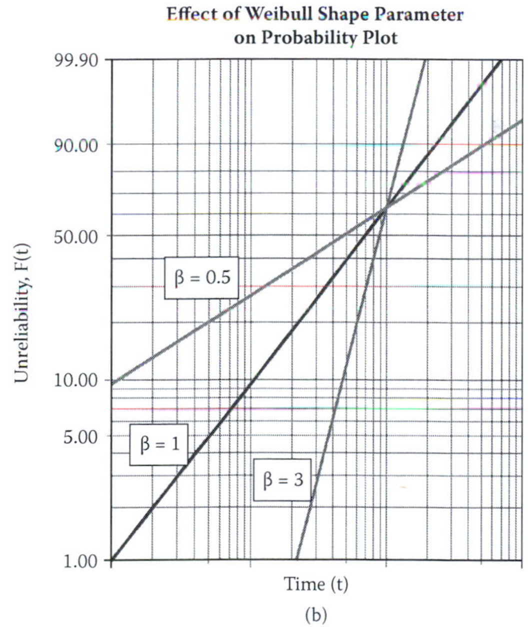

(b)

**FIGURE 6.13**
(a) Effect of shape parameter on Weibull probability density function. (Reprinted from ReliaSoft. With permission.) (b) Effect of shape parameter on unreliability. (Reprinted from ReliaSoft. With permission.)

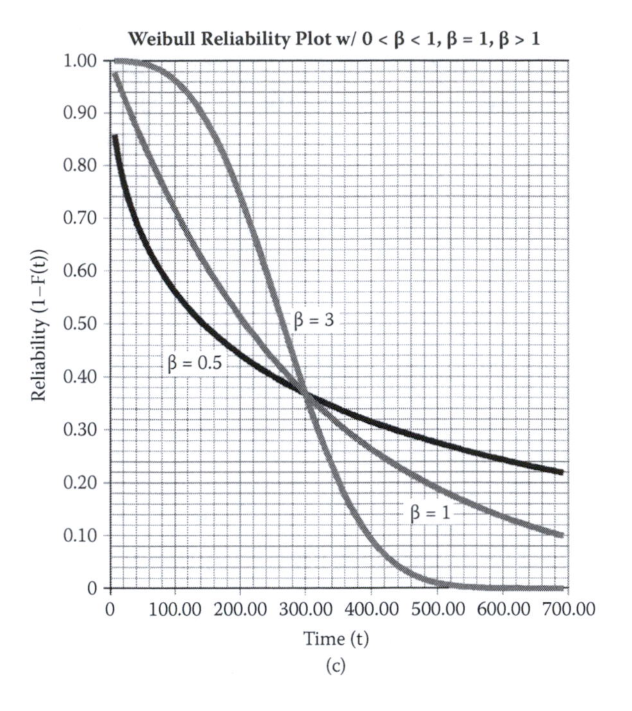

(c)

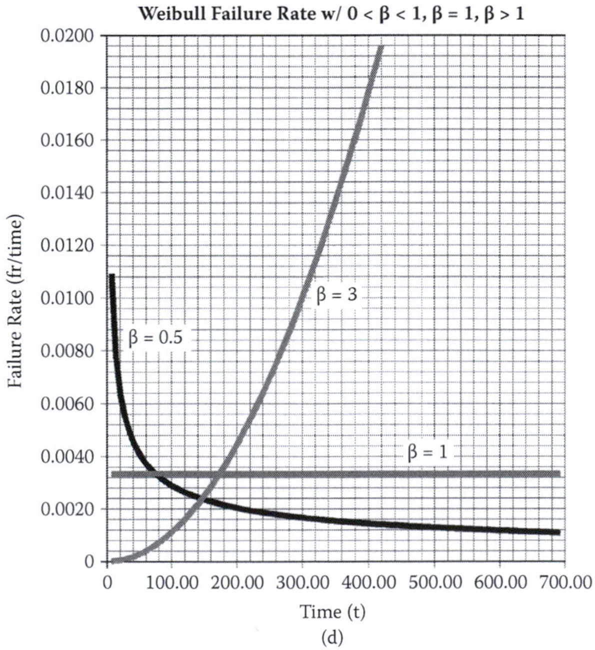

(d)

**FIGURE 6.13 (continued)**
(c) Effect of shape parameter on reliability. (Reprinted from ReliaSoft. With permission.) (d) Effect of shape parameter on Weibull failure rate. (Reprinted from ReliaSoft. With permission.)

and decrease, while the basic function shape remains unchanged, as does the initial location (the starting point or the origin) that is determined by the location parameter $t_0$. The shape parameter, β, is a dimensionless reference parameter. The effects of the shape parameter are demonstrated in Figure 6.13a to d. Figure 6.13a shows that the shape of a Weibull probability density function varies depending on the value of the shape parameter, where the probability density function, $f(t)$, is plotted against the variable time $t$. The profile of the Weibull probability density function looks very similar to that of a normal distribution when $\beta = 3$. However, the profiles of the Weibull probability density function are skewed and are drastically different from those of normal distributions when $\beta = 0.5$ or $\beta = 1$. In addition, the β value affects the slope of the regressed line when the cumulative density function, $F(t)$, is plotted on the Weibull probability paper, as shown in Figure 6.13b and c, where unreliability or reliability are plotted as a function of time. The x-axis can be just a simple logarithmic transformation of time or any other life dimension as shown in Figure 6.13b. Therefore, a numerical scale is not labeled, because it will not make any difference. The effect of Weibull shape parameter on unreliability will be exactly the same, independently from the numerical scale on the x-axis in Figure 6.13b. This can be readily observed by referring to Equation 6.13e for a two-parameter Weibull. The shape parameter plays a significant role in reliability analysis: it has a marked effect on the failure rate of a part whose properties follow the Weibull distribution. Statistical inferences can be made about a population's failure characteristics by considering whether the value of β is less than, equal to, or greater than 1. As illustrated by Figure 6.13d, populations with $\beta < 1$ exhibit a failure rate that decreases with time, populations with $\beta = 1$ have a constant failure rate, and populations with $\beta > 1$ have a failure rate that increases with time. A quick glance at the two-parameter Weibull, Equations 6.13a to e and 6.14, also shows the effects of β on reliability, which is one of the primary applications of Weibull analysis. The location parameter, $t_o$, defines the starting point or the origin where the event starts to evolve. It is also referred to as guaranteed life in reliability analysis. The probability of failure is zero until the time $t = t_o$, as illustrated by Equation 6.11.

## 6.5 Data Conformity and Acceptance

### 6.5.1 Dimension and Tolerance

The invertible fluctuation and the statistical distribution of most data often bring the following issues into discussion. The first one is the allowable tolerance. Today's advanced manufacturing technology, such as computer numerically controlled (CNC) machining, can produce very uniform parts with relatively constant dimensions, showing tighter tolerances than what the original design called for. A direct measurement of the OEM parts might produce

a very tight tolerance that makes its duplication either prohibitively expensive or technically impossible. Though truth in measurement is the basic principle of reverse engineering, an educated judgment call is often required to determine the allowable tolerance. Any relaxation of tolerance has to be justified based on either industrial standards, such as the customary best practice in the field, or documented specifications, such as the OEM's repair manual.

The standard tolerances of precision ball bearings are a good example of industrial standards widely acceptable for many reverse engineering applications. The Annular Bearing Engineering Committee (ABEC) of the American Bearing Manufacturers Association (ABMA) has published five primary precision grades, designated as ABEC 1, 3, 5, 7, and 9, respectively. The higher the grade number, the greater the precision of the bearing, and the finer the tolerance will be. For example, the tolerances on bearing bores between 35 and 50 mm range from +0.0000 to −0.0005 in. for ABEC grade 1, while from +0.00000 to −0.00010 in. for ABEC grade 9. In most reverse engineering applications, these standards provide an industrial guidance to make the necessary decisions on precision bearing dimensions and tolerances. The acceptable tolerance of a fillet radius is another example. It is a challenge to accurately measure a fillet radius bent on a thin plate. It is even more difficult to precisely determine the tolerance of the radius. Fortunately, some industrial standards have been established and listed in the engineering handbooks, such as the *ASM Handbook*, to specify the acceptable radius tolerance depending on the plate thickness. This information helps engineers make the appropriate decisions on fillet radius tolerance when necessary.

It is very important to consider the impact on the next higher or lower level assembly when making a tolerance adjustment. For example, the tolerance of a bolt hole diameter can directly affect the tightness of fastening, and that in turn can impact the bolt life cycle. The tolerance of the longitudinal length of a jet engine turbine blade can directly affect the clearance between this blade and the engine case that contains it. Even 1/10 of a 1-mm difference in tolerance can have a significant impact on the engine performance.

The second dilemma of data conformity in reverse engineering is that the measured data based on the OEM parts do not match those listed in the OEM engineering design drawing. Which dimension will prevail if the average diameter based on direct measurement is 0.49 mm, while the "unauthentic" design drawing lists the diameter as 0.50 mm? There are two options. In this situation, one can measure more parts and hope that the statistical average of the larger population will eventually converge to the value listed in the drawing. Alternatively, one can make an educated engineering judgment call based on experience and expertise.

### 6.5.2 Data Acceptance

From a deterministic perspective, quantitative engineering data are either acceptable or unacceptable. For instance, when the minimum low-cycle

fatigue (LCF) life of a part is defined as 20,000 cycles by a design criterion, if the LCF life obtained by a test is above 20,000 cycles, then this part is acceptable; if the test result is below 20,000 cycles, the part will be unacceptable. However, from the perspective of statistics, the decision of acceptance is more involved because it is based on a level of confidence along with the associated interval of confidence, instead of a singular data point.

To check the acceptance of a part made by reverse engineering, one can set the condition of acceptance of this part to demonstrate at a confidence level of 95%, a minimum LCF life of 10,000 ± 1,000 cycles, that is, with a confidence interval between 9,000 and 11,000 cycles—assuming that statistically the LCF life data will show a Weibull distribution, as they typically do. In this case, the test results of the sample parts have to meet the following condition for acceptance: show an LCF life of 10,000 ± 1,000 cycles at the confidence level of 95%. If twenty sample parts are tested and show a statistical LCF life of 10,100 ± 1,500 cycles at the confidence level of 95%, they are unacceptable because the lower LCF life boundary at 8,600 cycles based on testing is less than the allowed LCF life according to design criterion, 9,000 cycles. However, statistically, the confidence interval at the same confidence level will narrow when the test sample size increases. An unacceptable situation can theoretically become acceptable when more samples are tested because the confidence interval narrows as the sample size increases. In the above example, when the sample size increases from 20 to 50, the confidence interval might narrow to ±500 cycles at the same confidence level of 95%, provided the mean remains at the same value of 10,100 cycles. Therefore, the lower LCF life boundary based on the test results will increase to 9,600 cycles, which is higher than the minimum LCF life allowed by design, and becomes acceptable.

### 6.5.3 Source of Data

In reverse engineering, all the measured data or test results have to be confirmed with those of the OEM counterpart. Unfortunately, the baseline OEM counterpart data are usually not readily available for comparison. The design data are highly guided proprietary information. Most data of commercial parts have to be obtained by direct measurement or testing of the OEM part. The availability of the OEM part or the data affiliated with this part is vital to a reverse engineering project. Occasionally the new OEM part might be hard to obtain for reverse engineering analysis. Some special new parts can only be bought from the OEM itself, and the OEM might place a restriction on the purchase of this part. However, in most cases, new or preowned parts can be purchased from the open market. Furthermore, the Internet can also serve as an online parts locator. For example, http://www.ipls.com is a website that lists various aviation part inventories.

Most industrial specifications, such as SAE Aerospace Material Specifications (AMS) and ASTM standards, are available in the public domain. Other data depositories can also provide valuable information. The information handling

services company IHS, Inc. possesses millions of active and historical government (e.g., Department of Defense, FAA, and NASA) and industrial documents, including specifications, standards, drawings, regulations, forms, directives, and handbooks. Another example is the U.S. Defense Logistics Information Service, which creates, obtains, manages, and integrates logistics data from a variety of sources for dissemination through a due process.

Many alternative resources are available to provide the reference OEM data for comparison. The U.S. Government usually makes the data related to a part available for the public if this part is developed with public funding. Many military aircraft engines are developed with public funding, and a lot of commercial aircraft are direct derivatives of their military counterparts. For example, the civilian aircraft engines JT3D-7, CF34-3, CF6-6, JT4A-11, JT8D, JT9D-7R, and CFM-5-2A are very similar to their respective military counterparts: TF33, TF34, TF39, J75, J52, F105-PW, and F108. The derivative data obtained from the U.S. Government of a military aircraft engine identical to the civilian counterpart engine are usually deemed acceptable information to substantiate the engineering verification. However, a clear linkage between them has to be established. If the OEM adopts the same part number for both the military and civilian parts, such as some GE parts, the bridging will be relatively easy. Otherwise, an unambiguous linkage between the two parts has to be established before any comparison can be made. The OEM's illustrated parts catalog, maintenance and repair manuals can also provide valuable reference data for reverse engineering applications. However, although the allowed dimensions and tolerances listed in the maintenance and repair manuals might be appropriate for the used or alterated parts, reverse engineering is based on the dimensions and tolerances of a new part.

### 6.5.4 Statistical Regression and Relations between Mechanical Properties

It was noticed early on that the children of short parents have the tendency to be taller statistically, while those from tall parents will be shorter after proper adjustment for the effects of genes. Nature seems to have a tendency to converge to a common average in statistics. This convergence of data is the cornerstone of regression analysis. In reverse engineering, the regression methodology is used to find the best-fit value for a material property, such as low-cycle fatigue life. The regression methodology is also used to formulate a mathematical relationship between two material parameters, such as the quantitative relationship between tensile strength and hardness.

The benefit of being able to compare tensile strength using a simple nondestructive hardness test is obvious. However, engineers should also be aware of the uncertainty and potential errors that can result from this approach. The regression equations or models, such as Equation 3.25a to d, are established primarily based on statistical analysis of numerical data and often lack of knowledge of the underlying physical relations between the parameters.

As discussed in Chapter 3, hardness is a measurement of a material's resistance to plastic deformation. It is measured in various ways, for example, deformation resistance such as Brinell hardness and Vickers hardness, penetration depth such as Rockwell hardness, scratch resistance such as Mohs hardness, and rebound height such as scleroscope hardness. In general, an alloy with a higher hardness number will show better tensile strength in a qualitative sense. However, great caution is required to draw any quantitative tensile strength value based on a hardness number, or claim that two alloys will have the same tensile strength because they both show the same hardness number. The yield strength (YS) is a measurement of the stress a material can sustain at yielding. The ultimate tensile strength (UTS) is a measurement of the maximum stress a material can sustain before it fails. They both are measured in force per unit area, such as Newton per square meter (Pascal), or pound per square inch (psi). Besides Brinell hardness and Vickers hardness, most other hardness numbers are not measured in force per unit area. Even Brinell hardness and Vickers hardness are measured based on approximately indented surface areas that are different from the cross-sectional area used in tensile strength calculations. Now it should be clear that any quantitative relationship between hardness and tensile strength is only an approximation at best. Table 6.5 lists the hardness numbers of various cast irons and their corresponding tensile strengths and fatigue endurance limits (Dempsey, 2004). These data are also plotted in Figure 6.14. Qualitatively, a higher hardness number usually implies a stronger tensile strength and higher fatigue endurance limit. Quantitatively, only few empirical equations were ever validated for the relationship between hardness and tensile strength, or tensile strength and fatigue strength.

Another question often asked in reverse engineering is whether two alloys with the same tensile strength can claim to have equivalent fatigue resistance. From one perspective, these two material properties might be related because they both reflect a material's resistance to an external load. Numerous empirical relationships have been proposed between these two material properties,

**TABLE 6.5**

Hardness, Tensile Strength, and Fatigue Endurance Limit

| Material (Cast Iron ASTM No.) | Brinell Hardenss Number (BHN) | Ultimate Tensile Strength, MPa (ksi) | Fatigue Endurance Limit, MPa (ksi) |
|---|---|---|---|
| 20 | 156 | 151.7 (22) | 68.9 (10) |
| 25 | 174 | 179.3 (26) | 79.3 (11.5) |
| 30 | 201 | 213.7 (31) | 96.5 (14) |
| 35 | 212 | 251.7 (36.5) | 110.3 (16) |
| 40 | 235 | 293.0 (42.5) | 127.6 (18.5) |
| 50 | 262 | 362.0 (52.5) | 148.2 (21.5) |
| 60 | 302 | 430.9 (62.5) | 168.9 (24.5) |

*Source:* Dempsey, J. http://www.anvilfire.com, accessed April 24, 2009.

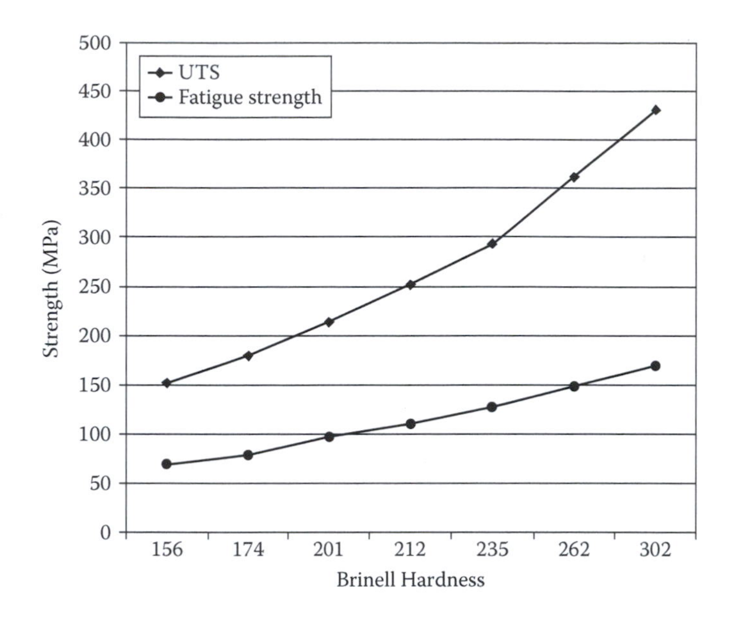


**FIGURE 6.14**
Brinell hardness vs. ultimate tensile strength and fatigue endurance limit.

and the ratio between fatigue endurance limit and ultimate tensile strength is referred to as the endurance ratio. Fatigue resistance measures a material's resistance to cyclic stress, while tensile strength reflects a material's resistance to monotonic stress. Fatigue strength is a material resistance to dynamic load, and tensile strength is a material's resistance to static load. In addition, the fatigue strength is much more sensitive to the surface finishing than the tensile strength is, and the effects of grain size on fatigue strength and tensile strength are also different. Figure 6.14 shows that both the ultimate tensile strength and the fatigue endurance limit of cast irons increase with increasing hardness. These relationships among hardness, ultimate tensile strength, and fatigue endurance provide a good qualitative reference, but the data usually cannot formulate a precise quantitative relationship among them.

By regression analysis, the fatigue endurance limits of most steels are approximately equal to 45 to 50% of the tensile strengths for smooth test specimens. However, the fatigue endurance limit might be only 20% of the tensile strength for a steel part with a badly corroded surface. The endurance ratio of Ti–6% Al–4% V titanium alloy spreads across a broad range from approximately 0.40 to 0.60 for smooth specimens (Bartlo, 1969), and the ratio is approximately 0.4 for a nickel-base superalloy Monet at 20°C. The correlation between fatigue endurance and tensile strength of aluminum alloys is known to be even less well defined. It is also worth noting that statistically, tensile strengths can be approximately represented by a normal distribution, while fatigue endurance limits usually fit the Weibull distribution better. The fatigue strength should never be derived from the tensile strength

by a simple quantitative relationship in a reverse engineering analysis without proper validation. These relations are usually material specific and also depend on other environmental and test conditions, such as temperature, loading speed, etc.

## 6.6 Data Report

Reverse engineering decodes what material the original part is made of and how it is manufactured. The deliverable of a reverse engineering project is a comprehensive design drawing that will allow an engineer to duplicate the original part. The data in a typical reverse engineering report should include the material composition, part dimensions, test results, manufacturing procedures, and all the associated specifications and standards. In contrast to a conventional machine design report, the data in a reverse engineering report should also include the baseline OEM part for comparison and conformity. Whenever possible, both the OEM specifications and the specifications call for the new parts to be attached to the reverse engineering report as part of the data package.

The test data report, including but not limited to the laboratory report, should be as comprehensive and accurate as the test samples represent. It should cover the sample source (e.g., as a purchased, used, or new part), part number, and testing methodology. Proper specifications and standards that guide the test, such as ASTM E92-82 (2003)e2, *Standard Test Method for Vickers Hardness of Metallic Materials*, or ASTM E8M-08, *Standard Test Methods for Tension Testing of Metallic Materials*, should be cited. The conclusion in the report must be based on the results from testing or analysis. A micrograph at adequate magnification showing recrystallization morphology should be included in the report to draw the conclusion that the part has a recrystallized microstructure. If only a micrograph at low magnification showing no visible recrystallization is attached to the report, or if no micrograph is attached, the conclusion of a recrystallized microstructure is based on "trust" instead of substantitation data. The relevant test conditions and environments, such as the temperature for a stress rupture test and the atmosphere for an oxidation test, should be recorded in detail. The calibration of the test instrument and its traceability should be verified. The estimated accuracy, if feasible, should be discussed. The report of false, fictitious, or fraudulent data may be punishable as a felony under federal statutes, including federal law, Title 18, Chapter 17.

It is standard practice in reverse engineering to present the dimensional measurements, such as the length, width, or radius, in statistical terms with defined means and standard deviations. These data usually would be directly referred to in machining or manufacturing. The test results can

also be reported as statistical data if sufficient tests of the same property have been performed, such as hardness numbers. However, if only limited tests have been conducted, the benefit of statistics might not be fully appreciated, and the exercise of statistical calculation might not be justified. If only three tensile tests are conducted, the utilization of a simple average tensile strength is more proper than a value of tensile strength from a reliability calculation. The report on alloy chemical composition usually only has a singular numerical percentage of each constituent element, primarily because often only one sample is used in the analysis. For critical parts, more samples might be required for the chemical composition analysis, so a report in a statistical format can be justified.

From the design and manufacturing perspectives, the average value is often used rather than the mean of a statistical data set to simplify the calculation in reverse engineering, unless otherwise justified. Theoretically speaking, the utilization of statistical data might be able to present a numerical percentage in reliability percentage. In reality, for most reverse engineering applications, the part reliability is so high that a semiquantitative conclusion based on average part strength and applied stress is often indifferent from the calculation based on statistics. However, statistics plays a much more prominent role in field management when a part fails and the development of an inspection schedule is required.

## References

Bartlo, L. J. 1969. Effect of microstructure on the fatigue properties of Ti-6Al-4V bar. In *Fatigue at high temperature,* ed. L.F. Coffin, 144–54. ASTM STP 459. Philadelphia: ASTM International.

Dempsey, J. http://www.anvilfire.com (accessed April 24, 2009).

# 7

# *Part Performance and System Compatibility*

The performance of a reverse engineered part compared to its OEM counterpart is vitally critical to the success of a reverse engineering project. The performance of these parts is usually evaluated based on three primary criteria: engineering functionality, marketability, and safety. From an engineering functionality perspective, part performance is judged based on its structural integrity and system compatibility. It is essential for a part to be free of structural defects. A structural defect can be quantitatively measured in terms of strength reduction from either material deficiencies or dimensional deviations. The performance of a commercial part reinvented using reverse engineering is also determined by the market. Many automotive spare parts produced by reverse engineering are evaluated primarily by the market acceptance. These parts are usually mass produced at a very competitive price. Marketability is the key to the success of these parts. However, for certain parts that affect public safety, engineering evaluation and market acceptance will only be part of the approval process. The assurance of public safety takes precedence. A regulatory government agency is usually entrusted by the people and Congress to evaluate and approve these parts to ensure their safety applications. The regulatory oversight is usually discretionary and often auditory in nature. Most reverse engineered parts for aviation applications are in this category. These parts are usually produced in limited quantity at a relatively high cost. This chapter will first focus on the technologies used to evaluate the reverse engineered parts, followed by a brief discussion on regulatory requirements. It is just a conceptual discussion on regulations in general instead of a comprehensive coverage on any specific certification process of a reverse engineered product.

The reliability of an OEM part depends on the degree of built-in redundancy based on the original design, and safety margin above the minimum performance requirements. However, the baseline in reverse engineering is beyond the minimum performance requirements and derived instead from the performance of the existing OEM part. For example, the minimum low-cycle fatigue life of an automobile shaft is defined as 200,000 cycles according to the design criterion. If the OEM shaft actually has a low-cycle fatigue life of 250,000 cycles, the performance requirement is raised to 250,000 cycles for a similar shaft reinvented by reverse engineering. The performance criteria in reverse engineering are based on the original part, instead of the original minimum design requirements.

Another unique quality requirement for a part produced using reverse engineering is that it must demonstrate cohesive system compatibility. Figure 7.1 shows an automobile engine that works properly only when all the parts are operating cohesively. This engine model is displayed at the SAE Automotive Headquarters in Troy, Michigan. A close collaborative interaction between the shaft, gears, and their surrounding parts to ensure proper system compatibility is mandatory. Any engine part produced using reverse engineering has to show satisfactory system compatibility before it can be installed as a substitute OEM part. An OEM component is part of a system that originated from the same invention idea, design concept, and manufacturing process. The interaction of any individual OEM part and its surroundings is usually fully harmonized during the design and manufacturing processes. However, the system compatibility of a substitute part later reinvented using reverse engineering has yet to be tested to demonstrate that it can fit in the existing system. Adverse effects on cohesive interaction might result from manufacturing variation, dimensional misfit, or configuration alteration. Figure 7.2 shows the electric system of a Toyota Camry car displayed at the Automotive Technology Center of Massachusetts Bay Community College. It is a real-life model with the head pointing to the left and the steering wheel in the center. It further demonstrates that all the functions in an automobile are integrated together through a network operating as a unit. System compatibility of every component is critical.

The system capability of a reverse engineered part has to be thoroughly diagnosed by its differential and integral characteristics relative to its surroundings.

**FIGURE 7.1**
An automobile engine system.

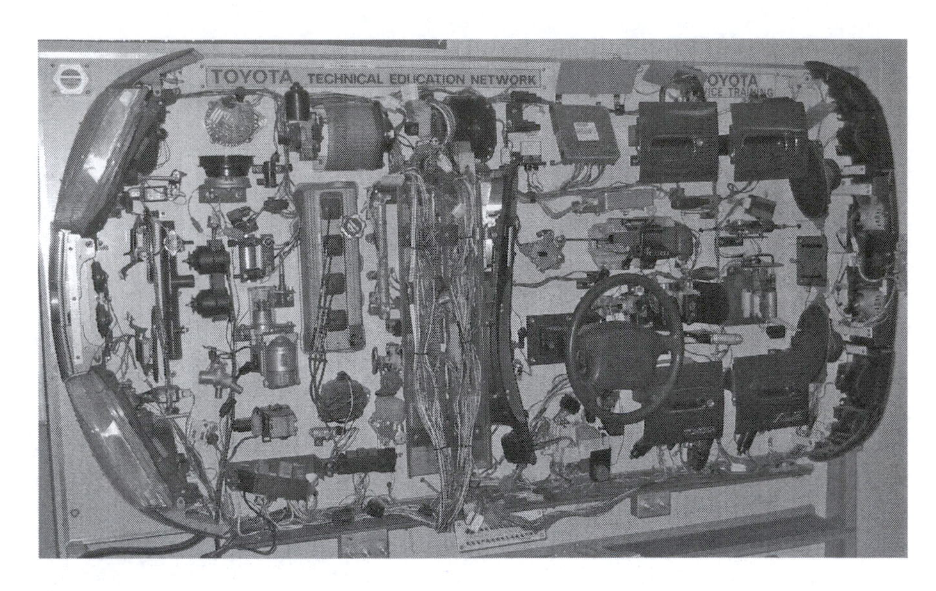


**FIGURE 7.2**
The electric system of a Toyota Camry car.

## 7.1 Performance Criteria

The performance of a part can be evaluated against either engineering criteria or regulatory requirements. The engineering criteria are usually based on scientific analyses and engineering tests, with a set of minimum performance requirements, such as room temperature tensile strength, cyclic load resistance, and tolerance to severe environmental attack. The regulatory requirements focus on safety and legality. Various federal regulatory agencies have published a variety of guidances to guide their respective industries. These two sets of performance requirements are usually tied closely to each other.

The following case exemplifies a typical part performance requirement and the challenge to reverse engineering. In the aviation industry, a life-limited part is usually designed based on its B.1 life. When a part is listed to have a life cycle of 15,000 hours, this usually implies that the OEM has claimed that only 1 of 1,000 parts (defined as the part's statistical B.1 life) will fail at 15,000 hours. The failure of a life-limited part is often defined as when a crack of 1/32 inch in length is detected. The OEM usually does not publish the methodology it uses to calculate the life cycle for this part, or the safety margin it integrates into this life prediction. A new part reinvented using reverse engineering might predict the same life cycle based on calculation with a different methodology. It is very difficult to claim the same safety margin for the new part, partially because two different methodologies are used and partially because the OEM's safety margin is never published. Furthermore,

in case this part fails and a safety management plan is required, two different field management plans might be needed. One is applicable to the OEM part, and another is specific for the part reinvented using reverse engineering due to the following reasons. Based on OEM's methodology and safety margin, it might conclude that there is a $10^{-7}$ chance the next failure will occur after 900 hours. Therefore, a mandatory inspection is required before 900 hours. However, based on another methodology and safety margin, it might conclude that the part reinvented using reverse engineering will have a $10^{-7}$ chance to fail after 800 hours. As a result, a field management plan has to require these parts to be inspected at 800, instead of 900, hours. This highlights that a part reinvented using reverse engineering often requires a new set of maintenance and repair manuals.

The aviation industries call most the aerospace-grade non-OEM parts Parts Manufacturer Approval (PMA) parts. PMA is a Federal Aviation Administration (FAA) design-and-production approval. It is a governmental certification system that marks a part that meets FAA regulatory performance requirements. A PMA part is an FAA-certificated part that satisfies FAA safety regulations and the fabrication inspection system. Many PMA parts are reinvented using reverse engineering.

The FAA has issued many orders, policies, advisories, and circulars to guide the aviation industries on PMA parts. The PMA industries also established various associations to advocate PMA parts and promote aftermarket business in recent years. The Modification and Replacement Parts Association (MARPA) was established in 2001. MARPA is a nonprofit trade association that advocates the interests of the PMA community. It works with its members and the FAA on PMA issues. The PMA-related subjects, from engineering challenges and regulatory requirements to business and marketing, are also being discussed in many open forums, most notably the Gorham International Conferences. In contrast to MARPA, which is an advocate of the PMA industry, the presentations at the Gorham International Conferences usually cover a wide range of perspectives.

Similar to aviation industries, rigorous government oversight through regulatory requirements and mandatory compliance on medical devices is also imposed on the medical device industries. In contrast, the spare parts and the aftermarket of automotive industries are more matured and less government regulated. Most of the oversight for reverse engineered automobile parts relies on the industry's own regulations.

### 7.1.1 Test and Analysis

When evaluating the performance of a mechanical part, there is nothing more critical to safety than mechanical strength. The part has to demonstrate that it has sufficient durability during normal operating conditions. Any premature part failure might cause safety concerns and could be costly.

Two methods are usually applied to show a part's durability and capability to properly perform its design functionality: test and analysis.

The most convincing way to demonstrate that the performance of a reverse engineered automobile engine valve is equivalent to its OEM counterpart is to conduct a side-by-side direct comparative test. However, in many cases, it is not practical to conduct this direct comparitive test. For example, the direct comparative test for an automobile crankshaft, as shown in Figure 7.3, or hypoid axel gear would be quite complicated and costly. A direct comparison of turbine engine shafts could also be prohibitively expensive. As a result, alternative analytical approaches are often considered whenever feasible.

A test provides direct evidence and a clear pass or fail. An analytic approach has inherent uncertainties and inaccuracies. It draws engineering conclusions based on indirect evidence and inference. One requirement of a jet engine high-pressure turbine blade is to show that the blade will have limited creep extension. Creep extension can affect the clearance between the blade tip and the turbine case and can cause engine stall, a serious safety concern in aviation. Let us assume that the OEM blade requires a creep extension less than 0.1 mm in 1,000 hours at the operating temperature 600°C. Both the PMA and OEM blades can be placed side by side to run an engine test at 600°C for 1,000 hours, and compare their respective creep extensions afterward. In case a direct comparative test is not feasible, engineers can still compare the creep resistance between the OEM and PMA blades by analysis. The PMA blade creep extension can be calculated based on creep theories by plugging in the relevant material parameters, such as diffusion coefficients and grain size. However, as discussed in Chapter 4, the mechanisms of creep are very complicated. The controlling parameter of creep rate varies depending on the applied stress and temperature. The effects of grain size on creep rate are also very complex. At low stress and high temperature, the creep rate is proportional to the applied stress and inversely proportional to the

**FIGURE 7.3**
Crankshaft in an automobile engine.

third power of grain size. At moderate stress and moderate temperature, the creep rate is proportional to the applied stress and inversely proportional to the square of grain size. The linear relationship between creep rate and applied stress will not hold in the high-stress and low-temperature region because the controlling mechanism of creep will shift from diffusion to sliding and dislocation movement. An accurate quantitative calculation of the creep extension requires a lot of substantiation data.

The theoretical calculation is often just an estimate, but a direct engine test can be prohibitively expensive. Should a direct comparative test be mandated to demonstrate equivalent high-pressure turbine blade performance? There are no standard answers to these types of frequently asked questions. Many factors affect the answers: part criticality, consequence of inferior performance, corporate knowledge, engineering expertise, accuracy of analysis, and costs. The best approach is always to reach a consensus among all stakeholders in advance. Most often, a combination of test and analysis approach is adopted for part performance evaluation.

Before an analytical calculation or laboratory test is conducted, the possible failure modes of the part should be thoroughly reviewed. The most vulnerable failure mode will determine the test method. Few mechanical parts fail due to static load such as tensile stress. However, the tensile stress plays a critical role in disk burst resistance under high-speed rotation. Tensile strength also represents a generic mechanical strength on many occasions. To conduct a tensile test and use the material's tensile properties to substantiate the equivalent performance of a reverse engineered part is usually acceptable as a first comparison. Other mechanical tests are also frequently required for critical parts and different failure modes.

Many mechanical parts are subject to cyclic stresses, and fatigue is their primary failure mode, such as the springs installed on the automobile piston engine and suspension, as shown in Figure 7.4a and b, respectively. These parts fail at a stress level well below the material's tensile strength. Because most materials lack well-defined quantitative relationships between their tensile and fatigue properties, fatigue tests are often required to obtain the accurate fatigue strength. Three challenges often face engineers when conducting fatigue tests: part size, part geometry, and test sample size. The size of some parts is relatively small; for example, the high-pressure turbine blades can be as small as only 2 to 3 cm in length. They are too small for standard fatigue test specimens. A fatigue test using subsize specimens should be consulted with all stakeholders, based on ratified reference and prior experience whenever feasible.

The material fatigue strength does not automatically reflect the part fatigue resistance to cyclic load, primarily because of the effect of part geometry. Irregular geometric shape or abrupt dimensional change such as the fillet radius raises local stress. These effects are quantitatively represented by the ratio between the local stress and average stress, defined as stress concentration factor. Since the effects of stress concentration factor on tensile and

(a)

(b)

**FIGURE 7.4**
(a) Valve springs in an automobile piston engine. (b) Suspension spring in an automobile.

fatigue properties are different, a fatigue stress concentration factor has to be used in fatigue analysis. This raises another frequently asked question in reverse engineering: whether a test of material coupon can be used to substitute the test of a part. In case it is decided to test material coupon instead of the part itself to evaluate the fatigue resistance, it is essential to assure that both the reverse engineered part and the OEM part have indistinguishable geometry, particularly dimension-wise. All the bending radii of the two parts should be identical or close enough to minimize any potential effects due to the discrepancy. These two parts should be manufactured by the same process. For example, both are either milled or grinded. Different manufacturing processes produce different surface finishings that can dramatically affect part fatigue resistance. Both parts should show identical or very similar microstructures. They both should have similar grain morphology such as grain size and grain texture. They should also show similar grain boundary configurations in terms of grain boundary network and precipitations along the grain boundary. Finally, both parts should have very similar amounts and distributions of strengthening precipitate, second phase, and, if any, inclusions.

Computer modeling and simulation are widely used in part performance evaluation. The accuracy and validity of modeling or simulation results depend on the software algorithm and its mathematic architecture. With the same input data, a three-dimensional computer modeling might predict a stress concentration factor that is five to ten times higher than a similar two-dimensional computer modeling's prediction. Generally speaking, a two-dimensional computer modeling is acceptable for a part with two-dimensional symmetry, such as a plate; however, a three-dimensional computer modeling is recommended for a part with three-dimensional features, such as a disk with bolt holes and fillets.

How many tests are required? Statistically, a high confidence level and a narrow confidence interval require a lot of tests. Fatigue properties in particular require tens, even hundreds, of tests to build a reliable database because of their inherent scattering nature. To establish a material specification or reverse engineer a critical part, these tests are mandatory. For a noncritical, non-life-limited part, a few fatigue tests are acceptable. Like many questions in reverse engineering, there are no cast-in-stone standard answers to this question. It is part specific, material specific, and also affected by many other factors, such as cost and marketing. A reasonable argument sustained by scientific data will usually prevail.

The immediate failure of a mechanical part soon after it is put in service, so-called infant death statistically, does not occur too often. The results of most field performances only come after a certain service period. The performance of a protective coating will not be fully tested until it is placed in the operating environment for a long period of time because the corrosion attack or even the stress corrosion cracking usually will not be observed until years after. Theoretical analyses on resistance to environment degradation,

as detailed in Chapter 4, can be applied whenever appropriate. Nonetheless, extrapolating a long-term effect based on short-term test data will always introduce some uncertainty. Though few reverse engineering projects mandate a real-life corrosion or stress corrosion cracking test, proper evaluation of a part's environmental resistance is critical, and the following section will focus on this subject.

### 7.1.2 Environmental Resistance Analysis

One of the biggest tasks facing evaluation of environmental resistance is the lengthy incubation period due to the slow progress of corrosive attacks and the difficulties in detecting the damage. To design an accelerated test providing short-term data for long-term performance prediction is a major endeavor when conducting part performance evaluation in reverse engineering. The observable environmental degradation of a part usually will not appear when the part is first placed in this environment. Accelerating corrosion tests often simulate the worst-case scenario, imposing a very aggressive corrosive environment. As a consequence, accelerated corrosion might occur at the onset, which is both contradictory to the real-life corrosion progress and unpredictable. Great caution has to be exercised to ensure that the corrosion mechanism is as similar as possible to the real-life case. Additional corrosion mechanisms can be inadvertently introduced in an accelerating test in the laboratory setting. For instance, using a metallic bolt to hold a test specimen in the laboratory test can create a galvanic couple and lead to spurious results.

Three of the most commonly used corrosion test settings are the laboratory, field, and service. Accelerating tests are difficult to conduct during field and service tests. Thus, most accelerating tests are in a laboratory setting. The results of these simulating tests heavily rely on the test specimen, corrosive medium, and test conditions. Corrosive behavior can be influenced by material microstructure and specimen surface finishing condition. The laboratory setting for an accelerating test should resemble the service conditions and environment as closely as possible to ensure similar corrosion mechanisms. If a part is manufactured by forging but the test coupon is prepared by machining, their different microstructures can result in different corrosive rates. The presence of scales or surface marks induced during test specimen preparation can also lead to different corrosion rates.

There are numerous factors that can affect corrosion rates. Several common factors and methods used for accelerating corrosion tests are listed below (Guthrie et al., 2002):

Aeration for immersion tests

Temperature

Electric current through the parts

Load for stress corrosion cracking tests
Acidity in immersion tests
Amounts of $NO_2$ and $SO_2$ for atmospheric tests
Relative humidity for atmospheric tests
Impingement velocity in erosion corrosion tests

The adjustment of test temperature is the most commonly used accelerating factor, and it usually has a noticeable effect for many corrosion mechanisms. Additionally, the adjustment of the corroding electrolyte and the ionic conductivity can also accelerate the corrosion rate.

A number of standards on corrosion characterization and test methods are published by ASTM and various other professional organizations. The corrosion rate can be measured by visual inspection, weight loss, or electrical or electrochemical resistance. For general corrosion tests of metals, ASTM Standards G31 and G50 provide guidance on immersion corrosion testing and atmospheric corrosion testing, respectively. For corrosion testing in freshwater, there are approximately forty ASTM standards providing guidance. For salt spray testing, ASTM Standard B117, *Standard Practice for Operating Salt Spray (Fog) Apparatus*, is one of the most widely referenced guidelines for continuous salt spray corrosion testing. It is an excellent reference document with a lot of instrumental information on the apparatus, procedures, and conditions required to create and maintain the salt spray test environment. However, ASTM B117 cautions engineers that the correlation between the corrosion observed in the laboratory tests and the corrosive damages observed in the natural environment is not always predictable. This standard further points out that the reproducibility of results in the salt spray exposure is highly dependent on the type of specimens tested and the evaluation criteria selected, as well as the control of the operating variables (ASTM, 2009a). Discrepant results have been observed when similar specimens are tested in different fog chambers, even though the testing conditions are nominally similar and within the specified ranges. Therefore, correlation and extrapolation should be considered only in cases where appropriate corroborating long-term atmospheric exposures have been conducted. ASTM F2129 provides guidance on the test method for conducting cyclic potentiodynamic polarization measurements to determine the corrosion susceptibility of small medical implant devices (ASTM, 2008). Many of these implant devices are potential candidates for reverse engineering. Table 7.1 summarizes some corrosion tests and their respective reference standards.

The interpretation of any test result in general, accelerated corrosion tests in particular, requires complete knowledge of material characteristics, specimen size and geometry, testing methods, and the environment. Any conclusion based on generalized material type, such as stainless steel, without specifying its metallurgical details in terms of composition and grain morphology will be error prone.

**TABLE 7.1**

ASTM Standards for Corrosion Tests

| Type of Test | Measurement and Application | ASTM Standard |
|---|---|---|
| Electrochemical | Potentiodynamic polarization resistance for general corrosion of metals | G59 |
| | Electrochemical impedance for general application | G106 |
| | Cyclic potentiodynamic polarization for localized corrosion susceptibility of iron-, nickel-, or cobalt-base alloys | G61 |
| | Galvanostatic measurement for repassivation potential of aluminum and its alloys | D6208 |
| Immersion | Laboratory immersion corrosion testing of metals | G31 |
| | Alternate immersion exposure of metals and alloys in neutral 3.5% sodium chloride solution | G44 |
| Salt spray | Relative corrosion resistance for specimens of metals and coated metals exposed in a simulated natural environment | B117 |
| | Modified salt spray (fog) testing simulating various variations of corrosive environments for ferrous and nonferrous metals; also organic and inorganic coatings | G85 |

The measurement of weight loss vs. time is the most convenient method to assess corrosion rate. For uniform general corrosion, the corrosion rate is usually proportional to the weight loss and inversely proportional to the area, exposure time, and density of the material. Unfortunately, this method fails to consider the effects of localized corrosion, such as pitting, crevice, and intergranular attack. The corrosive damages due to these localized corrosions can be severe, particularly in accelerated corrosion tests. The different rates of surface film or scale formation between accelerated and natural corrosion tests can further complicate the damage assessment and the correlation between their respective test results.

Figure 7.5 shows the corrosion at a bolt joint. Several forms of corrosive attack can occur simultaneously at a tightly joined connection, including crevice corrosion that is often observed in many mechanical joints by bolts, nuts, rivets, clamps, and other types of fasteners. Many of these mechanical components, such as the bots and nuts, are among the most popular parts for reverse engineering. Crevice corrosion is a localized form of corrosion usually associated with stagnant electrolyte chemistry and corrosive environment. The local electrochemical conditions and the details of narrow openings under washers or between the bolts and the flanges make the simulation of crevice corrosion in a laboratory setting very difficult. An accelerated test rarely produces reliable data for the life prediction of a part subject to crevice corrosion in a real-life service geometric configuration and environmental condition.

Pitting corrosion is highly localized. Many passive metals such as stainless steels and corrosion-resistant alloys will not corrode broadly, but are subject

**FIGURE 7.5** **(See color insert following p. 142.)**
Corrosion observed around a bolt joint.

to random pitting corrosion. Sometimes pitting corrosion can be quite small on the surface and very large below the surface, and cause unexpected catastrophic system failure. To simulate an accelerated test to quantitatively predict pitting corrosion behavior is very difficult, partially because it is not easy to quantify the rate of pitting corrosion.

The sensitivity of intergranular corrosion to heat treatment and the complexity of kinetics involved with segregation and precipitation that induce intergranular corrosion have made accelerated test of intergranular corrosion almost impossible. However, a proper selection of alloy in reverse engineering and careful control of the heat treatment process will help eliminate this type of corrosion.

The electrochemical process of corrosion is complex, and the corrosion rate depends on myriad physical and chemical parameters. Different degrees of corrosive attack are often observed at different locations on the same part seemingly exposed to the same corrosive environment. Figure 7.6 shows typical uniform general corrosion on a steel tube. Most areas show relatively uniform corrosion, while an unusually severe corrosive attack at one location has penetrated through the tube thickness. Multiple corrosion mechanisms can be activated; therefore, different failure modes predominate in accelerated and natural tests. This might require the use of different corrosion models in corrosion rate calculation and introduce different confidence levels in

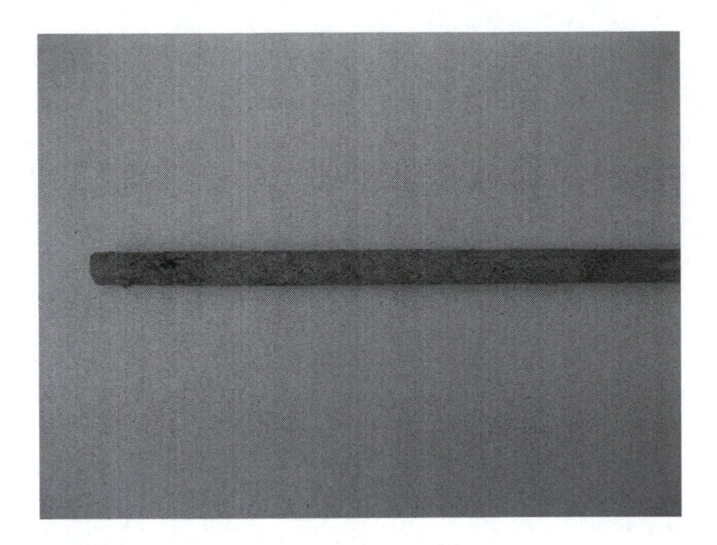

**FIGURE 7.6** (See color insert following p. 142.)
Uniform corrosion on a steel tube.

damage assessment. Despite that the application of accelerated corrosion tests presents an attractive technique for part performance evaluation in reverse engineering to expedite the test and potential cost savings, great caution has to be exercised to draw meaningful correlations and predictions.

Most accelerated corrosion tests cannot accurately predict the corrosion rate or part corrosion life. These accelerated tests should be used to provide qualitative insight into how materials will react in a corrosive environment, instead of being used as a quantitative prediction of part performance. Specific analytical method, test, and modeling are often applied for individual cases due to the complexity of these subjects. *ASM Handbooks* provide a comprehensive discussion on corrosion in a set of three volumes: 13A (ASM, 2003), 13B (ASM, 2005), and 13C (ASM, 2006). Volume 13A introduces the fundamental principles of corrosion mechanisms, their testing methods, and corrosion protection. Volumes 13B and 13C focus on specific materials, environments, and industries.

## 7.2 Methodology of Performance Evaluation

The performance of a part can be evaluated using either a conventional deterministic method based on a decisive test result or a probabilistic method based on statistics of probability. Regardless of the method chosen, an unambiguous pass/fail criterion should be clearly defined in advance. It is also advisable to reach a consensus on this criterion among all stakeholders

beforehand. The ideal pass/fail criterion should be simple, direct, and quantitative. It should require minimum subjective interpretation of the test result. However, an objective evaluation sometimes can be a daunting task, as exemplified by the following case. To ensure proper mechanical strength, the resistance to grain growth of the alloy of a high-pressure turbine blade above a specified temperature, say 1,200°C, is critical. However, engineers have yet to establish an accurate method to precisely determine the grain size. It is noted that no mechanical parts with a polycrystalline microstructure have ever been produced with one single uniform grain size throughout the part. Usually the grain size of a part is either reported as an average value or presented with a distribution profile. Furthermore, the parameter, either the diameter of the grain or the intercept length from one side to another of the grain, used to measure the average grain size varies depending on the method adopted. A linear one-dimensional method will produce an average grain size different from those measured with a two-dimensional or three-dimensional method due to the irregularity of the grain geometric shape and the location of section. Any measurement of grain size or the quantity of a second phase in the matrix is semiquantitative at best. Therefore, the OEM usually establishes a set of micrograph databases for comparison. If an alloy shows a finer grain size after being exposed above 1,200°C compared to this set of standard micrographs, it is deemed acceptable. Otherwise, the part will be disposed. Comparison between two sets of micrographs is very subjective. The conclusion also depends on the standards, which are most probably different between the OEM and other firms. The determination of whether two sets of micrographs have "equivalent" grain morphology is often a judgment call left to the engineer's discretion. Unfortunately, this type of visual evaluation is still being practiced in the aviation industry today.

Traditionally, the deterministic method has been the predominant method used in most machine design projects. When a shaft is tested for low-cycle fatigue endurance, if it passes the predetermined pass/fail criterion, this part and all its peers are deemed to meet the performance requirements. Sometimes, it is mandatory to have this type of unambiguous test. For a single-engine fighter jet, the turbine engine has to operate properly at all times. Any other engine performance standard, even just with 0.1% probability to fail, is not acceptable. Therefore, the deterministic method is still the preferred method to test engine performance. All the tested engines are expected to fully meet the performance requirements, and the only acceptable statistic is 100%. For an automobile engine, it might be acceptable when 99.9% of the engines meet the performance standards. In this case, probabilistic methods can play a more significant role in part evaluation.

### 7.2.1 Test Parameter

Many parameters affect a part's functionality. These parameters might be metallurgical, mechanical, or functional in nature. The metallurgical properties

relate part performance to its chemical composition, microstructure, and manufacturing process. Most metallurgical tests are conducted in a laboratory, although various portable X-ray detectors are available for on-site composition verification in the field today. The mechanical properties focus on the relationship between the load applied to a part and the strength to resist the externally applied load. These properties are based on the principles of elasticity and plasticity. The alloy microstructure is a delicate bridge between metallurgical and mechanical properties. The principles of metallurgy explain the evolution of an alloy microstructure, and this alloy microstructure in turn is used to explain the variations of mechanical characteristics. A metallurgical solution and aging treatment induces a semicohesive precipitation microstructure in an Al–4% Cu alloy, and this microstructure in turn enhances the tensile strength of the Al–4% Cu alloy through precipitation hardening, that is, an elastic-plastic strain-hardening mechanism. As a result, many metallurgical parameters, such as solution and aging temperatures and grain and precipitate sizes, are directly related to the mechanical parameters, such as tensile strength and fatigue endurance. Most mechanical properties are also evaluated in a laboratory setting. In contrast to metallurgical and mechanical properties that are tested on individual parts or subcomponents, such as corrosion resistance of an engine valve or the wear resistance of a piston engine cylinder, part functionality is usually measured in an assembled environment, in terms of engine thrust, engine noise, or exhaust gas pollution from the engine. Different parameters are evaluated by different tests. For example, the corrosion resistance of a fuel pump is evaluated by a metallurgical test, its wear or fatigue resistance is evaluated by a mechanical test, and the fuel flow rate is tested in a functionality test.

A full-scale machine test with the concerned part installed to directly demonstrate the part performance might not be feasible due to either technical or financial restraints. Alternative methods are usually used to indirectly prove that the part has met the required performance standards. These alternate methods range from comparative analysis based on theories, bench tests, to block tests. However, the first step for every method is to determine the parameters that need to be evaluated.

The following is a typical thought process to determine the parameters that have to be evaluated for the performance of a shaft. There are three critical considerations: mechanical load, operating environment, and system compatibility with the surrounding parts. The mechanical load that a shaft is subject to will determine its life cycles and therefore operation safety. Presumably this is a rotating shaft that transfers power from one component to another, and is supported with two bearings at both ends without any other radial load in between. In this simple configuration, the primary stresses this shaft is subject to are shear stress resulting from torque, and the tensile and compressive stresses resulting from bending during the rotation. If a theoretical analysis is used to evaluate the performance of this shaft, the shaft geometrical dimension has to be accurately measured

to quantitatively determine the cross-sectional area and moment of inertia, which are critical in the calculation of part strength and deflection. The geometrical and dimensional measurements can be done relatively easily with today's advanced metrological technology. Special attention needs to be paid to abrupt geometrical changes, such as a keyway, that will sharply raise the stress concentration factor and might have a drastic adverse effect on part performance. The surface finish has to be carefully examined because it can have a significant impact on fatigue life. Theoretically speaking, the material strength of this shaft should be directly measured from a sample cut from this shaft. The material strength heavily depends on the microstructure and other metallurgical characteristics, which in turn rely on the manufacturing and production process. The data extracted from mechanical or material reference books might not accurately reflect the true material strength despite having the same nominal chemical composition, similar heat treatment, and manufacturing process. It is always advisable to verify the material characteristic data extracted from any reference book before using it in performance evaluation.

A certain minimal number of tests are required to obtain a set of statistically sound tensile, compression, and shear strengths. The issue is whether it is feasible to actually conduct adequate tension, compression, and torsion tests with the specimens cut from the shaft due to the availability of the OEM part, the part size for a standard-size test coupon, and other factors. A compromise in reverse engineering is to conduct limited tests on the specimens directly cut from the part and compare the test results with the reference data to make sure the application of the reference data in this specific analysis is appropriate.

The tensile, compression, and shear strengths are the fundamental material properties that are used for material selection and part design to ensure proper static strength and acceptable deformation. The most probable failure mode for a power-transmitting rotating shaft is fatigue. It might be possible to theoretically derive fatigue strength from measured tensile strength, though this derivation is material specific and requires sufficient substantiation data.

When the material fatigue data are extracted from a reference book to calculate the part fatigue strength, proper modification of the reference data might be required. Many material fatigue strengths reported in a reference book are based on the test with specific $R = \sigma_{min}/\sigma_{max}$ ratio, usually –1, such as in a sinuous stress profile whereby the absolute value of the minimum stress in compression, $\sigma_{min}$, is equivalent to the maximum stress in tension, $\sigma_{max}$, but with the opposite sign. The real-life alternative cyclic stresses a rotating shaft is subject to rarely follow any uniform profile. The $R$ ratio in a service condition fluctuates widely instead of staying with a single value of –1. Therefore, the material fatigue strength cited from the reference database established with $R = -1$ has to be properly revised before applying it to the mathematical equation to calculate the fatigue life cycle of the rotating shaft. In case the sample test coupons are cut from the shaft for laboratory testing,

it is essential to ensure the samples are representative for the entire shaft. The sample size should meet the specification requirements, and multiple samples should be extracted from the part from various orientations and locations. The other alternatives are to conduct a bench test or block test to simulate the part operation conditions and ensure the part has the required fatigue strength. A real-life shaft test in a similar operating condition will obviously provide the most convincing proof of the part fatigue strength, and therefore the test is always recommended whenever it is feasible, particularly for a life-limited critical shaft.

It often requires more discussions and elaborations to determine what part parameters need to be tested in the postproduction evaluation on part functionality and system compatibility than what data need to be obtained to manufacture this part in reverse engineering. A hardness value might be the only test parameter required to demonstrate that a simple replacement part such as a retainer, bracket, or bushing will perform equivalently to the OEM part for its designed functionality. Additional tensile and fatigue tests are usually sufficient for typical bolts and nuts produced using reverse engineering. Corrosion, oxidation, and other environmental degradation resistance tests might also be in order if these bolts and nuts are used as fasteners for a machine operating in a corrosive or other severe environment. The elastic modulus and the spring constant most probably should be measured for a reverse engineered spring that will be used to adjust the engine valve operation in an automobile engine.

### 7.2.2 Test Plan

A more comprehensive test plan of part performance should be scheduled when the alloy composition or fabrication process cannot be fully verified in reverse engineering, as exemplified in the following cases. The laboratory analysis often identifies that the OEM part is made of an aluminum alloy having a composition within the range of a typical 2024 aluminum alloy, but cannot verify the exact composition range. It is not uncommon that the OEM modifies a commercial alloy with minor alloying elements to meet the specific performance requirement of the part. For instance, small variations in the amounts of copper and magnesium in a 2024 aluminum alloy can change the alloy tensile strength, which might be a specific design factor imposed by the OEM. A small amount of lithium added to a typical commercial 2024 aluminum alloy will increase its rigidity and tensile strength but decrease its ductility and reduce the density. The laboratory analysis might be able to detect these subtle alloy modifications, but still classifies it into the nominal 2024 aluminum alloy category. These minor discrepancies should trigger an inquiry into more performance tests.

Welding is one of the few fabrication processes that cannot be fully decoded using reverse engineering. Most coating processes are patented, and the details of these processes are proprietary information guarded

closely by the manufacturer. The detailed heat treatment schedule still cannot be completely extracted using reverse engineering despite the advancement in modern analytical technologies. A part joined by welding, coated with protective material or heat treated, should be carefully evaluated with a well-thought-out test plan.

Should a wear resistance test be conducted? Is a tensile test needed? Will a low-cycle fatigue test have to be conducted? There are no standard textbook answers for these questions. The performance of each part has to be evaluated independently based on its criticality, design functionality, and service environment. For example, a wear resistance test is not a commonly required test for most reverse engineered parts, but it is essential to verify the performance of a reverse engineered automobile brake disk, as shown in Figure 7.7a. The wear resistance test is also recommended for the automobile engine cylinder interior surface, as shown in Figure 7.7b. The burst resistance of a rotating compressor disk in an aircraft engine is dependent on the

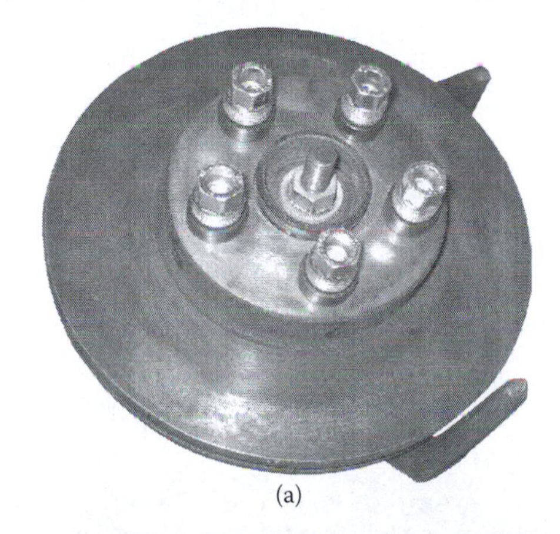

(a)

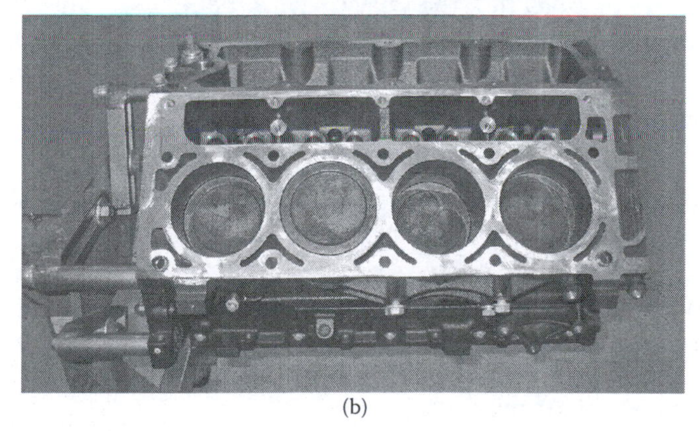

(b)

**FIGURE 7.7**
(a) Automobile brake disk. (b) Automobile engine cylinders.

tensile strength of the material used to manufacture this disk. Therefore, in the test plan, a tensile test might be required to obtain the tensile strength of the alloy if the disk burst resistance is evaluated based on theoretical calculation instead of a spin test. Figure 7.8 shows an automobile shaft. The low-cycle fatigue test is usually required for a rotating shaft that often fails by low-cycle fatigue. For a part that is subject to thermal fatigue or vibration, a high-cycle fatigue test should be conducted. Thermal fatigue usually results from combined mechanical and thermal stresses. The resistance to thermal fatigue is a critical performance requirement for the components operating at elevated temperatures, such as a combustion chamber, or in the high-pressure turbine blades of a jet engine. A test or analysis on thermal fatigue resistance on a part used for these applications might be necessary.

To demonstrate the satisfactory performance of a jet engine part manufactured using reverse engineering, five different levels of testing might be conducted. First, a material coupon test can be utilized to show that the part is made of the same material used by the OEM, and derivatively conclude that this part will have the equivalent mechanical strength of the OEM part. This level of testing is usually sufficient for relatively simple noncritical parts, such as the bolts that are used to fasten the fuel pump onto its base pad. As a second level of testing, a bench test is conducted, for example, to perform a bench test of a fuel pump to ensure that it will produce the same fuel flow rate as the OEM counterpart does. A third level of testing might be applied to a part that is assembled into the engine module for an engine block test. This fuel pump might be installed in the combustion module for the block test. At a still higher level, the combustion module installed with this fuel pump might be assembled with all other engine modules and the entire engine tested on a rig for an engine or rig test. Finally, the engine equipped with this fuel pump might be installed onto an aircraft for a flight test to test that the aircraft will operate properly. Whatever level of testing is chosen, it has to satisfy the customers' expectations and comply with regulatory requirements.

Some subtly different definitions of fatigue cycles are used for material sample coupon fatigue test and engine test in the aviation industry. The

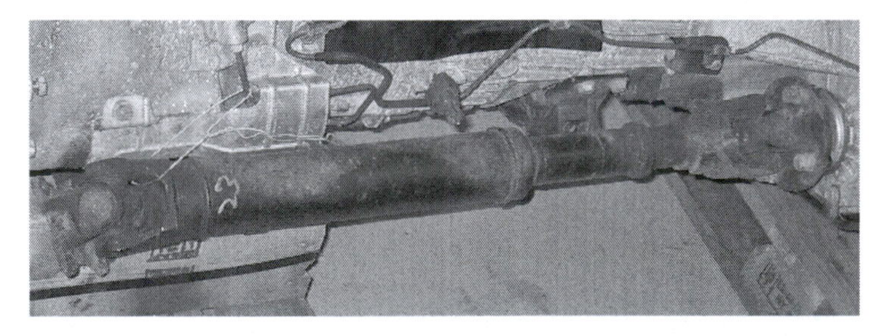

**FIGURE 7.8**
An automobile shaft.

fatigue endurance evaluation of an engine shaft is a good example to highlight this subtle difference in terms of fatigue cycles. Fatigue failure results from cyclic stresses, such as the alternate tension and compression stresses induced by a rotating shaft. One of the simplest fatigue test methods is to apply completely reversed flexural loading on a rotating specimen. The Moore rotating-beam machine is one of the most frequently used machines for the classic flexural stress fatigue test. In some flexure tests, the beam specimen is rotated and loaded by its own weight through gravitation; in others, a stationary flexure specimen is acted on by loads rotating around it. If the specimen is circular in cross section, the flexural stress will fluctuate sinusoidally. One single revolution will induce a complete reversing stress cycle that is counted as a complete cycle in the typical rotating fatigue test, usually at a constant rotating speed, on a sample material coupon. With the assumption that the specimen will remain a plane cross section, the maximum tension and compression stresses can be calculated with the simple flexural formula, Equation 7.1, where $\sigma$, $M$, $r$, and $I$ are the flexural stress, the moment about the neutral axis, the radius, and the area moment of inertia about the neutral axis, respectively. The flexural fatigue test can also conducted by applying cyclic load to a flat triangle specimen that fluctuates like a cantilever beam, as described in ASTM Specifications B593, *Standard Test Method for Bending Fatigue Testing for Copper-Alloy Spring Materials* (ASTM, 2009b). Fatigue tests are also frequently conducted by applying axial load to the specimen with reversal of stress. ASTM E466 details a fatigue test with a constant amplitude for metal (ASTM, 2007).

$$\sigma = \frac{Mr}{I} \tag{7.1}$$

Changes in rotation speed of a shaft will also induce stress variations. If the shaft rotation speed change pattern is a well-defined periodic event, a cyclic stress pattern can also be established. For an aircraft engine, the engine shaft rotation speed is relatively slow when the engine is in the idling state. The rotation speed rapidly increases at aircraft takeoff, as does the induced stress. The rotation speed and the stress remain more or less steady during aircraft cruise, and then both the rotation speed and stress come down during landing and engine shutoff. Similar rotation speed and stress cycles will repeat every time the aircraft takes off and lands. One single complete engine start-stop stress cycle for an aircraft engine consists of a flight cycle profile that includes starting the engine, accelerating to maximum thrust, decelerating, and stopping the engine. The fatigue life cycles for an aircraft engine life calculation are based on the start-stop flight cycles instead of the individual shaft revolution cycles. It can be very beneficial if engineers are aware of this subtle difference in determining what tests should be conducted and what parameters need to be measured during the test.

### 7.2.3 Probabilistic Analysis

The statistical uncertainties in applied load, material properties, and crack geometry have led to the development of probabilistic fracture mechanics in part life prediction. This probabilistic analysis calculates the probability of crack initiation and part failure with statistical models and reliability theories. There is no universal guidance for the selection of either deterministic fracture mechanics or probabilistic fracture mechanics in part performance evaluation. Deterministic fracture mechanics provide a definite pass or fail decision, and typically calculates the crack initiation time with lower bound conservative input. This methodology is often used to ascertain allowable crack lengths. Many machine design criteria define part failure at the observation of crack initiation. However, most mechanical parts show longer life expectancy than what is predicted solely based on crack initiation, and will remain in service continuously until a critical crack length has been reached. The safety of a cracked part is ensured by proper inspections. The inspection intervals calculated by deterministic fracture mechanics are usually very conservative. Probabilistic fracture mechanics might be used as an alternative to determine inspection intervals.

The application of probabilistic analysis in part life evaluation has the advantage of analyzing the repeatability of a property and the reproducibility of a function. This method provides the opportunity to analyze the part in a statistical sense of reliability. It is challenging to statistically evaluate part performance in a machine design project. It is even more difficult in reverse engineering to evaluate part performance using the probabilistic method, primarily because of limited part availability. There are two critical questions in applying the probabilistic method for reverse engineering. First, what level of randomness is required for the data source? Second, how many sample parts are required to demonstrate compliance with the statistical performance requirement and with an acceptable statistical confidence level?

It is essential to have a random data source for any meaningful statistical analysis. Unfortunately, in reverse engineering this fundamental requirement is more difficult to satisfy than in most other fields. The level of randomness is often a concern in many reverse engineering projects. Both economic and technical restrictions have imposed some limits on data randomness in reverse engineering. Practically speaking, most reverse engineering projects only have a limited number of parts to work with, partially due to fiscal consideration and partially due to part availability.

Many reverse engineering projects start with a few or even only one OEM part. It might already be a monumental task to get just that one patented ball bearing from the OEM that owns the proprietary data of the bearing and refuses to sell any new part to the company trying to reverse engineer it. To obtain multiple new OEM parts from a sole-source supplier to satisfy the statistical randomness requirement can be very difficult at times. A statistical requirement of multiple OEM parts might impose a prohibitively expensive

financial burden to reverse engineer a jet engine turbine disk that costs tens of thousands of dollars for just one piece. To reverse engineer a new replacement part using a used part that has ceased production is another example. It is not always practical or feasible to obtain multiple used parts or even just a second one.

Beyond the concerns about the statistical randomness of data, limited part availability also imposes a serious challenge to establish the baseline criteria for part performance evaluation. In a conventional machine design project, the performance requirements are usually developed by design criteria when the project first starts. In contrast, the baseline used to evaluate the part performance in reverse engineering is established by testing the OEM part. The scenario that an OEM part exceeds the minimal design criteria does occur from time to time. Statistically, it requires multiple OEM parts to establish the standard for reverse engineering. Unfortunately, the establishment of baseline requirements for performance evaluation in reverse engineering is often not rigorously pursued.

After the establishment of performance standards for comparison, statistical analysis in the probabilistic method can be another task. Chapter 6 presented an introductory discussion on the statistics about how many test sample parts are required to demonstrate compliance with the statistical performance requirement and within an acceptable confidence level. More details on this subject can be found in statistics reference books.

The aforementioned issues and discussions once again highlight the challenges facing reverse engineering today. The fiscal, business, and technical restrictions often require engineers to make judgment calls based on the best available data in reverse engineering. A consensus among all stakeholders on the test method and pass/fail criterion for part performance evaluation is advisable, particularly for life-limited, critical, or expensive parts. In case a regulatory requirement has to be satisfied, it is mandatory to fully understand the expectations on part performance before a reverse engineering project starts.

## 7.3 System Compatibility

System compatibility is essential in all mechanical systems. In our daily life we are aware that it does not meet the system compatibility requirement if we add lead-containing gasoline into an automobile that uses only lead-free gasoline. In aviation, special aviation-grade fuels are required for aircraft. The use of alcohol-based fuels can cause serious performance degradation and fuel system component damage. Therefore, the usage of non-aviation-grade fuel is not system compatible and strictly prohibited in most cases. System compatibility requirement is not always clear-cut. For example, will

a steel oil pump satisfy the system compatibility requirement when it is used to replace an aluminum oil pump?

Figure 7.9 shows a cutaway view of an automobile engine that is displayed at SAE Automotive Headquarters in Troy, Michigan. It illustrates the details of the valve spring, engine cylinder, air inlet, and gas exhaustion systems. There are hundreds of parts in this engine system. To replace any OEM part with a new reverse engineered part, it is critical to ensure that the new part has acceptable system compatibility in this engine assembly.

The rigorousness of system compatibility requirement for a part heavily depends on the part functionality and the system configuration. In a typical internal combustion gasoline engine used in automobiles, most interactive parts directly contact each other. A minor dimensional variation of an automobile engine part usually has a tolerable impact on system compatibility. Therefore, small variations in part dimensions are often acceptable when an automobile part is reinvented using reverse engineering. This can reduce the costs and increase the marketability of the part, and still with acceptable performance. In contrast, there are fewer direct contacts between a rotating part and its stationary surroundings in a modern turbine jet engine. A close clearance between this part and its surroundings can be vitally critical. One-tenth of 1 mm variation in dimension can have a significant impact on system compatibility of a turbine jet engine. Figure 7.10 shows a cutaway view of the combustion and turbine sections of a modern J79 jet engine, which is used to power the F-4 and F-104 fighter aircraft. It is exhibited in the MTU-Museum, Munich, Germany. The clearance between the

**FIGURE 7.9** **(See color insert following p. 142.)**
A cutaway view of an automobile engine.

**FIGURE 7.10** **(See color insert following p. 142.)**
A cutaway view of the combustion and turbine sections of a jet engine.

turbine rotor blade tip and its stationary case has a significant influence on engine performance. The clearances between the blade tips and the shroud are actively controlled with precision to minimize the leakage passage over the blade tips for efficient cruise flight, yet clearance can be adjusted during other periods of the flight so as to avoid blade tip rubbing. An active clearance control system is often integrated in the engine design to ensure proper operation. Engine stall, in-flight engine shutdown, or a serious incident can occur due to improper tip clearance alignment. To reverse engineer a jet engine rotor blade, precise dimension control is critical to satisfy the system compatibility requirement. Other reverse engineered parts, such as fuel nozzles and combustion canisters used in the combustion section, also need to satisfy the system compatibility requirements of the engine configuration.

### 7.3.1 Functionality

Part dimension is just one obvious parameter that can affect system compatibility. If a part shows an altered geometric form, then how well this part can physically fit in the system will immediately cause concern. System compatibility evaluation goes beyond just geometric form and physically fitting in. Engineers are more interested in how well a new part can be integrated into the system to perform the design function.

Material characteristics such as density, mechanical strength, and environmental degradation resistance all play a role in system compatibility. There are approximately 2 to 3 million fasteners in a large jet aircraft. A slight density variation of the material used for these fasteners can change the total

weight of the system, shift the gravity center of the system, change the resonance frequency of the system, and induce undesirable vibration during operation that might even lead to premature failure. There are hundreds of various airfoils in a modern jet engine, as illustrated in Figure 7.11. A slight variation in these airfoils can result in similar concerns as well. There are also hundreds of cooling holes drilled in a single turbine vane or blade that operates at elevated temperatures. Figure 7.12 shows the cooling holes on the high-pressure turbine vanes installed on a CF6 engine that is used to power many commercial aircraft, such as B747, B767, A300, and DC10. The turbine inlet temperature can reach approximately 1,300°C, and the turbine

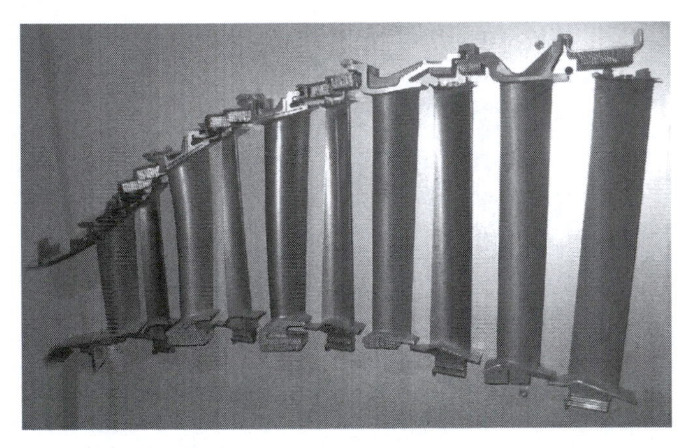

**FIGURE 7.11**
Various airfoils in a jet engine.

**FIGURE 7.12**
Cooling holes on high-pressure turbine vanes.

outlet temperature can still be as high as 800°C. A small alteration in the diameter of cooling holes directly affects the cooling effect, and therefore the engine operation. A minor deviation from the original design, such as the total number of holes or their configurations, can potentially change the net system mass and weight.

The system compatibility is not only determined by the part itself, but also often depends on the adjacent neighboring part. A slight variation in alloy composition might only have a negligible effect on alloy strength. However, this part can show an inferior corrosion resistance when it is in contact with another adjacent part because many corrosion activities depend on the relative electrochemical potential difference between the two contact parts. A variation in manufacturing process from casting to forging, or different heat treatments, can alter material mechanical properties such as wear resistance or yield strength. After installation to the system, this part can be in contact with another part. If the original design is to make this part having a weaker wear resistance and be replaced during the scheduled maintenance, an improved hardness can have an adverse effect on the life cycle of the mating part that is in contact with this part and is not scheduled for replacement during the service.

In the aviation industry, a vane or rotor blade with better mechanical strength often deems desirable. However, when a new vane or blade is reinvented using reverse engineering, it is critical that its system compatibility is thoroughly evaluated. A stronger vane or blade usually also has a higher hardness and better wear resistance. When this vane or blade is installed into a disk, it might wear out the disk before the disk's predicted life cycle based on OEM's original design. Figure 7.13 illustrates the vane assembly at the rim of a disk. In many disk designs, the low-cycle fatigue at the joint slot where the vane is installed is one of its life-limiting locations. The replacement of worn-out vanes is routinely carried out during the maintenance, while the replacement of a disk is very costly and less frequent. In case the vane is broken away, the dynamic impact of this "improved" vane might cause more damage to the surrounding parts than a normal vane does because it is stronger and possesses higher dynamic energy. It might even result in an uncontained failure if it breaks through the case that was originally design to have sufficient capacity to arrest the weaker broken vane.

Meeting the fatigue endurance or dimension tolerance requirements of a rotating shaft can be directly demonstrated by either analysis or test. However, the proof of a satisfactory indirect impact of a reverse engineered part to a system sometimes can be very challenging. The fracture of a rotating shaft may not immediately stop the rotation of its surrounding parts, such as a rotating seal attached to this shaft, and cease the operation of the machine. Instead, the rotating speeds of the remaining shaft and the adjacent parts might even spike up during the transit period immediately after the breakdown of this shaft and before the machine stops operation. The post-shaft-fracture effects during the transit period might cause more damage to

**FIGURE 7.13**
Rotor disk assembly.

the machine and impose more serious safety concerns than the broken shaft itself. The OEM often has the advantage of prior field experience to make necessary design integration to minimize the potential damage in this event. This is invaluable corporate knowledge and surely proprietary information rarely available in the public domain. In other words, a machine might be designed with a subtly built-in safety mechanism to prevent the shaft from ever reaching its critical rotating speed, and automatically stop the machine in the event that the shaft starts to over-speed. This can be done through some software control or a hardware mechanism that is not directly related to the mechanical strength of a rotating shaft. As a result, the most critical design safety factor is not the mechanical strength of the shaft because it most probably will not fail. Instead, the most critical design criterion of this reverse engineered part has become that of demonstrating its system compatibility with the safety control mechanism.

The significance of any property to a part performance depends on the part's design functionality. The mechanical properties are important functional factors to a wire or cable that is used to control a mechanical mechanism. For example, the mechanical strength is critical to a rope that is attached to a pulley to open a door gate. However, to replace a cotton or polymeric cable with a metal cable is not always system compatible. The metallic cable most probably will be stronger and last longer, but it might also impose some adverse effect on the surrounding parts, such as damaging the pulley surface or the adjacent frame. On the other hand, the mechanical strength is not essential for a harness cable or an electric cable under normal conditions. In another example, to replace an aircraft seat with an automobile seat might improve passenger comfort, but it is most probably unacceptable from

the perspective of system compatibility of an aircraft in terms of weight and flammability resistance.

### 7.3.2 Interchangeability

The importance of system compatibility has long been recognized in many industries, and the automobile aftermarket industry in particular. Most replacement parts in maintenance and repair are manufactured by suppliers other than the OEM in today's automobile industry. However, in the medical device industry and aviation industry, only selected and closely controlled parts are allowed in the aftermarket to ensure good system compatibility.

Many component parts are manufactured in accordance with the industrial standards for good system compatibility and part interchangeability. The most notable example are the standards for the bolts, nuts, and other fasteners. The diameter, type of thread, and length are all well standardized. Tables 7.2 and 7.3 summarize a few sample standards of the metric screw threads and the unified screw threads, respectively. Both systems adopt the

**TABLE 7.2**

Standards of Metric Screw Threads

| | Coarse Threads | | Fine Threads | |
|---|---|---|---|---|
| Major Diameter (mm) | Pitch (mm) | Minor Diameter (mm) | Pitch (mm) | Minor Diameter (mm) |
| 3 | 0.5 | 2.39 | — | — |
| 5 | 0.8 | 4.02 | — | — |
| 10 | 1.5 | 8.16 | 1.25 | 8.47 |
| 14 | 2 | 11.6 | 1.5 | 12.2 |
| 20 | 2.5 | 16.9 | 1.5 | 18.2 |
| 24 | 3 | 20.3 | 2 | 21.6 |

**TABLE 7.3**

Standards of Unified Screw Threads

| | | Coarse Threads | | Fine Threads | |
|---|---|---|---|---|---|
| Size | Major Diameter (in.) | Thread per Inch | Minor Diameter (in.) | Thread per Inch | Minor Diameter (in.) |
| 1/4 | 0.2500 | 20 | 0.1887 | 28 | 0.2062 |
| 3/8 | 0.3750 | 16 | 0.2983 | 24 | 0.3239 |
| 1/2 | 0.5000 | 13 | 0.4056 | 20 | 0.4387 |
| 3/4 | 0.7500 | 10 | 0.6273 | 16 | 0.6733 |
| 1 | 1.000 | 8 | 0.8466 | 12 | 0.8978 |

*Source:* ASME/ANSI Standard B.1.1.

standard terminology. The major diameter is the linear dimension between the thread crests, and the minor diameter is the linear dimension between the thread roots. The pitch is the linear distance between two adjacent threads along the axial direction. The metric screw threads are widely used internationally. They are in compliance with the ISO standards. The profiles and specifications of metric screw threads are detailed in the American Society of Mechanical Engineers (ASME) and American National Standards Institute (ANSI) Standard B1.13M (ASME, 2005). The unified screw threads are primarily used in the United States, and the details are discussed in the ASME and ANSI Standard B1.1. (ASME, 2003). The metric screw threads are specified based on the major diameter and the pitch. For instance, M10x1.5 is the designation of a metric screw thread with a major diameter of 10 mm and a pitch of 1.5 mm. The unified screw threads are specified based on the major diameter and the number of threads per inch. For example, "1/4 in-20" is the designation of a unified screw thread with a major diameter of ¼ in. and having 20 threads per inch in the axial direction. A newly designed machine is expected to adopt these standard fasteners instead of inventing its own fasteners whenever feasible. Another example is automobile tires. Most automobile tires are manufactured by tire companies instead of automobile OEMs. The manufacturing of these tires for both new cars and the aftermarket will follow a set of industrial standards to meet the interchangeability requirements for a variety of automobile models.

Numerous other off-the-shelf spare parts, such as tubing, bearings, and gears, are also manufactured with acceptable system compatibility in terms of fit, form, and function in the automobile and other industries. Usually they are readily available unless other regulatory requirements need to be satisfied.

System compatibility in hardware refers to seamless interaction and harmonized operation between the parts to perform the design function regardless of the system and part vendors. System compatibility in software usually refers to data interchangeability. It implies that one document file compiled by one system is also readable and operable in another system. Today's engineering exercise requires system compatibility in both hardware and software. Preferably the computer-aided design and manufacturing software systems utilized in reverse engineering to reinvent a part are compatible with the software systems used by the OEM. This software interchangeability helps data interaction for future maintenance and repair despite that the hardware part is the final product.

### 7.3.3 Cumulative Effect

It is usually allowed and acceptable in most cases that a minor alteration is introduced to a mechanical part. When the same or similar change is applied to the same part multiple times, or to multiple parts that are installed on the same machine, the resultant cumulative effects of this relatively minor change can accumulate and cross the critical threshold. It might be within

the design tolerance that a single seat out of 200 on an aircraft is replaced with a 5 kg heavier seat produced using reverse engineering. However, when all 200 seats are replaced by the similarly heavier seats, the center of gravity of the airplane might shift. This can create an unsafe condition and fail to meet the system compatibility requirement. It is worth noting that even in the most regulated aviation industry, whenever a PMA part is approved, there is virtually no regulation to prevent it from being installed multiple times in the same OEM product. Theoretically speaking, a PMA part is an approved aeronautic product that can be used to replace or substitute the relevant OEM counterpart with minor restrictions.

The cumulative effect on cooling a cylindrical engine is another example. All engines operate at an elevated temperature, and sufficient cooling is essential for proper engine operation and efficiency. A radial aircraft engine is an assembly of multiple sets of piston cylinders. Figure 7.14 shows an engine cylinder with cooling fins around the external surface. The replacements of these vintage radial engines are often reproduced using reverse engineering because of their lack of engineering support from the OEM. The total cooling effect depends on the heat transfer coefficient of the aluminum alloy used to cast these engines, the total number of fins, the size and surface area of the fins, and other geometric factors, such as the spacing between two

**FIGURE 7.14**
An engine cylinder with cooling fins.

adjacent fins. A variation in alloy composition, cast processing, a fin's shape, and spacing can all change the heat transfer efficiency and subsequently the cooling effect. If only one engine cylinder is replaced with a slightly altered replacement reproduced using reverse engineering, the difference in cooling efficiency of the entire engine assembly might still lie within the OEM design limit and satisfy the system compatibility requirement. However, when multiple engine cylinders are replaced, the cumulative effects of cooling efficiency could fall outside that limit, and fail to meet the system compatibility requirement.

## 7.4 Case Studies

### 7.4.1 Fastener Evaluation

The market for mechanical fasteners is worth billions of dollars. The fastener design, production, and reinvention are the subject of many projects. Fastener failure investigation and prevention is one of the most pressing concerns in industry. Tens of thousands of fasteners are used to join parts together in machines and constructions. Figure 7.15a and b shows arrays of fasteners utilized in a traditional metal bridge and a modern highway structure, respectively.

The nuts and bolts are among the most frequently reverse engineered parts in aviation industries. Despite the fact that these parts are usually standard off-the-shelf parts in many industries, such as the automobile and construction industries, the aerospace-grade fasteners are subject to specific requirements and possess several key elements to be good reverse engineering candidates: vast quantity in application, engineering maturity in design and production, and critical components in structure integrity. The interest in reverse engineering fasteners in the aviation industry is further augmented by the regulatory requirements. The part used in an aircraft has to be approved by a government agency for flight safety. There are tens of millions of fasteners in a large jet airplane, and these aerospace-grade fasteners have to comply with their respective design specifications for specific functions. They are of high quality and subject to rigorous quality control. The quality requirements boost their prices up to a hundred times higher than those of the fasteners used in the automobile, construction, and household industries. Beyond the fiscal potential, the fasteners also play a critical role in aviation safety. The 1988 Aloha Airlines Flight 243 accident started when a small section of the Boeing 737 airplane roof ruptured during the flight at an altitude of 7,300 m (24,000 ft) in the middle of the ocean. The crack grew larger and eventually tore off about 5.5 m (18 ft) of the cabin skin and structure, consisting of almost the entire top half

(a)

(b)

**FIGURE 7.15**
(a) A bolt-jointed metal bridge. (b) Bolts in a modern highway structure.

of the aircraft fuselage, extending from just behind the cockpit to the forewing area. The most probable root cause of this accident was attributed to metal fatigue exacerbated by crevice corrosion around the rivets that were used for the fuselage lap joints. This accident prompted numerous actions, and also led to the integration of damage tolerance into new aircraft design.

The aircraft and its components designed with the integration of damage tolerance will tolerate a certain amount of damage and still remain intact until the damage is detected and repaired.

Figure 7.16 shows the different sections of a turbojet engine Pratt & Whitney (PW) J57 exhibited at the New England Air Museum in Windsor Locks, Connecticut. This engine has sixteen two-spool stages of compressors. It was one of the first turbojet engines utilizing axial compressors. In contrast to the centrifugal compressor, the compressed air through an axial compressor flows parallel to the engine axial direction. The PW J57 turbojet engine and its variants, such as JT3 series engines, have been used to power many military and civilian aircraft, such as B-52 bomber and Boeing 707 aircraft. The front fan, the subsequent low-pressure and high-pressure compressors, the middle section of the combustion chamber, the turbine, and the final exhaust section are all bolted together by numerous side-by-side fasteners.

Figure 7.17 is a photo of the front section of an F-86F fighter. Figure 7.18 shows the attachment of an aircraft engine to a commercial jet through a pylon. Both illustrate that the airframes of modern aircraft are fastened together with thousands of rivets. The regulatory certification requirements, the market demands in quantity, and the attractive profit margins result in a very unique commercial market for aerospace-grade fasteners reproduced using reverse engineering. The interchangeability requirement is one of the elements most fasteners are expected to satisfy. However, different designs and configurations of various engine and aircraft models also require these fasteners to meet specific standards for system compatibility. The temperature profile from the front fan section of a jet engine through various compressor

**FIGURE 7.16**
Sections of a turbojet engine jointed together by bolts.

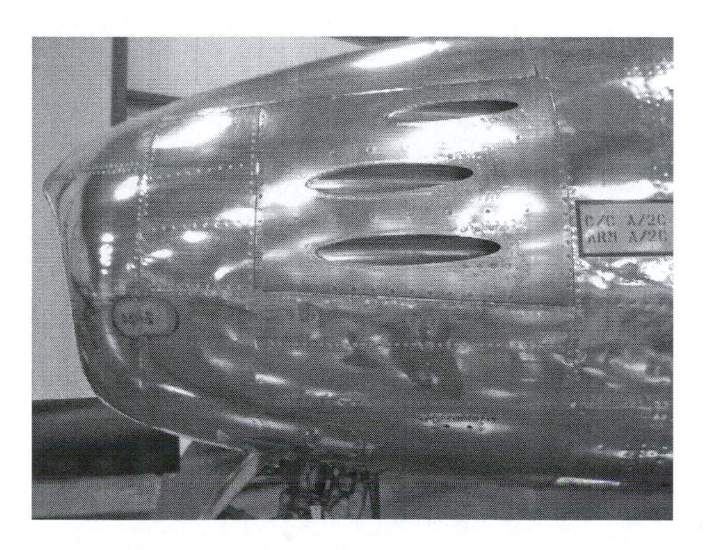

**FIGURE 7.17**
The front section of an F-86F fighter.

**FIGURE 7.18**
Rivets on a pylon.

sections, the combustion chamber, and the turbine section might change dramatically from the ambient temperature rising to as high as 1,500°C and then cooling down to 300°C, depending on the engine and aircraft models. The required strength, durability, and other characteristics also vary from section to section where they are used. The fasteners utilized to bolt a turbojet engine of various models are usually made of stainless steel, titanium alloy, or nickel-base superalloy with different levels of temperature resistance. The

tensile strength, fatigue strength, and creep resistance are usually the important criteria in the evaluation of these fasteners. In contrast, the rivets used to fasten the airframe require different properties. The high-temperature properties are not the determining factor in the selection of these rivets. Instead, the shear strength of these rivets is more critical. Various governing codes published for specific industries should be complied with when applying riveted connections in boilers, bridges, buildings, and other structures that might have potential safety concerns. For instance, the Boiler Construction Code of the American Society of Mechanical Engineers needs to be complied with to build a boiler. The following are the typical failure modes of a rivet. These failure modes shed some light on what properties and characteristics have to be evaluated in testing these rivets when they are produced using reverse engineering. As illustrated in Figure 7.19a, the primary load most rivet joints are subject to is shear force. However, the shear strength of the rivet is not the only material property that has to be specified. Figure 7.19b to g illustrates various possible failure modes of riveted joints. Figure 7.19b

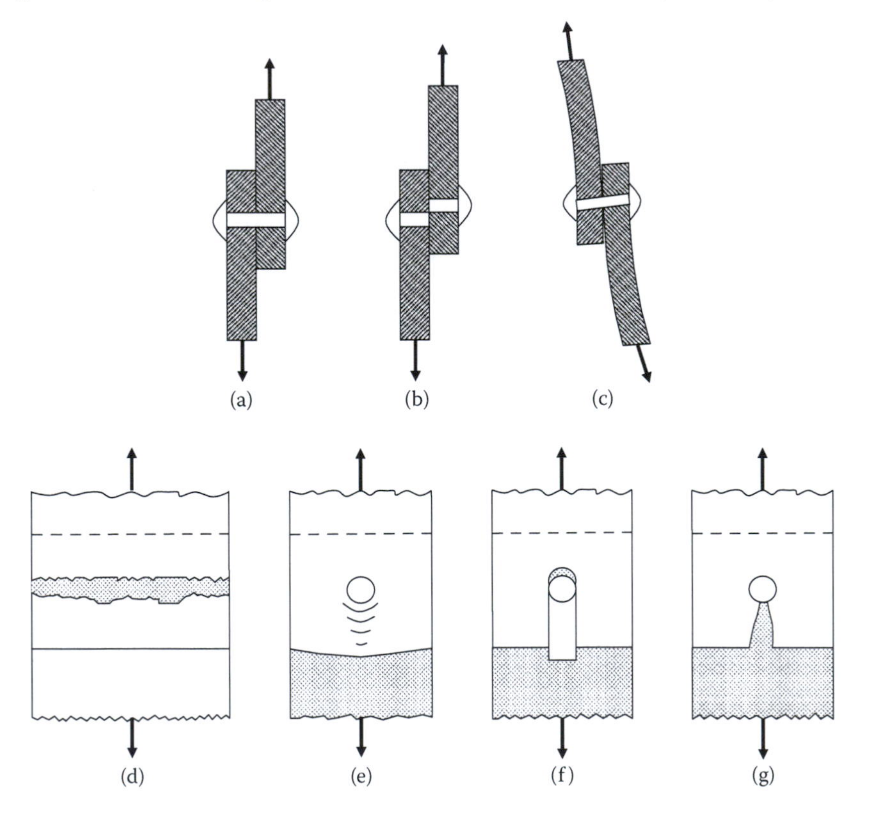


**FIGURE 7.19**
Schematics of rivet joint and failure modes: (a) riveted connection, (b) failure by pure shear, (c) bending, (d) cracking due to tension, (e) crushing, (f) shearing, and (g) tearing.

is a schematic of the configuration when the riveted connection fails by pure shear stress. The simple criterion to prevent this type of failure is to ensure that the shear strength of the rivet is larger than the applied shear stress, as calculated by Equation 7.2, where $\tau$, $F$, and $A$ are the shear stress, applied shear force, and cross-sectional area of the rivet, respectively.

$$\tau = \frac{F}{A} \tag{7.2}$$

In most machine designs, the cross-sectional area is calculated based on the nominal diameter of the rivet, even though that diameter often expands after installation. The ultimate tensile strength and yield strength in tension are often readily available in most mechanical engineering handbooks or material databases. Due to lack of data on yield strength in shear and shear strength, they are usually calculated by Equation 7.3a and b, where $\sigma_{sy}$, $\sigma_y$, $\sigma_{su}$, $\sigma_u$ are the yield strength in shear, yield strength in tension, shear strength, and ultimate tensile strength, respectively:

$$\sigma_{sy} \approx 0.58\sigma_y \tag{7.3a}$$

$$\sigma_{su} \approx 0.62\sigma_u \tag{7.3b}$$

Equation 7.3a is an approximation based on the theory of maximum distortion energy for ductile materials. The maximum distortion energy theory postulates that a given material only has a limited capacity to absorb the energy of distortion. A material will start yielding when the externally imposed distortion energy reaches the maximum capacity that the material can absorb. The distortion energy tends to change the shape of the material, but not the volume. Engineering materials can withstand enormous hydrostatic stress and volume change, and still result in no yield. Despite the enormous hydrostatic pressure of a deep sea, a sunken ship can remain intact for years. The maximum distortion energy theory predicts that the yield strength in shear is approximately equal to 58% the yield strength in tension. Equation 7.3b is based on studies on steel bolts (Fisher and Struik, 1974). Figure 7.19c is the schematic showing a rivet connection failure due to bending. Because of the complex configuration of a rivet connection and the effect of stress concentration, the calculation of bending stress by Equation 7.1 is usually only an approximation. The bending moment, $M$, can be approximately calculated by the shear force $F$ as $M = Ft/2$, where $t$ is the total thickness of the connecting parts. Since a rivet connection seldom fails due to bending, the consideration of bending is usually used for the improvement of safety margin, or system compatibility. The close contact and tight squeeze between the rivet and the component can cause crushing and subsequent failure of either the rivet or the component, as sketched in Figure 7.19e. The failure of crushing is caused

by bearing stress. Considering the cylindrical contact surface between the rivet and the component, the uncertainty of the stress distribution profile, and the effect of the stress concentration, a precise calculation of bearing stress is very challenging. This brings to light the fact that the failure of a riveted connection goes beyond the mechanical strength of the rivet. It also depends on the interaction between the rivet and the component. A stronger rivet does not always prevent riveted parts from failing. Two more possible failure modes are further illustrated in Figure 7.19f and g, where the connected component, instead of the rivet, might fail first, due to shearing and tearing when the rivet location is too close to the edge. The type of rivet and the total number of rivets used in a part design are usually the design parameters integrated with the type of components and the locations of rivet holes. The application of replacement rivets has to be thoroughly evaluated to ensure proper system compatibility.

Catastrophic disasters caused by bolt failures also occur in building structures. Kemper Arena is a huge indoor arena in Kansas City, Missouri, built in 1974. On June 4, 1979, a major storm with 110 km/h (70 mph) winds and heavy rains caused a large portion of Kemper Arena's roof to collapse. Similar to the above-mentioned Aloha accident, bolt fatigue failure was one of the reasons for the collapse. The bolts used on the Kemper Arena roof were the ASTM A490 bolts. A490 bolts are heat-treated, high-strength structural bolts made of alloy steel. However, they are of relatively low ductility and susceptible to brittle fracture. These bolts can only sustain high tensile stress under static conditions. When the A490 bolts are subject to dynamic load such as what the Kemper Arena was subject to on that stormy day, they fail at a much lower stress. The specifications and applications of A490 bolts are detailed in the *Specification for Structural Joints Using ASTM A325 or A490 Bolts*, published by the Research Council on Structural Connections (RCSC, 2004). The nominal tensile strength of an A490 bolt is 779.1 MPa (113 ksi); while the maximum tensile stress for fatigue loading should not exceed 372.3 MPa (54 ksi) for a fatigue life of 20,000 cycles. The allowed stress level reduces to 337.8 MPa (49 ksi) for a fatigue life between 20,000 and 500,000 cycles, and further reduces to 262 MPa (38 ksi) for a fatigue life higher than 500,000 cycles. Although the total bolts used for the arena construction were enormous, once a few of the bolts gave way, that triggered a domino effect and led to the subsequent cascading failure of the roof. It is often recommended to conduct a thorough evaluation prior to using A490 bolts in a structure.

### 7.4.2 Door Stairs

Many medium and small airplanes are equipped with boarding door stairs as shown in Figure 7.20. They are usually equipped with one or two side handles that function as hand holders for the passengers during boarding and deplaning. A small airplane usually has only one side handle, which also helps attach the door stairs to the airplane. Instead of rigid steel bars,

**FIGURE 7.20**
Door stairs.

flexible metal cables or even fabric cables can be used for these side handles in small aircraft. Though the door stairs are not as critical a component as the oil tank or propeller, they are a necessity to the aircraft and do open and close regularly. Their frequent use has sometimes led to breakage of the side handles or cables, particularly when they are designed to carry part of the load of the door stairs.

There are two basic requirements for these side handles: strength and flexibility. The replacement of a side handle or cable with a reverse engineered substitute is relatively risk-free and requires little performance evaluation as long as these two basic requirements are met. This invites a variety of creative ways to replace the side handles or cables. The basic fit, form, and function requirements of reverse engineering might be easily satisfied in most cases, but the system compatibility with the surroundings can cause unexpected concern. When a metal link chain is used to replace the original handle, it might have sufficient strength to carry the load, and proper flexibility for door opening and retraction operations, but it might fail the system compatibility requirement and damage the surrounding airframe.

## 7.5 Regulatory Certification of Part Performance

In some cases it is a legal mandate to have a part certificated by a government agency before it is allowed to enter the marketplace for public usage. The certification process is a regulatory instead of an engineering evaluation

process. In a certification process, the fatigue life cycle of a shaft can be estimated with two-dimensional computer modeling despite the shaft is actually subject to three-dimensional stresses. The estimated life cycle is not accurate. However, this inaccuracy can be compensated for by a higher safety margin. The life cycle estimated with two-dimensional modeling might not be deemed as accurate engineering data when documented in an engineering handbook. However, one-third of this estimated life cycle, that is, with a safety margin of 3, could be accepted as valid substantiation data in a certification process. A certification process focuses on approving a part for public usage. It is more relevant to the commercialization of the reverse engineered parts than the parts' engineering merits.

The scientific theories and engineering principles are solely based on the experimental data, and do not change until new test data prove otherwise. However, the standards to meet for certification are often the minimum standards, which are revised periodically and can be affected by many non-scientific factors. The revision of certification procedures and regulations is a legal and governmental process that is subject to the influence of public psychology and opinion. Historically the revisions of FAA regulations are often prompted by aviation accidents or incidents, and rarely driven by new scientific discoveries.

With respect to airworthiness, Title 49 of the United States Code (USC), "Transportation," Section 40113, designates the FAA as the regulatory entity with the duties and powers to take actions, as appropriate, to conduct investigations, prescribing regulations, standards, and procedures, and issuing orders. Section 44701 further elaborates this delegation that the FAA shall promote safety flight of civil aircraft in air commerce by prescribing minimum standards required in the interest of safety for appliances and for the design, material, construction, quality of work, and performance of aircraft, aircraft engines, and propellers. As a means of surveillance for the flight public, Section 44704 entrusts the FAA to issue a type certificate for an aircraft, aircraft engine, or propeller, or for an appliance when the aircraft, aircraft engine, propeller, or appliance is properly designed and manufactured, performs properly, and meets the regulations and minimum standards prescribed in Section 44701. A supplemental type certificate can be issued later, when applicable, for a change to a previously type-certificated aircraft, aircraft engine, propeller, or appliance. The oversight of engineering design of an aircraft, aircraft engine, propeller, or appliance is carried out through the type certification process, while their quality production showing compliance to the type-certificated specifications is regulated through the issuance of product certification. Section 44704 ensures the safe operation of an aeronautic product such as an aircraft through the process of airworthiness certification. The registered owner of an aircraft has to apply for an airworthiness certificate for the aircraft to demonstrate that the aircraft conforms to its type certificate and is in good condition for safe operation. Besides the aviation programs, 49 USC also covers the rail programs, the

motor vehicle and driver programs, and other government agencies. Each government agency is entrusted with its respective oversight and regulatory responsibilities.

In compliance with 49 USC, the organization and operation of the FAA are detailed in Title 14 of the Code of Federal Regulations, "Aeronautics and Space." Most FAA surveillance practice and guidance materials are based on this Code of Federal Regulations.

There is no government mandate to regulate the general reverse engineering practice. In the mostly self-regulated industries such as the automotive industry, the burden of approving the parts produced using reverse engineering usually falls to the respective business associations, such as the Certified Automotive Parts Association (CAPA). However, in aviation industries, the federal government still bears most of the regulatory responsibilities. The application of reverse engineering to the type-certificated parts in aviation industry requires government approval, particularly for safety-critical parts. A few FAA orders and policies have been published to provide instructions on reverse engineering. The following information is in the public domain, and is only presented here as sample regulatory guidance on reverse engineering.

FAA Order 8110.42, *Parts Manufacturer Approval Procedures*, was first issued on August 4, 1995. Revision A of this order was issued on March 31, 1999. Both versions contain the same statement on reverse engineering in paragraph 9.c (2) (g) (FAA, 1995):

> Reverse Engineering. Special care should be taken in evaluating "identicality" based upon "reverse engineering." The process of "reverse engineering" is one way to develop the design of a part. However, "reverse engineering" a part will not normally produce a design that is identical to a type certificated part. While an applicant could establish the use of identical materials and dimensions, it is unlikely that a showing could be made that the tolerances, processes, and manufacturing specifications were identical. If the design can not be approved by identicality then the test and computation method should be used. The applicant must show that its design complies with the applicable regulations. The extent of substantiating data required by the FAA should take into account the degree to which the design is identical.

This is the only statement on reverse engineering in the original and Revision A of this order. Revisions B and C were issued on September 9, 2005, and June 23, 2008, respectively. More statements on reverse engineering were included in them.

A certificated product simply means the product is in compliance with the regulatory requirements effective at the time when the product is certificated. The certification requirements of the same product can change many times during the life cycle of the product. The certification requirements for reverse engineering many years after this product has been introduced

into the market often depend on the service history of the product and the consensus of all stakeholders.

## References

American Society of Mechanical Engineers. 2003. *2003 unified inch screw threads (UN and UNR thread form).* Standard B1.1. New York: ASME.

American Society of Mechanical Engineers. 2005. *2005 metric screw threads: M profile.* Standard B1.13M. New York: ASME.

ASTM. 2007. *Standard practice for conducting force controlled constant amplitude axial fatigue tests of metallic materials.* ASTM E466-07. West Conshohocken, PA: ASTM International.

ASTM. 2008. *Standard test method for conducting cyclic potentiodynamic polarization measurements to determine the corrosion susceptibility of small implant devices.* ASTM F2129-08. West Conshohocken, PA: ASTM International.

ASTM. 2009a. *Standard practice for operating salt spray (fog) apparatus.* ASTM B117-09. West Conshohocken, PA: ASTM International.

ASTM. 2009b. *Standard test method for bending fatigue testing for copper-alloy spring materials.* ASTM B593-96(2009)e1. West Conshohocken, PA: ASTM International.

Federal Aviation Administration. 1995. *Parts Manufacturer Approval procedures.* Order 8110.42. Washington, DC: FAA.

Fisher, J. W., and Struik, J. H. 1974. *Guide to design criteria for bolted and riveted joints.* New York: Wiley.

Guthrie, J., Battat, B., and Grethlein, C. 2002. Accelerated corrosion testing. *Mater. Ease* 6:11–15. http://ammtiac.alionscience.com/pdf/2002MaterialEase19.pdf (assessed June 30, 2009).

Research Council on Structural Connections (RCSC). 2004. *Specification for structural joints using ASTM A325 or A490 bolts.* Chicago: American Institute of Steel Construction.

# 8

# *Acceptance and Legality*

The progress of technology is incremental. With a few exceptions, such as the accidental discovery of the X-ray, most scientific discoveries and engineering inventions are based on prior scientific principles or existing subjects. Most breakthrough innovations are based on accumulated evolutions. The inventions of the automobile and aircraft revolutionized the transportation system. However, these inventions were not accomplished overnight. Instead, they resulted from the persistent efforts of thousands of engineers over many decades. Reverse engineering, when properly utilized, will accelerate later inventions from earlier discoveries. Adequate legal acceptance of reverse engineering is crucial for the continued discovery and reaping of these technical and social benefits.

Figure 8.1 shows an early aircraft turbine disk and blade assembly, a turbine wheel for the BMW 109-003 E1 engine, with air-cooled hollow blades welded onto the disk rim. A modern low-pressure turbine disk and blade assembly designed for the PW4084 engine is illustrated in Figure 8.2a. Compared to the simple welded joint of the 109-003 E1 engine, the PW4084 has improved efficiency and functionality. Figure 8.2b is a close-up view of PW4084's low-pressure turbine assembly, and it shows a much more complex configuration. Both assemblies are exhibited in the MTU-Museum in Munich, Germany. The blades on both disk assemblies, though made of different alloys and by different processes, show similar airfoil profiles that were designed based on the same principles of aerodynamics. The inventors of each generation of the disk assembly deserve recognition of their respective contributions and protection of their intellectual properties. However, that protection should never impede further advancement of new turbine disk assembly design.

## 8.1 Legality of Reverse Engineering

Reverse engineering is used to duplicate the original design, or to create a new model that improves an existing product. It also enriches the design and manufacturing processes by enabling compatibility and interoperability between products. Reverse engineering uses scientific analyses to discern the know-how embedded in an existing product. It enhances market competition. The impact of a reverse engineered product depends on the individual part and

**FIGURE 8.1**
Turbine wheel.

the industry. The technical analysis for reverse engineering can be very expensive. It also often requires a long marketing time to introduce a new product. These factors allow the OEM to recoup original research and development costs and establish a strong market share before the reverse engineered product enters the market.

The legality of reverse engineering by an individual to understand the design and functionality of an invention has rarely been challenged, as long as the know-how knowledge thereby obtained is not transferred to other people and is not the cause of any market-destructive appropriations. However, the utilization of the know-how in commercial applications can trigger serious legal concerns regarding the infringement of inventor's intellectual properties. As a result, most legal guidance focuses on regulating post–reverse engineering activities in commercial applications, such as making competitive products based on reverse engineering.

### 8.1.1 Legal Definition of Reverse Engineering

The U.S. Supreme Court stressed the importance of reverse engineering, characterizing it as an "essential part of innovation," likely to yield variation on products that may further advance the technology (Supreme Court, 1989). The standard legal definition of reverse engineering, from *Kewanee Oil Co. v. Bicron Corp.* (Supreme Court, 1974), is "starting with the known product and working backwards to divine the process which aided in its development or manufacture." In the field of science and technology, reverse engineering is to uncover the knowledge of know-how.

**FIGURE 8.2**
(a) Low-pressure turbine assembly. (b) Close-up view of low-pressure turbine assembly details.

In his treatise, Pooley emphasized that the "fundamental purpose of reverse engineering is discovery, albeit of a path already taken." He also identified six reasons for reverse engineering: learning, changing or repairing a product, providing a related service, developing a compatible product, creating a clone of the product, and improving the product (Pooley, 1999). From the legal perspective, the definition of reverse engineering was further broadened to the process of extracting knowledge or know-how from a human-made artifact. A human-made artifact refers to an object that embodies knowledge or

know-how previously discovered by other people. Hence, the engineering required to uncover the knowledge is "reverse" engineering (Samuelson and Scotchmer, 2002).

### 8.1.2 Legal Precedents on Reverse Engineering

Despite the potential scientific and social benefits of limiting a monopoly due to copyright, the U.S. Federal Circuit Court has upheld the U.S. Copyright Act not to preempt any contractual constraints prohibiting reverse engineering. Private parties are free to contractually forego the limited ability to reverse engineer a software product under the exemption of the Copyright Act (Federal Circuit Court, 2003). However, in the absence of a contractual agreement, the Fifth Circuit Court ruled that a state law prohibiting all copying of a computer program was preempted by the federal copyright law (Fifth Circuit Court, 1998). Besides including "non–reverse engineering" clauses in the license agreement, OEMs and inventors might also adopt more broad defensive strategies to protect the business interest reaped from their products. The most commonly used protective contract terms are long-term maintenance contracts with exclusive OEM parts only, and restrictive warranties that will be null and void when a non-OEM part is used in the machinery. Technically, the OEM might also tightly integrate separate parts together into a single system to make reverse engineering individual parts difficult and costly. In other cases, reverse engineering is thwarted by the sheer complexity of the product. It can be prohibitively expensive to decode the design details, or where the process is so time-consuming that it will miss a rapidly shifting market opportunity.

In the last few decades, the advancement of reverse engineering has been a turbulent journey due to legal challenges. In the 1970s and 1980s, some states forbade the use of a direct molding process to reverse engineer. These state laws prohibit the use of an existing product, such as a boat hull, as a "plug" for a direct molding process to manufacture identical products. The plug molding process does not aim to understand the design details; it essentially copies the original product. Therefore, this process will not likely lead to any further invention based on the know-how thereby obtained, one of the propelling benefits of reverse engineering. The supporters of the anti-plug law argued that the plug molding process will undermine incentives for further innovation and harm the industries. The opposition side argued that these laws were economically driven and lobbied for by special interest groups that impede fair competition and technology advancement. In 1989 the U.S. Supreme Court struck down the antiplug state laws that were enacted in twelve states, including Florida and California. The court pointed out that the antiplug law "prohibits the entire public from engaging in a form of reverse engineering of a product in the public domain." It went on to state that "where an item in general circulation is unprotected by a patent (such as a boat hull), reproduction of a functional attribute is legitimate competitive

activity" (Supreme Court, 1989). Nonetheless, further legal protection of an original design of a useful article such as a vessel hull or deck was enacted by the U.S. Congress in 1998. The term *design* is defined below in Title 17 of the United States Code (USC), "Designs Protected," Section 1301:

> A design is "original" if it is the result of the designer's creative endeavor that provides a distinguishable variation over prior work pertaining to similar articles which is more than merely trivial and has not been copied from another source.

This (new) law has effectively banned any reverse engineering by either plug molding or other molding on "designed" vessel hulls.

Since the late 1970s, a series of court decisions was made on reverse engineering semiconductor chips and software programs with mixed support for reverse engineering. First, the semiconductor industry successfully obtained legislation to protect chip layouts from reverse engineering to clone chips. Then, the legality of decompilation, a widely used process for reverse engineering of software, was challenged. In the *1992 Sega Enterprises Ltd. v. Accolade Inc.* case, the U.S. Ninth Circuit Court of Appeals ruled that decompilation was an acceptable reverse engineering practice for achieving interoperability (Ninth Circuit Court, 1992). In the 2003 *Bowers v. Baystate Tech. Inc.* case, the U.S. Federal Circuit Court of Appeals upheld the enforceability of licenses forbidding reverse engineering of software and other digital information. Title 17 USC Section 1201 further restricts reverse engineering the technical protections for copyrighted digital works. It only allows for the bypassing of technical controls and the making of such tools when necessary to achieve interoperability among programs. Except in very limited circumstances, this law deligitimizes reverse engineering in commercial applications, and it also outlaws the manufacture or distribution of tools for such reverse engineering, as well as the disclosure of information obtained in the course of lawful reverse engineering (Samuelson and Scotchmer, 2002). The paradox of reverse engineering is still evolving in the engineering and legal communities.

## 8.2 Patent

Intellectual properties are protected through patent, copyright, trade secret, trademark, service mark, and mask work. Patents protect new, useful, and nonobvious inventions. Copyright provides protection for an expression fixed in a tangible medium. Patent and copyright infringement are the primary legal concerns in reverse engineering. Trade secret law protects confidential commercial information and knowledge. Trademarks and service marks protect characteristic marks distinguishing products and services from others. Mask works protect images that are used to create the layers of semiconductor chips.

The U.S. Patent and Trademark Office (USPTO) defines a patent as a property right granted by the U.S. government to an inventor "to exclude others from making, using, offering for sale, or selling the invention throughout the United States or importing the invention into the United States" for a limited time in exchange for public disclosure of the invention when the patent is granted.

Patent rights are territorial. The USPTO reviews and approves U.S. patents according to U.S. patent laws, regulations, policies, and procedures. U.S. patents are only enforceable in the United States. The term of a U.S. patent is 20 years from the filing date; and from the time of patent grant, the inventor has the exclusive rights to make, use, and sell the invention. The length of patent term varies in different countries. In the United States, a patent application can be filed up to 1 year after the date of first sale or publication of the invention. These prefiling disclosures, although protected in the United States, may preclude patenting in other countries. However, it is universal that an invention will be in the public domain when the term expires. The following discussions only apply to U.S. patents. U.S. patent laws are detailed in 35 USC, "Patents." The relevant patent rules are detailed in Title 37 of the Code of Federal Regulations (CFR), "Patents, Trademarks and Copyrights." Reverse engineering rights and patent laws are mutually exclusive of each other because most, if not all, technical details of a patented invention have already been disclosed to the public and require no more engineering to discover them. The issue is how much legal room an individual with little or no knowledge about the disclosed information has to study or re-create the same or similar invention. For noncommercial purposes, an individual can apply reverse engineering to disassemble a rightfully obtained product to study the product without infringing any patent rights. Proper legal consultation is strongly recommended before applying the information learned through reverse engineering to reinvent a similar product or improve it.

A U.S. patent is a personal property and is only granted to the inventor. When two or more individuals work together to make an invention, if each had a share in the ideas forming the invention, regardless of their respective levels of contribution, they are joint inventors and the patent will be issued to them jointly. In case of a joint invention, each inventor owns the same share of the patent unless otherwise agreed. On the other hand, if one of these individuals has provided all the ideas of the invention, and the other has only followed instructions in making it, the individual who contributed the ideas is the sole inventor and the patent will only be granted to this individual alone. For example, the patent is only granted to the engineer who designed and invented a medical device, but not to the individual who manufactured this device following the design drawing given to him by the inventor, or the employer who financially sponsored the testing of the invention. The design drawing is the outcome of an invention to create a new part. The patent ownership may be transferred through employment agreement, and/or by an express assignment, wherein ownership is. Despite the fact that the

inventor owns the patent, the employer often benefits from the patent license. Universities may patent a U.S. government-funded invention, but the U.S. government has a royalty-free license under the Bayh-Dole Act. A patent can be sold, traded, licensed, rented, or mortgaged. Any infringement of a patent due to improper reverse engineering practice can be enforced through the legal system to impose an injunction or to recover damages.

The types of patents include utility, design, and plant patents. The utility patent covers apparatus, method and composition of matter. Most engineering inventions are under utility patent protection. The design patent protects an ornamental design, and the plant patent applies to the reproduction of a plant. A patent holder can pursue someone for patent infringement even if that person did not copy or know about the patent in question. It is always advisable to consult with legal professionals whenever an issue related to patent infringement arises in a reverse engineering project.

The patent application is a long and complex legal endeavor requiring thorough examination and public disclosure. The patent application fees and maintenance costs can also be expensive, usually in the tens of thousands of dollars. The provisional patent application provides a simple provision period for a potentially enforceable patent. A provisional patent application can be filed in a much simpler format, and is widely used by inventors prior to their filing of utility patents for their new inventions. The provisional patent application will establish an official filing date and proof of invention. It allows subsequent public disclosure, such as publication, marketing, or sales, without loss of potential patent rights. The USPTO neither examines nor formally publishes a provisional patent application. A provisional patent application will automatically expire in 1 year. However, a later filed non-provisional patent application is entitled to the priority date of the earlier provisional application provided the provisional application fairly describes the claimed invention; enables its manufacture, application or practice; and describes the inventor's "best made" of the invention. The provisional patent application can effectively extend the patent protection period one additional year, from 20 to 21 years.

An individual must file a patent application in each country in which he or she would like to secure patent protection. Therefore, multiple patent applications to several countries are usually required to have proper international protection coverage, although the procedures of Patent Cooperation Treaty may be utilized to mitigate the costs of filing the patent application in multiple countries. As a general rule, no new disclosure can be added after filing, with a few exceptions, and a new application has to be filed for any additional disclosures. A nonprovisional utility patent application to the USPTO usually requires detailed specifications and drawings that are also part of the key engineering elements used to judge whether the patent is infringed by a reverse engineering process later in court when a legal challenge is raised. The specification is a written description of the invention and of the manner and process of making and using the same. The specification must be in such full, clear, concise, and exact terms as to enable any person skilled in the art

or science to which the invention pertains to make and use the same. For example, a computer program listing may be submitted as part of the specification as set forth in 37 CFR § 1.96. Drawings are often necessary in a patent application for understanding of the subject matter sought to be patented, and when included, the drawings must show every feature of the invention as specified in the corresponding claims. A comprehensive engineering drawing in a utility patent on a mechanical part is vitally essential. Omission of a drawing may cause a patent application to be considered incomplete. Even granted, the lack of detailed drawings may impede the patent from obtaining protection against a reverse engineered part.

A patent claim defines the scope of patent protection; it is a property that embodies the right. The claim or claims in a patent must particularly point out and distinctly claim the subject matter that the patent applicant regards as the invention. The predictability and reproducibility are two critical elements in the evaluation of originality of a new invention. These criteria from time to time restrict the patentability of some chemical and biological inventions because minor variations can make the results either unpredictable or irreproducible. The predictability and reproducibility are also two critical elements in reverse engineering, particularly for material chemical composition identification and manufacturing process verification. The wording of the claims directly affects whether a patent will be granted, and also has significant impact later on any patent infringement lawsuits involving reverse engineering. To pursue the maximum protection, the claims written in a patent application are usually as broad as possible. For example, the word *fastener* instead of *bolt* is used. Similarly, *polymeric material* instead of *polymer foam*, and *adhesive medium* instead of *glue* are preferred wordings in a patent claim. The mixing of these legal and engineering terminologies often causes some confusion in reverse engineering. The usage of *fastener* instead of *bolt* might broaden the legal protection of an invention, but the term *fastener* covers many other mechanical parts beyond just bolt, such as rivet. In engineering terms, a bolt is very different from a rivet in terms of required mechanical strength, geometric shape, and manufacturing process. It is arguable to claim that patent protection over a unique bolt design can automatically extend to a rivet simply by adopting the more generic word *fastener* in the claim.

To avoid patent infringement, a reverse engineering process should properly conduct a thorough and comprehensive patent search at the beginning to identify any potential risks and pitfalls. One strategy is to modify instead of copy the OEM part as much as feasible in reverse engineering. When the patent in question requires three elements, try to utilize only two instead of three of them in the reverse engineered part. This will significantly reduce the risk of patent infringement. Understandably, this poses a challenge to duplicate an "equivalent or better" substitute part reinvented by reverse engineering for the OEM counterpart. Obviously, a "better" rather than an "identical" part is the option to take from the perspective of patent defense.

The legal endeavor of a patent infringement fight is often mixed with emotion, technology, legal maneuvers, politics, and even personality. Beyond monetary rewards and legal obligations, the pride of two rival inventors may also be at stake. In 2001, NTP, Inc., a patent-holding company, brought a patent infringement lawsuit against Research in Motion Limited (RIM), who makes the BlackBerry mobile email system. During the course of a 5-year litigation, thousands of pages of documentation and expert opinions were filed calling the validity of NTP's patents into question. RIM claimed that it just codified the technology that was already widely in use by RIM and countless others. The jury at a U.S. District Court found the patent was valid and RIM willfully infringed NTP's patent. The judge instructed RIM to pay NTP $53 million in damages, $4.5 million in legal fees, plus 8.55% royalty, and issued an injunction ordering RIM to cease and desist from infringing the patents. This would shut down the BlackBerry systems in the United States. Numerous charges, countercharges, and appeals were filed. Eventually RIM settled its BlackBerry patent dispute with NTP and agreed to pay NTP a total of US$612.5 million in 2006. This case exemplifies a real-life legal saga on patent infringement.

## 8.3 Copyrights

### 8.3.1 Copyright Codes

Copyrights protect an expression fixed in a tangible medium, such as writing, painting, or sculpting. In contrast to patents, under the U.S. copyright laws all works are automatically given copyright protection the moment they are created. Most of the literary, pictorial, graphic, audiovisual, or music works, such as books, art, sculptures, photographs, motion pictures, songs, video games, and computer software, that do not fall into the public domain are potentially copyrightable. A copyright permission is clearly required when an image is photocopied from a book. However, how many words can be excerpted from a book before a copyright permission becomes mandatory is not clearly defined. It is usually recommended to get copyright permission when a quote or series of short quotes total several hundred words or more from a book. Data are not copyrightable; only the format in which the data are presented or published can be copyrighted. To reproduce a table in its entirety, including the format, might infringe on the author's copyright if the table's unique artistic expression is copied. Simply using the data listed in the table and reformatting them in a different table does not infringe any copyright because copyright does not prevent use of the underlying ideas, concepts, systems, or processes. A list of data or the raw data in a database is factual information that does not show originality. However, a drawing or diagram based on these data shows schematic expression, and therefore it is copyrightable. A passage of a song is

copyrightable, but short phrases and slogans are not. To reverse engineer the hood of a truck might not infringe the copyright, but copying the decoration or image on the hood will infringe the copyright.

Copyrights protect the author's right to reproduce, prepare derivatives of, distribute or display publicly, and perform publicly his or her works. A re-creation based on the underlying idea always plays a critical role in the progress of arts and science—it is legal and often encouraged—but a direct derivative work on a copyrighted art without permission is not allowed. The factual data, such as a part failure rate, and the scientific theory, such as the theory of fracture mechanics, are examples of underlying ideas and concepts that are not copyrightable. In contrast, a fictional character in a novel is usually copyrightable. Although copyright registration is not required, it is highly advisable because a copyright registration must be filed before it can be used to sue and claim damages. A person is guilty of copyright infringement if this individual has access to and copies the work, but a person cannot be sued for copyright infringement if he or she independently created the same piece of work. However, proof of independent creation of a work is not a sufficient defense for patent infringement.

The evolution of legal process over copyright and patent protection often progresses in parallel with the advancement of technology and demand for business protection. Software used to be considered copyrightable, but was difficult to patent, partially because the patent applicant had to demonstrate distinctive hardware effects by the software to secure a patent. Nonetheless, the copyright protection on software is relatively weak because reverse engineering can often decode it. Today more and more software developers are seeking patent protection for their inventions.

There are distinctive differences between the ownership of copyright and patent, as detailed in 17 USC Section 201, "Ownership of Copyright." The authors of a joint work are the copyright owners. The ownership is attributed to the owner's "originality" contribution to the work; simply providing some ideas to the work does not qualify a person to hold the copyright. The employee owns the copyright, though not the patent. In the case of a work made for hire, the employer or other person for whom the work was prepared is considered the author of the work. In contrast to patent ownership, the individual, company, or institute that funded or commissioned the work is the copyright owner. The employer can request an employee to sign an agreement to give all the copyrights of the work to the employer, no matter when, where, or how the work is done, even if part of the work is done at the employee's home after regular work hours.

The term of *copyright* is also different from patent. For works created after January 1, 1978, the term for a single author is the life of the author plus 70 years; for joint authors, it is 70 years after the last surviving author's death. For works made for hire where the employer holds the copyright, the term will be 95 years from first publication or 120 years from the year of creation,

**TABLE 8.1**

Copyright Terms

| Work | | Copyright Term |
|---|---|---|
| Works originally created on or after January 1, 1978 | A work by an individual author | Life of the author plus an additional 70 years after the author's death |
| | A joint work prepared by two or more authors who did not work for hire | 70 years after the last surviving author's death |
| | Works made for hire, and for anonymous and pseudonymous works | 95 years from publication or 120 years from creation, whichever is shorter |
| Works originally created and published or registered before January 1, 1978 | | Subject to the law in effect before 1978 and subsequent revisions, renewable for a total term of protection of 95 years |
| Works originally created before January 1, 1978, but not published or registered by that date | | Same as for the works created on or after January 1, 1978: the life-plus-70 or 95/120-year terms will apply to them as well; the law provides that in no case will the term of copyright for works in this category expire before December 31, 2002, and for works published on or before December 31, 2002, the term of copyright will not expire before December 31, 2047 |

*Source:* Data from http://www.uspto.gov/smallbusiness/copyrights/faq.html, assessed January 10, 2010.

whichever comes first. Table 8.1 summarizes the copyright terms for various types of works created in different time periods.

The purpose of copyright protection extends beyond just protecting and rewarding the authors and creators. Much copyright protection legislation encourages authors to share their creative works with society and help promote the progress of science. The mere fact that a work is copyrighted does not mean that every element of the work may be protected, as illustrated by the following excerpts.

> The copyright holder has a property interest in preventing others from reaping the fruits of his labor, not in preventing the authors and thinkers of the future from making use of, or building upon, his advances. The process of creation is often an incremental one, and advances building on past developments are far more common than radical new concepts. See *Lewis Galoob Toys, Inc. v. Nintendo*, No. 91-16205, slip op. at 5843 [22 USPQ2d 1857] (9th Cir. May 21, 1992). Where the infringement is small in relation to the new work created, the fair user is profiting largely from his own creative efforts rather than free-riding on another's work. A prohibition on all copying whatsoever would stifle the free flow of ideas without serving any legitimate interest of the copyright holder. *New Kids*

> *on the Block v. News Am. Publishing*, Nos. 90-56219, 90-56258, slip op. at 4 n.6, 1992 WL 171570 [23 USPQ2d 1534] (9th Cir. July 24, 1992). See also Harper & Row, 471 U.S. at 549.

The 1976 U.S. Copyright Act sets fair limits on the scope of copyright protection. The copyright protection based on 17 USC does not extend to any idea, process, or method of operation regardless of the form in which it is presented. The Copyright Act permits a party in rightful possession of a work to undertake necessary efforts, including disassembly of a hardware component or decompilation of a software program, to gain an understanding of the unprotected functional element of this work and the work's ideas, processes, and methods of operation. To protect processes or methods of operation, a creator must look to patent laws based on 35 USC. To separate the protectable expression from the unprotectable ideas, concepts, facts, processes, and methods of operation is a challenge in reverse engineering because the delineating boundaries between these two areas are not always clearly defined.

Fair use is distinguished from other uses of copyrighted work in 17 USC, "Copyrights," Chapter 1, "Subject Matter and Scope of Copyright," Section 107, "Limitation on Exclusive Rights: Fair Use." It states that

> the fair use of a copyrighted work, including such use by reproduction in copies or phonorecords or by any other means . . . , for purposes such as criticism, comment, news reporting, teaching (including multiple copies for classroom use), scholarship, or research, is not an infringement of copyright. In determining whether the use made of a work in any particular case is a fair use the factors to be considered shall include—
>
> (1) the purpose and character of the use, including whether such use is of a commercial nature or is for nonprofit educational purposes;
> (2) the nature of the copyrighted work;
> (3) the amount and substantiality of the portion used in relation to the copyrighted work as a whole; and
> (4) the effect of the use upon the potential market for or value of the copyrighted work.
>
> The fact that a work is unpublished shall not itself bar a finding of fair use if such finding is made upon consideration of all the above factors.

The teaching of and research in reverse engineering in an educational institute have much more freedom to use copyrighted work than practicing reverse engineering in a company for commercial products. Personal home use, such as off-air video- or audiotaping, is a common example of fair use.

The subject of reverse engineering is discussed in the 17 USC, Chapter 9, "Protection of Semiconductor Chip Product," Section 906, "Limitation on Exclusive Rights: Reverse Engineering; First Sale." It states that

> it is not an infringement of the exclusive rights of the owner of a mask work for — (1) a person to reproduce the mask work solely for the

> purpose of teaching, analyzing, or evaluating the concepts or techniques embodied in the mask work or the circuitry, logic flow, or organization of components used in the mask work; or (2) a person who performs the analysis or evaluation descried in paragraph (1) to incorporate the results of such conduct in an original work which is made to be distributed.

The Digital Millennium Copyright Act (DMCA) anticircumvention rules and 17 USC Section 1201 prohibit any person from "circumvent[ing] a technological measure that effectively controls access to a work protected" via reverse engineering in general. However, as excerpted below, it enumerates exceptions to the general prohibitions and permits reverse engineering for specific purposes such as achieving interoperability. The legitimacy of reverse engineering in this case is determined by purpose or necessity. Though reverse engineering a component is usually deemed legal in principle, the cultural effects and economic impact are sometimes applied as part of the criteria for the determination of legality of reverse engineering, particularly in the fair use arena.

> (f) Reverse engineering
>
> (1) . . . a person who has lawfully obtained the right to use a copy of a computer program may circumvent a technological measure that effectively controls access to a particular portion of that program for the sole purpose of identifying and analyzing those elements of the program that are necessary to achieve interoperability of an independently created computer program with other programs, and that have not previously been readily available to the person engaging in the circumvention, to the extent any such acts of identification and analysis do not constitute infringement. . . .
>
> (3) The information . . . and the means . . . may be made available to others if the person . . . provides such information or means solely for the purpose of enabling interoperability of an independently created computer program with other programs, and to the extent that doing so does not constitute infringement under this title or violate applicable law other than this section.
>
> (4) . . . the term "interoperability" means the ability of computer programs to exchange information, and of such programs mutually to use the information which has been exchanged.

The privilege of reverse engineering for interoperability allowed as an exception in the DMCA does not extend to free disclosure of the information obtained. Unless the sole purpose of the disclosure is to accomplish interoperability, information disclosure is subject to stringent restriction under the DMCA. However, the rights of free speech inherited in the First Amendment of U.S. Constitution may override economic and other considerations.

Though the above USC citations primarily focus on software and information technologies, the basic principles also apply to hardware and mechanical components. If a patented product is not sold with restrictive licenses,

a bona fide purchaser can reverse engineer it. Nonetheless, a competitive product produced by reverse engineering may still infringe the patent itself. Patent infringement does not require proof of copying. The inherent distinctions between hardware and software also result in noticeable differences in the legal codes and regulations that guide reverse engineering. To effectively enforce the DMCA anticircumvention rules and block the readily available, inexpensive software copying tools, the DMCA goes beyond reverse engineering itself and the post–reverse engineering data perforation. It has broad antitool provisions to target the techniques, primarily software programs, used for reverse engineering. In contrast, there are few legal rules, if any, that ban mechanical tools or analytical instruments, such as a scanning electron microscope, used in reverse engineering hardware.

The reverse engineering of a mechanical part encompasses the decoding or disassembling of the object part into the design details right up to the development of a new product. The decoding or disassembling process has to be reconciled with existing copyright laws. Making a duplicate of the OEM part could potentially constitute a breach of the exclusive rights of the copyright holder. However, copyright only protects the specific expression of an idea, not the idea itself. The reproduction of an OEM part might be allowed in "certain special cases," provided that such reproduction does not conflict with normal exploitation of the work and does not unreasonably prejudice the legitimate interests of the inventor. If reverse engineering through the decoding process is a necessary process to gain the nonprotected elements of an idea, it can be justified as a special case. If the data obtained by reverse engineering are only used to re-create an independent "new" part, not to impede the OEM's right to continue its supply of the original part to the market, the concerns of "normal exploitation" and "unreasonably prejudicing the legitimate interest of the inventor" might be voided. The general consensus is that reverse engineering is permitted under U.S. copyright and patent laws.

### 8.3.2 Legal Precedents on Copyrights

Reverse engineering for reinvention and copyright protection on creation both have been embraced by society for centuries. Reverse engineering was not a serious legal challenge for copyright laws until the late twentieth century, when computer software and other information technology products became the subject matter of copyright laws. The know-how of the artistic and literary works the copyright laws traditionally protected is usually apparent on the surface and needs little engineering to further decode. The underlying function of a software program goes beyond the aesthetic expression of a product, and to understand the software operation is an engineering task. Reverse engineering is therefore needed to decode the functionality of the product; it might infringe the privileges protected by copyright laws.

The conceptual separation between aesthetic and functional elements of an industrial product has made reverse engineering legally acceptable, as

evidenced by the 1987 court decision in *Brandir International, Inc. v. Cascade Pacific Lumber Co.* (Second Circuit Court, 1987). To distinguish the fine line between protectable "works of applied art" and "industrial designs not subject to copyright protection," the court ruled that the bicycle rack at issue is not copyrightable. Considering the fusion between function and aesthetics, the following argument prevailed in this case. Despite that the bicycle rack was derived in part from the "works of art," its final form was essentially a product of industrial design primarily for utilitarian function. The aesthetic elements cannot be conceptually separated from the utilitarian elements. The inseparability of these two elements made the bicycle rack not copyrightable. In analogy, it will be an interesting legal challenge to decide when a "designed" streamlined automobile body will be deemed a copyrightable work of art, and under what circumstance it will be viewed as an industrial design not subject to copyright protection.

Most legal precedents on reverse engineering are related to software and information technologies because today the majority of legal challenges related to reverse engineering are in these fields. The vulnerability of information products to reverse engineering is primarily attributed to two factors. First, modern technologies make software decoding relatively easy in this digital era. Second, the reverse engineered products potentially have devastating market impacts on the OEM products. However, imposing excessive legal restrictions on reverse engineering through legislation could detrimentally thwart technology advancement. The effect of limitations on reverse engineering brought by the DMCA has yet to be analyzed. The debate of both the merits and the damages of reverse engineering will certainly continue in the legal community. The following cases will help engineers sense the legal boundaries of reverse engineering.

Decompilation is allowed for the purpose of achieving interoperability in software reverse engineering, provided the final code is not substantially similar in expression to the original code. The illegality of decompilation can certainly protect software from one of the most effective reverse engineering methods. At the same time, it will also tilt the delicate balance between intellectual property assets protection and fair market competition. The legality of reverse engineering is therefore often not the issue; the key factor lies in the degree of similarity.

In the case of *Microsoft Corp. v. Shuuwa System Trading K.K.*, the defendant (Shuuwa) decompiled the plaintiff's (Microsoft) basic interpreter into a form of source code and then published the results in a book that was commercially distributed. The book at issue is entitled *PC-8001 Basic Source Program Listings* and listed Shuuwa System Trading K.K. as the publisher. The Tokyo District Court found that this constituted infringement of the plaintiff's copyright, primarily because the defendant reproduced the plaintiff's entire source code in its published manual.

The U.S. federal court decision in the *Saga Enterprises Ltd. v. Accolade, Inc.* case affirmed that decompilation of a software program was fair use. Two

conditions need to be satisfied for a software decompilation to be seen as fair use. First, "there is a legitimate reason to gain an understanding of the unprotected functional elements of the program." Second, there is "no other means of access to the unprotected elements."

Sega's video game cartridges had a security system; Accolade disassembled its code by reverse engineering, and integrated it into their competing products. Among other things, the U.S. Court of Appeals, Ninth Circuit, was asked to determine whether the Copyright Act permits persons who are neither copyright holders nor licensees to disassemble a copyrighted computer program in order to gain an understanding of the unprotected functional elements of the program. In light of the public policies underlying the Copyright Act, the court concluded in its decision (Ninth Circuit Court, 1992) that when the person seeking the understanding has a legitimate reason for doing so, and when no other means of access to the unprotected elements exist, such disassembly is, as a matter of law, a fair use of the copyrighted work. The court believed that the two conditions of fair use had been satisfied in this case. The decompilation of the specific software program to gain understanding of the functional requirements for compatibility was a legitimate reason. The object code could only be understood by decompilation, and there was no other means to do so.

Few intellectual property assets protection laws explicitly mandate technology advancement as a perquisite for reverse engineering. However, in the *Sega v. Accolade* case, the development of a "new" code based on the alleged infringed program seemed to play a significant role in the court decision. The forward engineering requirement in a reverse engineered product presents an incentive for follow-up innovation and healthy competition for the market. The 1984 Semiconductors Chip Protection Act (SCPA) permits reverse engineering of chip circuitry, and even reuse of the knowledge learned in reverse engineering, in a new original chip design. The SCPA encourages engineers to learn from earlier inventions to propel ahead new designs. At the same time, it prohibits the copying of OEM designs for selling cloned products, and provides fair protection for the legitimate interests of the original inventors.

The U.S. Government is not precluded from receiving and holding copyrights transferred to it by assignment, bequest, or otherwise, but a work of the U.S. Government is generally available in the public domain in accordance with 17 USC Section 105. There is a wealth of information in the public domain for reverse engineering applications. This is not an invitation to misappropriate protectable expression. No individuals or companies can use reverse engineering as an excuse to commercially exploit or otherwise misappropriate protected expression.

In the *Atari Games Corp. v. Nintendo of America, Inc.* case, the acceptance of reverse engineering was not an issue. However, the court decision pointed out that the fair use reproductions of a computer program must not exceed what is necessary to understand the unprotected elements of the work. It

is acknowledged that an individual cannot even observe, let alone understand, the object code on Nintendo's chip without reverse engineering. Atari engineers chemically etched off the top layer from Nintendo's chip surface to reveal the original object code. Through microscopic examination of the "peeled" chip, they transcribed the original object code into a list of 1s and 0s and keyed in these data to a computer. The computer then analyzed the object code to further reveal the program's functions. This reverse engineering process is qualified as a fair use by the court (Federal Circuit Court, 1992).

Nonetheless, the court found Atari liable in this case. In the conclusion of the court decision, it noted the substantial similarity between the two codes. The court also noted that Atari improperly obtained a copy of the Nintendo source code from the Copyright Office to help them decompile the Nintendo code. The court decision pointed out that "copies or reproductions of deposited articles retained under the control of the Copyright Office shall be authorized or furnished only under the conditions specified by the Copyright Office regulations." Although the source of substantiation data or authenticity is not the primary concern in many certification processes of a reverse engineered part, during litigation the source of authenticity can play a pivotal role, as exemplified in this case.

The legality of reverse engineering is well established. The courts have rarely weighed the practice of reverse engineering itself in their decisions as long as the original products and information are obtained by fair and honest means, such as through purchasing from the open market. A perceived wrongful acquisition, use, or disclosure of information, such as trespassing or deceit, or breach of a license agreement, can have serious adverse effects on both the judge and jury decisions, as demonstrated in the *Atari v. Nintendo* and *RIM v. NTP* cases. This adds another dimension of complexity to the reverse engineering arena.

International trade and globalization have made the legal issues of reverse engineering international affairs. Different countries have established different patent and copyright laws based on their respective legal systems and needs. The Japanese copyright law perceives computer software as an economic asset meant to contribute to the development of industrial economy as opposed to the development of culture. Since reverse engineering plays a major role in software development, the application of reverse engineering to software development is relatively liberally regarded as a lawful act in Japan. The European Community (EC) also allows for decompilation in restricted circumstances. On May 14, 1993, the EC issued the European Council Directive on Legal Protection of Computer Programs. It allows a rightful possessor of software to reverse engineer, that is, decompile it to achieve interoperability only, but not for the development, production, or marketing of a computer program substantially similar to the original program. Any contract or license terms forbidding software decompilation for interoperability are null and void.

## 8.4 Trade Secret

Trade secret is the information companies or individuals keep secret to give them an advantage over their competitors. The formula for the soft drink Coca-Cola is one of the most quoted trade secrets. It is also a very successfully protected trade secret. There is no report that any reverse engineering has ever fully decoded the Coca-Cola formula in terms of complete predictability and reproducibility. Other trade secrets include specific manufacturing processes, unique design drawings, special stock-picking formulae, and favorite customer lists. There is no filing required to establish trade secret protection. However, certain actions and procedures are expected to be taken to ensure that information is not disclosed to the public.

The trade secret protection is governed by state laws that vary from state to state. The term for trade secret is forever, as long as it remains confidential without public disclosure. Therefore, reverse engineering is the principle method to expose a trade secret in the public domain. The trade secret will be lost instantly if disclosed by either reverse engineering or other means. In contrast to patent and copyright laws that protect intellectual properties by restricting reverse engineering, trade secret laws expose inventions to reverse engineering with little protection.

Patent and trade secret are mutually exclusive. The same invention cannot hold a patent and claim trade secret protection at the same time. Trade secrets and trademarks are also inherently exclusive of each other. Trademarks are used to increase consumer recognition, while trade secrets are used to keep the underlying information away from the marketplace and the public. Trademark protects words, names, symbols, sounds, or colors that distinguish goods and services from those manufactured or sold by others, and indicate the source of the goods. In other words, a reinvented product via reverse engineering might be able to duplicate the substance inside of the product, but is not allowed to use the trademark of the product. Unlike patents, trademarks can be renewed forever, as long as they are being used in commerce. Both state common laws and federal registration protection provide coverage for trademarks.

The U.S. Copyright Office allows dual protection of copyright and trade secret for copyrightable material, particularly computer software. For example, with a computer program for a security control system, the program itself can be protected by copyright, and the programming algorithm is also protectable as a trade secret. The Copyright Office allows the author to obtain copyright protection on the code, without disclosing the underlying algorithm. Reverse engineering of such a program will focus on the trade secret algorithm.

The U.S. Patent and Trademark Office defines a service mark as "a word, name, symbol or device that is to indicate the source of the services and to distinguish them from the services of others." A service mark is very similar

to a trademark, except that it identifies and distinguishes the source of a service rather than a product.

### 8.4.1 Case Study of Reverse Engineering a Trade Secret

The following case study of reverse engineering a trade secret is based on the U.S. Ninth Circuit Court of Appeals report on *Chicago Lock Co. v. Fanberg* (Ninth Circuit Court, 1982).

The Chicago Lock Company (hereafter referred to as "the Company") manufactured and sold a highly secured tubular lock under the registered trademark "Ace." These locks are primarily used for maximum security applications, such as burglar alarms. The distinctive feature of Ace locks is the secrecy and difficulty of reproduction associated with their keys. The keys to Ace locks are stamped "Do Not Duplicate," and the Company will only sell a duplicate key to a bona fide lock owner. Nonetheless, a proficient locksmith would be able to "pick" the lock, decipher the tumbler configuration via reverse engineering, and grind a duplicate tubular key. Through the years, several locksmiths have accumulated substantial key code data, albeit noncommercially and on an *ad hoc* basis.

Fanberg and his father gathered these data and compiled the serial number–key code correlations into a book entitled *A-Advanced Locksmith's Tubular Lock Codes* without the permission of the Company. This book would allow an individual to duplicate the keys if the serial number of the lock was given. In 1976 and 1977, the Fanbergs advertised and sold this book for $49.95. On December 2, 1976, the Company filed a three-count complaint against the Fanbergs for trademark infringement under 15 USC § 1051 et seq., federal unfair competition under 15 USC § 1125(a), and California common law unfair competition under former California Civil Code § 3369. In the District Court, the Fanbergs won the federal claims. However, they lost the state law claim on the ground that the confidential key code data were a "valuable business or trade secret-type asset" of the Company, and that the Fanbergs' publication of their compilation of these codes so undermined the Company's policy as to constitute "common law unfair competition in the form of an unfair business practice within the meaning of Section 3369 of the Civil Code of the State of California."

An appeal was filed to the U.S. Ninth Circuit Court of Appeals. The Ninth Circuit Court agreed with the Fanbergs' argument that the District Court erroneously concluded that they were liable under Section 3369 of the Civil Code of the State of California for acquiring the appellee's trade secret through improper means, and on this basis the Ninth Circuit Court reversed the District Court.

The Ninth Circuit Court pointed out: "A trade secret does not offer protection against discovery by fair and honest means such as by independent invention, accidental disclosure or by so-called reverse engineering, that is, starting with the known product and working backward to divine the process.

(Sinclair, 42 Cal. App. 3d at 226, 116 Cal. Rptr. at 661)." The court believed that "it is the employment of improper means to procure the trade secret, rather than mere copying or use, which is the basis of liability.... the Fanbergs bought and examined a number of locks on their own, their reverse engineering (or deciphering) of the key codes and publication thereof would not have been use of 'improper means.' Similarly, the Fanbergs' claimed use of computer programs in generating a portion of the key code-serial number correlations here at issue must also be characterized as proper reverse engineering."

The Ninth Circuit Court further concluded that the Fanbergs' procurement of other locksmiths' reverse engineering data is also a "fair business practice" because they did not intentionally induce the locksmiths to disclose the trade secrets in breach of the locksmiths' duty to the company of nondisclosure.

In this case study, the Circuit pointed out that "a lock purchaser's own reverse-engineering of his own lock, and subsequent publication of the serial number-key code correlation, is an example of the independent invention and reverse engineering expressly allowed by trade secret doctrine. See Sinclair, 42 Cal. App.3d at 226, 116 Cal. Rptr. at 661." It is lawful and a fair business practice to decode a security key configuration via reverse engineering. It is fair competition.

## 8.5 Third-Party Materials

There are a lot of engineering data, computer codes, and literature belonging to a third party that can be used without infringing any copyrights or patents. These third-party materials fall into four categories: fair use, open source, creative commons, and public domain.

Fair use is defined in 17 USC Section 107 and has been discussed in previous sections. In the arena of open source, original work can be used free of infringement. This model of sharing information has gained increasing popularity with the wide use of the Internet. In 1998, Netscape released its code as open source under the name of Mozilla and marked a milestone for open source. Open source allows for concurrent use of different agendas and approaches in production.

Creative Commons is a nonprofit organization that offers flexible copyright licenses for creative works via reverse engineering or other methods. Creative commons licenses provide a flexible range of protections and freedoms for authors, artists, and educators. Creative commons are built upon the "all rights reserved" concept of traditional copyright, but move forward in a voluntary "some rights reserved" approach.

A license is a contract of permission from the copyright owner to use all or part of a copyrighted work for a particular purpose during a specified period. It is not uncommon to have certain restriction clauses included in a

license agreement to prohibit systematically copying, printing, or downloading substantial portions of the content. License agreements also often prohibit publicly exposing the information, such as posting to a public website. Some license agreements expressly prohibit reverse engineering. The legality of these restrictions on reverse engineering has been challenged in court. Most license terms, even though they are overwhelmingly restrictive, are upheld in court as long as the stakeholders had willingly agreed upon them.

A creative work is considered to be in the public domain if it is not protected by copyright, and it may be freely used by anyone. To take a photo of a building in open space requires no permission because the displayed art is in the public domain. However, taking a photo of a person requires specific permission. Most printed materials of the U.S., Canadian, and British governments are in the public domain and require no permission. Nonetheless, many government-sponsored agencies do copyright their materials, and therefore permission is required to use them.

The expiration of the term of copyright is one of the most common reasons for a work to become unprotected. The terms of copyright depend on the nature and publication dates of the work, as summarized in Table 8.1.

Orphan copyrights refer to the copyrights of the works that are still protected but no rights holder can be located. It is a general consensus that orphan works should be made available for reverse engineering as a last resort, after a diligent search has failed to locate the copyright owner. Unfortunately, the legal guidelines for what constitutes a diligent search are yet to be established. Those utilizing orphan copyrights in reverse engineering should prepare reasonable financial compensation and remedies should the copyright owner be found later.

## References

Federal Circuit Court. 1992. *Atari Games Corp. v. Nintendo of Am. Inc.*, 975 F.2d 832, 842–44.

Federal Circuit Court. 2003. *Bowers v. Baystate Tech. Inc.*, 320 F.3d 1317, 1323, 1325.

Fifth Circuit Court. 1988. *Vault Corp. v. Quaid Software, Ltd.*, 847 F.2d 255, 270.

Ninth Circuit Court. 1982. *Chicago Lock Co. v. Fanberg*. 676 F.2d 400.

Ninth Circuit Court. 1992. *Sega Enterprises Ltd. v. Accolade Inc.*, 977 F.2d 1510.

Pooley, J. H. A. 1997. Trade secrets. *Law Journal Press (looseleaf updated)*. 5.02[2]:5–17.

Samuelson, P., and Scotchmer, S. 2002. The law and economics of reverse engineering. *Yale Law J.* 111:1575–663.

Second Circuit Court. 1987. *Brandir International, Inc. v. Cascade Pacific Lumber Co.*, 834 F.2d 1141.

Supreme Court. 1974. *Kewanee Oil Co. v. Bicron Corp.*, 416 U.S. 470.

Supreme Court. 1989. *Bonito Boats, Inc. v. Thunder Craft Boats, Inc.*, 489 U.S. 141, 160–61, 164.

# Appendix A: Symbols and Nomenclature

| | |
|---|---|
| 2N | number of load reversals to failure (N = number of cycles to failure) |
| $a$ | crack length, displacement |
| $a_o$ | atomic distance. |
| A | cross-sectional area, stress ratio between the amplitude of alternating stress and mean stresses, cross-sectional area |
| $A_o$ | original cross-sectional area |
| b | fatigue strength exponent |
| $c$ | crack length, distance from the neutral axis, fatigue ductility exponent |
| $C$ | empirical constant, Larson-Miller constant |
| CDF(X) | cumulative density function, probability distribution function |
| $d$ | average grain size, spacing between two atomic planes, grain size |
| $d_x$ | diameter of bore hole |
| $d_y$ | diameter of rod |
| D | diameter of indenter ball |
| $D(x)$ | distribution function |
| $D_b$ | grain boundary diffusivity |
| $D_i$ | diameter of indention |
| $D_l$ | lattice diffusivity |
| $da/dn$ | fatigue crack propagation (or growth) rate |
| e | initial indention |
| E | Young's modulus, modulus of elasticity |
| $E_i$ | potential energy of electron in atomic shell, i = 0, 1, 2, 3 |
| erf | error function |
| $erf^{-1}$ | inverse error function |
| F | applied force, indenting force |
| F(X) | distribution function, cumulative distribution function |
| $F_1$ | minor load |
| $F_2$ | major load |
| G | shear modulus |
| $h$ | height, indention depth |
| $H_B$ | Brinell hardness number |
| $I$ | moment of inertia of cross-sectional area |
| $J$ | polar moment of inertia |
| $k$ | Boltzmann's constant ($13.8 \times 10^{-24}$ J/K), material parameter, number of standard deviation |
| $k_t$ | stress concentration factor |
| $K$ | stress intensity factor |
| $K_c$ | fracture toughness |
| $\Delta K$ | range of stress intensity factor |
| L | specimen length at the moment |
| $L_f$ | final specimen length |

| | |
|---|---|
| $L_o$ | original specimen length |
| $m$ | Walker exponent |
| $M$ | bending moment |
| MPa | macro pascal |
| n | sample size, nano, number of cycles, integer representing the diffraction order |
| $n_1$ | number of cycles the component is exposed under stress $\sigma_1$ |
| $n_i$ | number of cycles the component is exposed under stress $\sigma_i$ |
| $N$ | number of cycles to failure, population size, sample size |
| $N_1$ | number of cycles to failure under stress $\sigma_1$ |
| $N_f$ | fatigue life |
| $N_i$ | number of cycle to failure under stress $\sigma_i$ |
| nm | nanometer |
| $p$ | pressure, pica, empirical constant |
| P | (total) load, force, Larson-Miller parameter |
| $P(x)$ | probability function, cumulative probability function |
| Pa | pascal |
| PDF(x) | probability density function |
| $q$ | plastic-constraint factor |
| $r$ | radius |
| $r_x$ | length of the plastic zone in the x direction |
| $R$ | stress ratio between minimum and maximum stresses |
| $Ra$ | roughness average, two-dimensional measurement |
| $S$ | applied stress |
| $Sa$ | roughness average, three-dimensional measurement |
| SI | International System of Units (SI/Systeme International) |
| t | thickness, time |
| T | torque, time, temperature |
| x | exponent constant |
| $x_i$ | value for individual variate |
| X | variate |
| $\delta$ | grain boundary width |
| $\Delta K_{th}$ | critical threshold value |
| $\Delta L$ | change in length |
| $\varepsilon$ | creep deformation |
| $\dot{\varepsilon}$ | creep rate |
| $\varepsilon_e$ | engineering (normal) strain |
| $\varepsilon_f$ | true fracture strain |
| $\varepsilon_f'$ | fatigue ductility coefficient |
| $\varepsilon_o$ | initial strain |
| $\varepsilon_t$ | true or natural strain |
| $\varepsilon_x$ | lateral strain |
| $\varepsilon_z$ | strain in the thickness direction |
| $\Delta\varepsilon$ | total strain range |
| $\frac{\Delta\varepsilon_e}{2}$ | elastic strain amplitude |

| | |
|---|---|
| $\Delta\varepsilon_p$ | plastic strain range |
| $\frac{\Delta\varepsilon_p}{2}$ | plastic strain amplitude |
| $\phi$ | the angle between the tensile axis and the normal to the slip plane |
| $\Phi(X)$ | cumulative distribution function |
| $\gamma$ | shear strain |
| $\gamma_s$ | surface energy |
| $\lambda$ | the angle between the tensile axis and slip direction, electron wavelength |
| $\mu$ | micro, statistic mean, original distribution mean |
| $\mu_{safety}$ | the mean value of the safety margin distribution |
| $\mu_x$ | mean of part strength |
| $\mu_{\bar{x}}$ | sampling distribution mean |
| $\mu_y$ | mean of applied stress |
| $\theta$ | angle between the incident electron and the scattering plane |
| $\rho t$ | radius of curvature at crack tip |
| $\mu_z$ | mean of safety margin |
| $\mu_{safety}$ | stress of safety margin |
| $\mu_{strength}$ | strength of a part |
| $\mu_{stress}$ | stress applied to a part |
| $\sigma$ | standard deviation, tensile stress, stress, normal stress, applied tensile stress |
| $\sigma_1$ | the algebraically largest principal stress |
| $\sigma_3$ | the algebraically smallest principal stress |
| $\sigma^2$ | variance |
| $\sigma_a$ | stress amplitude, amplitude of alternating stress, alternating stress |
| $\sigma_b$ | bending stress |
| $\sigma_e$ | engineering (normal) stress, fatigue endurance limit |
| $\sigma_{eff,\max}$ | effective maximum stress |
| $\sigma_{eff,\max}/2$ | effective alternating stress |
| $\sigma_f$ | fracture strength, fracture stress |
| $\sigma_f'$ | fatigue strength coefficient |
| $\sigma_h$ | hoop stress |
| $\sigma_m$ | mean stress |
| $\sigma_{\max}$ | maximum stress, theoretical cohesive strength in tension |
| $\sigma_{\min}$ | minimum stress |
| $\sigma_{nom}$ | normal stress |
| $\sigma_o$ | constant material parameter |
| $\sigma_{R,\max}$ | algebraic maximum stress at a specific $R$ ratio |
| $\sigma_s$ | shear strength |
| $\sigma_{safety}$ | standard deviation of safety margin distribution |
| $\sigma_{sampling}$ | standard deviation for sampling distribution |
| $\sigma_{sy}$ | yield strength in shear |
| $\sigma_t$ | true stress |
| $\sigma_u$ | ultimate tensile strength |
| $\sigma_{ut}$ | ultimate tensile strength |

$\sigma_{Walker}$ Walker equivalent stress
$\sigma_x$ standard deviation of strength, transverse elastic stress
$\sigma_{x,\max}$ maximum stress in the x direction
$\sigma_y$ longitudinal stress, standard deviation of stress, yield strength, alternating stress range, stress range
$\sigma_{yc}$ yield strength in compression
$\sigma_{yield}$ yield strength
$\sigma_{y,\max}$ maximum stress in the y direction
$\sigma_{yt}$ yield strength in tension
$\sigma_z$ standard deviation of safety of margin
$\tau$ shear stress
$\nu$ Poisson's ratio
$\tau_{crss}$ critical resolved shear stress
$\tau_{\max}$ maximum shear strength
$\tau_t$ torsion stress
$\Omega$ atomic volume

# Appendix B: Acronyms and Abbreviations

| | |
|---|---|
| 2D | Two-dimensional |
| 3D | Three-dimensional |
| A | Ampere for electric current |
| A2LA | The American Association for Laboratory Accreditation |
| AAS | Atomic-absorption spectroscopy |
| ABEC | The Annual Bearing Engineering Committee |
| ABMA | American Bearing Manufacturers Association |
| ABS | Acrylonitrile butadiene styrene |
| AC | Advisory Circular |
| ACO | Aircraft Certification Office (FAA) |
| AES | Atomic emission spectroscopy |
| AMS | Aerospace material specification |
| ASME | American Society of Mechanical Engineers |
| ANSI | American National Standards Institute |
| ASCII | American Standard Code for Information Interchange |
| ASQ | American Society for Quality |
| ASTM | ASTM International, originally known as the American Society for Testing and Materials |
| ATOS | Advanced TOpometric Sensor |
| BCC | Body-centered cubic |
| BHN | Brinell hardness number |
| CAD | Computer-aided design |
| CAE | Computer-aided engineering |
| CAM | Computer-aided manufacturing |
| CAPA | The Certified Automotive Parts Association |
| CCD | Charge-coupled device |
| CE | Carbon equivalent |
| CFD | Computational fluid dynamics |
| CFR | Code of Federal Regulations |
| CMM | Coordinate measuring machine |
| CNC | Computer numerically-controlled |
| COS | Continued operational safety |
| CSS | Cross-sectional scanning |
| DIN | Deutsches Institut für Normung in German, and the German Institute for Standardization in English |
| DMCA | Digital Millennium Copyright Act |
| DMLS | Direct metal laser sintering |
| DPH | Diamond pyramid hardness |
| DXF | Drawing exchange, drawing interchange |
| EASA | European Aviation Safety Agency |
| EC | European Community |

| | |
|---|---|
| EDM | Electrical discharge machining |
| EDS | Energy dispersive X-ray spectrometry |
| EDX | Energy dispersive X-ray analysis |
| EDXA | Energy dispersive X-ray analysis |
| EELS | Electron energy loss spectroscopy |
| ELI | Extra-low-interstitial |
| EPMA | Electron probe microanalysis, electron probe microanalyzer |
| FAA | Federal Aviation Administration |
| FCC | Face-centered cubic |
| GD&T | Geometric dimensioning and tolerancing |
| GE | General Electric |
| GEAE | General Electric Aircraft Engine |
| GPA | Grade point average |
| GTAW | Gas tungsten arc welding |
| HCF | High-cycle fatigue |
| HCP | Hexagonal close-packed |
| HP | Horse power |
| HPT | High pressure turbine |
| HVOF | High velocity oxyfuel |
| IAF | International Accreditation Forum |
| ICP-AES | Inductively coupled plasma atomic emission spectrometry |
| ICP-MS | Inductively coupled plasma mass spectrometry |
| ICP-OES | Inductively coupled plasma optical emission spectrometry |
| IEC | International Electrotechnical Commission |
| IGES | Initial Graphics Exchange Specification |
| ISO | International Organization for Standardization |
| JIS | Japanese Industrial Standards |
| kph | Kilometers per hour |
| ksi | Kilo pounds per square inch |
| kv | Kilo voltage |
| LCD | Liquid crystal display |
| LCF | Low-cycle fatigue |
| LEES | Low energy electron spectroscopy |
| LLP | Life limited parts |
| MARPA | Modification and Replacement Parts Association |
| MIL-HDBK-5 | Military Handbook 5 |
| MIT | Massachusetts Institute of Technology |
| mm | Millimeter |
| MMPDS | Metallic Materials Properties Development and Standardization |
| mph | Miles per hour |
| NADCAP | National Aerospace and Defense Contractors Accreditation Program |
| NAP | National Accreditation Program |
| NASA | National Aeronautics and Space Administration |
| NIMS | National Institute for Materials Science |
| NIST | National Institute of Standards and Technology |

| | |
|---|---|
| nm | Nanometer |
| NRC | The National Research Council |
| NRL | Naval Research Laboratory |
| NURBS | Non-uniform rational B-Spline |
| OEM | Original equipment manufacturer |
| Pa | Pascal, Newton per square meter |
| PMA | Parts manufacturer approval, premarket approval |
| psi | Pound per square inch |
| PVD | Physical vapor deposition |
| PW | Pratt Whitney |
| QMS | Quality management system |
| Ra | Roughness average |
| RAB | Registrar Accreditation Board |
| RE | Reverse engineering |
| RIM | Research in Motion Limited |
| RoHS | Restriction of Hazardous Substances |
| RR | Rolls Royce |
| SAE | Society of Automotive Engineers |
| SCC | Stress corrosion crack |
| SCPA | Semiconductors Chip Protection Act |
| SEM | Scanning electron microscope, scanning electron microscopy |
| SI | International System of Units |
| SLA | Stereolithography |
| SLS | Selective laser sintering |
| SME | Society of Manufacturing Engineers |
| SS | Stainless steel |
| SST | Solution Support Technology |
| STC | Supplemental type certificate |
| STEP | Standard for the Exchange of Product data |
| STL | Standard triangulation language, Stereolithography |
| TEM | transmission electron microscope, transmission electron microscopy |
| TIG | Tungsten inert gas |
| USAF | U.S. Air Force |
| USC | United States Code |
| USPTO | United States Patent and Trademark Office |
| UTS | Ultimate tensile strength |
| VAR | Vacuum arc remelting |
| VIM | Vacuum induction melting |
| WDS | Wavelength dispersive spectrometry, wavelength dispersive X-ray spectroscopy |
| WDX | Wavelength dispersive X-ray analysis |
| Wt% | Weight percentage |
| YS | Yield strength |
| μm | Micro meter |

# Index

**M**

## S

## Y



## Z